Polarized Electron/Polarized Photon Physics

PHYSICS OF ATOMS AND MOLECULES

Series Editors

P. G. Burke, *The Queen's University of Belfast, Northern Ireland*
H. Kleinpoppen, *Atomic Physics Laboratory, University of Stirling, Scotland*

Editorial Advisory Board

R. B. Bernstein *(New York, U.S.A.)*
J. C. Cohen-Tannoudji *(Paris, France)*
R. W. Crompton *(Canberra, Australia)*
Y. N. Demkov *(St. Petersburg, Russia)*
C. J. Joachain *(Brussels, Belgium)*

W. E. Lamb, Jr. *(Tucson, U.S.A.)*
P.-O. Löwdin *(Gainesville, U.S.A.)*
H. O. Lutz *(Bielefeld, Germany)*
M. Standage *(Brisbane, Australia)*
K. Takayanagi *(Tokyo, Japan)*

Recent volumes in this series:

ATOMIC PHOTOEFFECT
M. Ya. Amusia

ATOMIC SPECTRA AND COLLISIONS IN EXTERNAL FIELDS
Edited by K. T. Taylor, M. H. Nayfeh, and C. W. Clark

ATOMS AND LIGHT: INTERACTIONS
John N. Dodd

ELECTRON COLLISIONS WITH MOLECULES, CLUSTERS, AND SURFACES
Edited by H. Ehrhardt and L. A. Morgan

ELECTRON–MOLECULE SCATTERING AND PHOTOIONIZATION
Edited by P. G. Burke and J. B. West

THE HANLE EFFECT AND LEVEL-CROSSING SPECTROSCOPY
Edited by Giovanni Moruzzi and Franco Strumia

INTRODUCTION TO THE THEORY OF LASER–ATOM INTERACTIONS,
Second Edition
Marvin H. Mittleman

INTRODUCTION TO THE THEORY OF X-RAY AND ELECTRONIC SPECTRA
OF FREE ATOMS
Romas Karazija

MOLECULAR PROCESSES IN SPACE
Edited by Tsutomu Watanabe, Isao Shimamura, Mikio Shimizu, and Yukikazu Itikawa

POLARIZATION BREMSSTRAHLUNG
Edited by V. N. Tsytovich and I. M. Ojringel

POLARIZED ELECTRON/POLARIZED PHOTON PHYSICS
Edited by Hans Kleinpoppen and W.R. Newell

THEORY OF ELECTRON–ATOM COLLISIONS, Part 1: Potential Scattering
Philip G. Burke and Charles J. Joachain

A Continuation Order Plan is available for this series. A continuation order will bring delivery of each new volume immediately upon publication. Volumes are billed only upon actual shipment. For further information please contact the publisher.

Polarized Electron/Polarized Photon Physics

Edited by

Hans Kleinpoppen
University of Stirling
Stirling, Scotland

and

W. R. Newell
University of London
London, England

Plenum Press • New York and London

Library of Congress Cataloging-in-Publication Data

```
Polarized electron/polarized photon physics / edited by Hans
  Kleinpoppen and W.R. Newell.
      p.   cm. -- (Physics of atoms and molecules)
    "Proceedings of two United Kingdom Engineering and Physical
  Science Research Committee workshops on Polarized Electron/Polarized
  Photon Physics, held September 22-23, 1993, and April 15-16, 1994,
  in York, England"--T.p. verso.
    Includes bibliographical references and index.
    ISBN 0-306-45131-X
    1. Electrons--Polarization--Congresses.  2. Photons--Polarization-
  -Congresses.   I. Kleinpoppen, H. (Hans)  II. Newell, W. R. (William
  Robert), 1943-     .  III. Workshop on Polarized Electron/Polarized
  Photon Physics (1993 : York, England)  IV. Workshop on Polarized
  Electron/Polarized Photon Physics (1994 : York, England)  V. Series.
  QC793.5.E628P64  1995
  523.7'2112--dc20                                            95-37115
                                                                  CIP
```

Proceedings of two United Kingdom Engineering and Physical Science Research Committee workshops on Polarized Electron/Polarized Photon Physics, held September 22–23, 1993, and April 15–16, 1994, in York, England

ISBN 0-306-45131-X

©1995 Plenum Press, New York
A Division of Plenum Publishing Corporation
233 Spring Street, New York, N.Y. 10013

10 9 8 7 6 5 4 3 2 1

All rights reserved

No part of this book may be reproduced, stored in a retrieval system, or transmitted in any form or by any means, electronic, mechanical, photocopying, microfilming, recording, or otherwise, without written permission from the Publisher

Printed in the United States of America

PREFACE

The EPSRC (Engineering and Physical Science Research Committee of the U.K.) suggested two Workshops (York University, 22–23 September, 1993 and 15–16 April, 1994) for possible development of polarized electron/photon physics as targeted areas of research. The remit of these meetings included identifying research groups and their activities in polarized electron/polarized photon physics, listing relevant existing facilities (particularly *electron spin sources and polarimeters*), possible joint projects between research groups in the U.K., recognizing future needs of projects for research of the highest scientific merit and referring to international comparisons of these research activities. Although very diverse but interconnected, the areas of research presented at the Workshops embrace atomic, molecular, surface, and solid state physics. *In more detail these areas covered:* electron spin correlations and photon polarization correlations in atomic and molecular collisions and photoionization, electron spin effects in scanning tunneling microscopy, surface and interface magnetism from X-ray scattering and polarized Auger electrons (including analysis of domain structures in solids and surfaces), polarized electrons from multiphoton ionization, quasi-atomic effects in solid state physics, dichroism in molecular and surface processes, Faraday rotation and high-field magneto-optics and polarization effects in simultaneous higher order electron–photon excitations.

It is obvious from the spectrum of research fields presented at the Workshops that physicists of primarily two communities, namely those studying electron and photon spin interactions with *gaseous atomic and molecular targets* and those using *condensed matter targets* for their studies, interacted very closely with each other. This kind of *interdisciplinary communication* between various groups and the expertise available in polarized electron/polarized photon physics has generally been highly valuable to the participants in the Workshops, which involved exchange of knowledge in technology, recognizing similarities between the above subfields, and learning about new developments from each other. It has been possible to exchange experience between the research scientists in these fields and joint expertise should help newcomers to start building up facilities which will certainly be needed to a larger extent for future research in this country and overseas. Subfields of special interest in research are: quasi-atomic effects in solid state physics, (particularly for joint projects of atomic and solid state physicists); polarized electrons in multi-photon ionization and continuum processes; Faraday rotation and high-field magneto-optical processes; electron spin and polarized photon analysis of what is normally called "complete" atomic collision experiments (particularly in connection with one-electron atoms, i.e., atomic hydrogen and alkali atoms for which severe discrepancies exist between theory and experiments!); development of low energy circularly polarized synchrotron radiation and the availability of advice and access to efficient Mott detection of spin polarized electrons. While most of the existing and future projects for using polarized electrons and polarized light are basic fundamental research projects, applications of the magneto-optical Kerr effect, of spin polarized scanning tunneling microscopy, of circularly polarized X rays (e.g.

in synchrotron experiments) give information on images of surfaces and in the development of novel materials that have great potential in *wealth creation for society*.

Summarizing, the two Workshops held have achieved and/or established the following:

1) Close contact and possible future informal cooperation between researchers from atomic, molecular, surface and solid state physics and electron optics. This particularly refers to an exchange of expertise and knowledge of various groups for research in the above subfields.

2) An awareness that there is a kind of "correlation" between the various disciplines or research areas in applying polarized electrons and polarized photons.

3) In the period between the two Workshops researchers from various places (e.g., Brighton and Daresbury) were stimulated by discussions and exchange of knowledge so that they were able to present improved new results on electron spin polarimetries at the last Workshop.

Clearly modern research activities in given subfields highly benefit from exchange and stimulation of knowledge for which the two Workshops at York University provided the base. Members of the ad hoc Committee for organizing the Workshops were:

Prof. J.P. Connerade, I.C. London;
Prof. B.L. Gyorffy, Bristol University;
Prof. H. Kleinpoppen (Chairman), Stirling University;
Prof. J.A.D. Matthew, York University;
Dr W.R. Newell, University College London.

As editors we are pleased to recognize the work of the authors of this Proceedings of the Workshops.

Hans Kleinpoppen
W.R. Newell

CONTENTS

Hans Kleinpoppen
INTRODUCTORY REMARKS ON POLARIZED ELECTRON/
POLARIZED PHOTON PHYSICS   1

W. Raith, G. Baum, P. Baum, L. Grau,
B. Leuer, R. Niemeyer and M. Tondera
MEASUREMENT OF EXCHANGE AND SPIN-ORBIT
INTERACTION EFFECTS IN ELASTIC ELECTRON-
CESIUM SCATTERING   23

H.R.J. Walters
THEORY OF ELECTRON SPIN-EFFECTS IN
ELECTRON-ATOM SCATTERING   37

Albert Crowe
CORRELATIONS STUDIES OF ELECTRON
IMPACT EXCITATION - PAST, PRESENT, FUTURE   61

Jurgen Beyer
POLARIZATION CORRELATIONS FROM ELECTRON
IMPACT EXCITATION OF EARTH ALKALINE ATOMS   81

E.A. Seddon, I.W. Kirkman and F.M. Quinn
ELECTRON POLARIMETRY ON THE SYNCHROTRON
RADIATION SOURCE (SRS)   95

D. Murray Campbell
SOURCES AND DETECTORS OF POLARIZED
ELECTRONS   107

E.A. Seddon
ELECTRON SPIN POLARIMETRY INSTRUMENTATION   121
SURVEY: 1994

Fredrik Schedin, Ranald Warburton
and Geoff Thornton
A BOLT-ON SOURCE OF SPIN POLARISED ELECTRONS
FOR STUDIES OF SURFACE MAGNETISM   133

M. Hardiman, I.R.M. Wardell, M.S. Bhella,
M. Whitehouse-Yeo, P. Gendrier,
C.J. Harland, G. Roussel, C-K Lo,
S. Lis, D. König and J. Agernon
SPIN POLARIZED ELECTRON DETECTORS FOR
SURFACE MAGNETISM   147

Muhammad Afzal Chaudhry and Hans Kleinpoppen
RESONANCE LINE RADIATION FROM SPIN POLARIZED
SODIUM AND POTASSIUM ATOMS   159

Gaetana Laricchia
STUDIES WITH POLARIZED POSITRONS   169

W.R. Newell
POLARIZATION EFFECTS IN
SIMULTANEOUS ELECTRON PHOTON EXCITATION   177

Alan J. Duncan and Zahoor A. Sheikh
POLARIZATION CORRELATION AND COHERENCE
LENGTHS OF TWO-PHOTON RADIATION   187

D.G. Thompson
SOME THEORETICAL AND COMPUTATIONAL
RESULTS FOR ELECTRON MOLECULE SCATTERING   197

N.J. Mason
SPIN DEPENDENT ELECTRON SCATTERING FROM
ORIENTED MOLECULES: AN EXPERIMENTAL APPRAISAL   209

J.B. West
PHOTOIONISATION AND FLUORESCENCE OF CALCIUM
AND STRONTIUM IN THEIR P-D GIANT RESONANCE REGIONS   225

J.P. Connerade
FARADAY ROTATION IN THE UV/VUV AND HIGH
FIELD MAGNETO-OPTICS   235

Natalia E. Karapanagioti, George Droungas
and Jean-Patrick Connerade
STEPWISE MULTIPHOTON STUDY OF YTTERBIUM   247

J.A.D. Matthew
SPIN-POLARISED ELECTRON SCATTERING
IN CONDENSED MATTER - AN ATOMIC APPROACH   261

J.A.C. Bland
MAGNETO-OPTIC KERR EFFECT STUDIES OF
ULTRATHIN MAGNETIC STRUCTURES   269

G. van der Laan
MAGNETIC X-RAY DICHROISM. AN EFFECTIVE WAY
TO STUDY THE SPIN AND ORBITAL MAGNETIZATION
IN MAGNETIC MATERIALS   295

Andrew J. Rollason
X-RAY MAGNETIC SCATTERING   309

T-H. Shen
SPIN-POLARISED SCANNING TUNNELLING
MICROSCOPY AND RELEVANT TECHNIQUES -
A SURVEY OF PRESENT STATUS   329

SERIES PUBLICATIONS   343

INDEX   347

Polarized Electron/Polarized Photon Physics

INTRODUCTORY REMARKS ON POLARIZED ELECTRON/ POLARIZED PHOTON PHYSICS

Hans Kleinpoppen

Unit of Atomic Physics
University of Stirling
Stirling FK9 4LA, Scotland

INTRODUCTION

Spin effects are widespread phenomena in many areas of physics: in elementary particle, nuclear, atomic, molecular, surface, solid state and astrophysics and also in electron optics. However, in these Proceedings we restrict ourselves and exclude particle and nuclear physics. A variety of theoretical and experimental data are reported in these Proceedings in connection with applying polarized electrons and polarized photons; some of the investigations presented are based upon joint applications of polarized electrons, polarized photons, polarized atoms, aligned molecules and layers at surfaces. Apart from applications of spin-polarized electrons in elctron optics the topics in these Proceedings can be divided into two parts:

(1)   Collisions and spectroscopic interactions of polarized electrons or polarized photons with targets of free atoms and molecules, the atoms may be polarized and the molecules aligned (see Fig.1);

(2)   collisions of polarized electrons or polarized photons and spectroscopic interactions with surfaces and solids (see Fig. 2).

EXAMPLES OF COLLISIONS AND SPECTROSCOPIC INTERACTIONS OF (POLARIZED) ELECTRONS OR POLARIZED PHOTONS WITH TARGETS OF FREE (POLARIZED) ATOMS OR (ALIGNED) MOLECULES

Fig.1 illustrates the first scheme introduced. A collimated beam of incoming particles collides with a target of free atoms or molecules; the projectiles may consist of unpolarized or polarized particles such electrons $e^-(\uparrow)$, positrons $e^+(\uparrow)$, atoms $A(\uparrow)$, aligned molecules M or polarized photons $h\nu$. The outgoing particles B_1, B_2, ... leaving the target can also be composed of such particles. The geometry of joint angular correlations and polarization correlations between scattered particles (which may give information on angular correlations) may contain interesting and important characteristics of physical quantities of the collisional or spectroscopic interactions involved. Such information on the physical processes involved can quantum mechanically be "complete", i.e., it represents an optimum knowledge of the physics extracted from the experiment.

Polarized Electron/Polarized Photon Physics
Edited by Hans Kleinpoppen and W. R. Newell, Plenum Press, New York, 1995

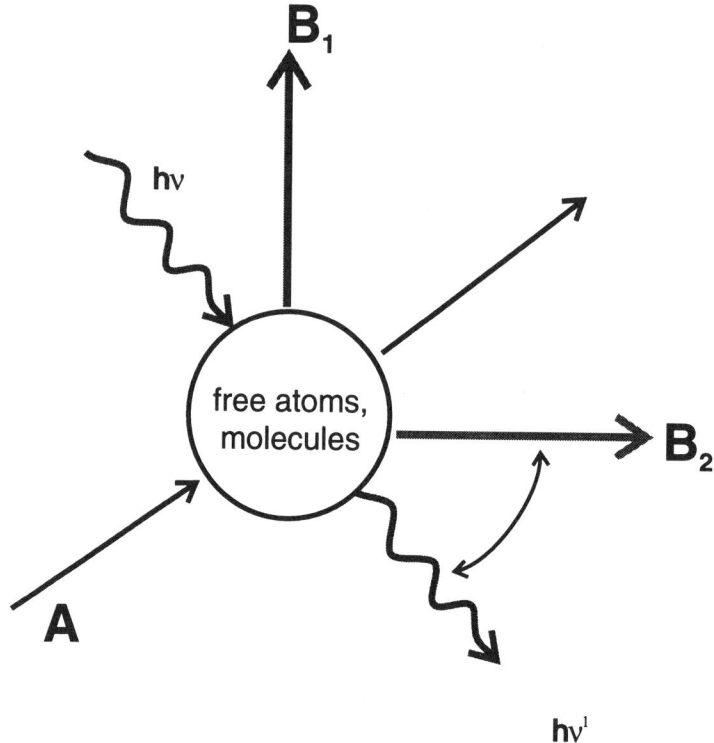

Fig.1. General scheme for scattering processes with polarized projectiles A colliding with a target of free atoms or molecules; the projectiles may consist of spin polarized electrons $e^-(\uparrow)$, positrons $e^+(\uparrow)$, atoms $A(\uparrow)$, ions $Io(\uparrow)$, or polarized photons $h\nu(\uparrow)$; the target may also consist of polarized atoms or aligned molecules. The outgoing particles B_1, B_2 can be spin-polarized and the photons polarized. Polar (θ) and azimuthal angles (ϕ) of the scattered particles are measured with reference to the direction of the incoming particles and the scattering plane; they describe geometric angular correlations between the detected particles after the collision.

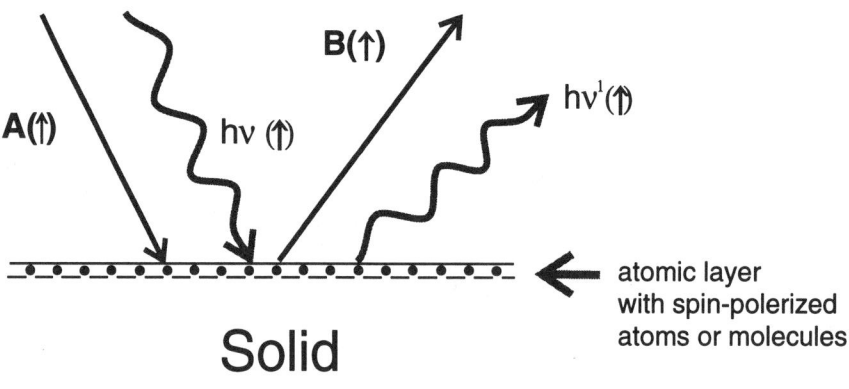

Fig.2. Same scheme as in Fig.1 but the target shall consist of condensed matter with a spin-polarized atomic or an aligned molecular layer (also called polarized "adatoms" or aligned "admolecules"). The incoming particles may penetrate deeper into the target or even pass through it.

In collision processes such "complete" experiments can be approached by studies of spin-dependent particle-particle or photon-particle coincidence experiments (including combinations of both types of experiments). Such complete experiments are highly complicated in general, they remain goals of worthwhile studies in the (hopefully!) near future.

"Complete" atomic collision experiments have revolutionized atomic collision physics since the beginning of the 70's. First proposals for complete experiments were made by several authors[1-3] but I particularly like to refer to a kind of scheme outlined by U.Fano[1]. For simplicity we consider electron-atom collision processes. Their differential cross section σ depends on the atom, the primary electron energy E and the scattering angles θ and ϕ, i.e. $\sigma = \sigma(E,\theta,\phi)$. However, this information on the scattering process is limited; a complete information is contained in the state vector $|\psi\rangle = \Sigma f_m \psi_m$ describing the collision to which the differential cross section can be normalized to its bracket by the relation $\sigma = C\langle\psi|\psi\rangle = C|\psi|^2$. Fano's suggestion[1] how to obtain the state vector from experiment is as follows: prepare initially the collision particles into a pure state called $|\psi_{in}\rangle$. Because of the linearity of the Schrödinger equation which includes a collisional interaction potential U the initial pure state $|\psi_{in}\rangle$ will be transferred into another pure state $|\psi_{out}\rangle$ according to the scheme

$|\psi_{in}\rangle = |P\rangle|A\rangle$ $\qquad\qquad$ $|\psi_{out}\rangle = |...\rangle...|...\rangle$
$\qquad\qquad\qquad$ linear operator
state vector $\qquad\quad$ for the Hamiltonian $\qquad$ state vector
before collision $\qquad$ with interaction $\qquad\qquad$ after collision
$\qquad\qquad\qquad$ potential U(t)
$\qquad\qquad\qquad$ acting during the
$\qquad\qquad\qquad$ collision process

$|P\rangle$, $|A\rangle$, $|...\rangle$ are state vectors for the projectile, the target atoms and the collisional products, respectively.

The question arises what are the experimental tools for this procedure ? So far particle-photon coincidences and/or spin reactions are being or have been applied from which quantum mechanically complete and optimum information can be extracted.

Particle-Photon Coincidence Experiments

Probably the simplest example for complete experiments is the excitation of $^1P_1$-states of helium by electron impact: $e + He \rightarrow He(n^1P_1) + e'$. The relevant state vectors before and after the excitation are

$|\psi_{in}\rangle = |e\rangle|He(1^1S_o)\rangle$, $\qquad$ (before excitation) $\rightarrow$
$|\psi_{out}\rangle = |e'\rangle|He(n^1P_1)\rangle = |e'\rangle\Sigma f_m \psi_m(^1P_1)$, (after excitation) $\rightarrow$

whereby $|e\rangle$ and $|e'\rangle$ are the state vectors of the incoming and outgoing electron; we note that their spin components are symmetric with regard to the scattering plane, it therefore can be neglected in the state vectors $|e\rangle$ and $|e'\rangle$. The state vector $|He(n^1P_1)\rangle$ for the excited helium atom is the sum of the magnetic substates ψ_m of the $^1P_1$-state with the complex amplitudes f_1, f_o and f_{-1}. Because of planar symmetry of the scattering process $f_1 = -f_{-1}$ and the differential cross section can be expressed by $\sigma(E,\theta) = |f_o|^2 + 2|f_1|^2 = \sigma_o + 2\sigma_1$ with σ_o and σ_1 as partial differential cross sections for exciting the magnetic sublevels $m = 0$ and $m = \pm 1$. By allowing a fixed phase difference $\chi = \chi_o - \chi_1$ between the two amplitude, the quantities $|f_o|$, $|f_1|$ and χ or $\lambda = \sigma_o/\sigma = |f_o|^2/\sigma$, χ and σ represent full sets of data for a "complete" type of experiment. While the differential cross section is made up by an incoherent superposition of the magnetic sub-cross sections the electron-photon coincidence rates are due to coherent superpositions of contributions of data from the magnetic substates. By setting f_0 real, f_1 may have a phase χ so that $f_1 = |f_1|e^{i\chi}$; defining the so-called λ-parameter by $\lambda = |f_0|^2/\sigma$ the angular correlation becomes

$$N_{e,\gamma} = \lambda\sin^2\theta_\gamma + (1-\lambda)0\cos^2\theta_\gamma - \sqrt{\lambda(1-\lambda)}\sin\theta_\gamma\cos\theta_e\cos\chi$$

for a fixed electron scattering θ_e angle and observing the photons in the scattering plane at an angle θ_γ. An equivalent expression that makes the interference effect more transparent is given by

$$N_{e,\gamma} = 1/\sigma |\int_0 \sin\theta_\gamma - \sqrt{2}\int_1 \cos\theta_\gamma|^2$$

The physical picture following from this descripton of the $^1S \to {}^1P$ excitation implies that two coherent oscillators are excited parallel to the z-axis (amplitude f_0 in Fig.3) parallel to the x-axis (amplitude $\sqrt{2}f_1$ in Fig.3) in the scattering plane.

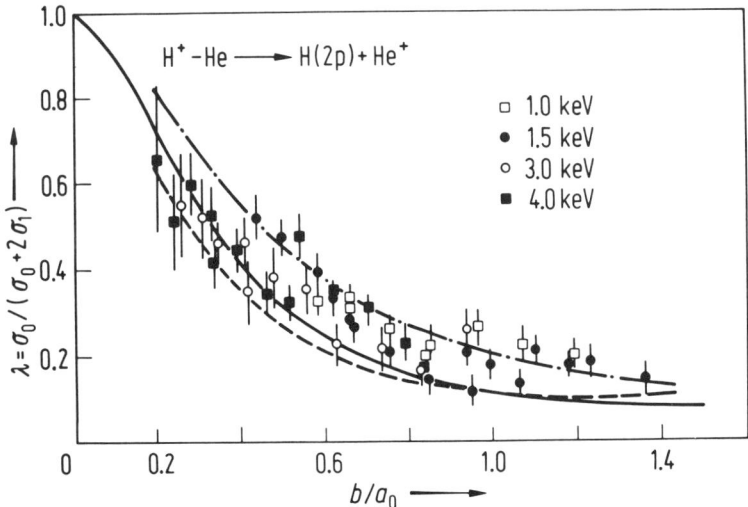

Fig.3. λ-parameter for the relative population of the H(2P, m = o)-state as a function of the impact parameter (in atomic units) of the charge exchange process $H^+ + He \to H(2P) + He^+$ at various energies. Experimental data with error bars after Hippler et al[21]; theoretical predictions by Macek and Wang[22] (full curve) and Fritsch (private communication) for 1 keV (dotted curve) and 4 keV (dotted curves including points).

Standage and Kleinpoppen[4] tested the *model of coherent impact excitation* of the He 3^1P_1 state as follows. The pure state vector $|\psi(^1P_1)>$ that is represented by a linear superposition of the magnetic substates $|\ell m_\ell>$ will "decay" into a 2^1S_0 state with the emission of a photon $h\nu$. If the decay process conserves the coherence during the photon transition into the lower state both the atom and the photon should be in pure states. The decay process can then be described by the reaction

$$|^1P_1> \to |^1S_0> |h\nu>$$

A critical test of these assumptions can be based upon a study of the state of the photon from the $^1P_1 \to {}^1S_0$ transition. Being in a pure state the photon radiation is expected to be completely coherent. The coherence properties of the photon radiation can be measured by means of the classical Stokes parameters. Criteria for coherence of the photon radiation (Born and Wolf[12]) are the vector polarisation

$$|\mathbf{P}| = (|P_1|^2 + |P_2|^2 + |P_3|^2)^{\frac{1}{2}}$$

and the coherence correlation factor

$$\mu = \frac{P_2 - iP_3}{\sqrt{1 - P_1^2}}$$

with $|\mu|$ as degree of coherence and ß as effective phase between two orthogonal light vectors. Complete coherence is valid for $|\mathbf{P}| = |\mu| = 1$. Such a complete coherence of the excitation-deexcitation $1^1S_o \rightarrow 3^1P_1 \rightarrow 2^1S_o$ of helium could be verified by measuring the Stokes parameters of the helium $\lambda = 5016$ Å line radiation[4].

Another method of interpreting the angular correlation data is based upon the anisotropic population of the magnetic substates of the excited state. Fano and Macek[5a] connected the anisotropic population of the magnetic substates with alignment tensors (A^{Col}) and an orientation vector (O^{Col}) of the excited atom, which can be calculated from λ and χ or from the Stokes parameters P_1, P_2 and P_3 (Standage and Kleinpoppen[4]).

A third way of parameterizing results of coherence and correlation analysis of atomic excitation in planar scattering experiments consists in representing the data in terms of the charge cloud distribution of the excited atom. By using the state vector $|\psi>$ extracted from the rate of the electronic charge $e = \int |\psi|^2 e \, d\tau$ can be represented by its angular dependence $|\psi|^2 d\tau$ at volume element $d\tau$ which provides typical pictures for charge distributions and the orientation angle γ of the electron charge in the excited state. We note the planar symmetry of the charge clouds instead of the usual cylindrical symmetries of charge clouds in atomic spectroscopy.

As pointed out by Csanak and Cartwright[11] the whole angular behaviour of the orientation or equivalently the orbital angular momentum $<L_\perp>$ transferred into the excited state is a *structure* due to a quantum mechanical interference effect among the electron partial waves without any background. This is a remarkable kind of a "pure" *Ramsauer-Townsend* effect without background as seen in the structure of the orbital angular momentum transfer $<L_\perp>$ in the electron impact excitation of $^1P_1$-states. A further interesting aspect with regard to the analysis of extracting the dynamics from "complete" collision experiments is based upon finding out to what degree or proportion the various partial electron waves contribute to the excitation process.

It could be shown by Fano and collaborators[8,9] that a partial wave of the incoming electrons with angular momentum $\ell_a - 1$ is a "parity favoured" while $\ell_b + 1$ is a "parity unfavoured" case in He(1^1P_1) excitation at 80 eV.

We would like, however, to emphasise that the most fundamental result extracted from such electron-photon coincidence experiments is the detection of the **coherent nature of the excitation process** as most clearly described by the interference between the magnetic sublevel amplitudes f_o and f_1. It is worth remembering that **coherent excitation of atoms by photon excitation** was already detected in the middle of the twenties (i.e. in the Hanle effect[10]) by measuring the de-polarization of resonance fluorescence radiation in small magnetic fields; however, coherent electron impact excitation was not detected before about the middle of the 70's.

A considerable number of studies on complete experiments by applying electron-photon coincidences have been reported and summarized. The reader will note the relevant contributions of H.-J. Beyer and A. Crowe in these Proceedings where relevant references can be found.

Other "complete" atomic collision experiments in the area of ion-atom and atom-atom collisions followed the coincidence measurements of electron impact excitation of atoms. Again they are based upon **coherent excitation** which allows extraction of complex scattering amplitudes from particle-photon angular or spin and photon polarization correlations (see, e.g., refs. 20 and 21). A simple example for such

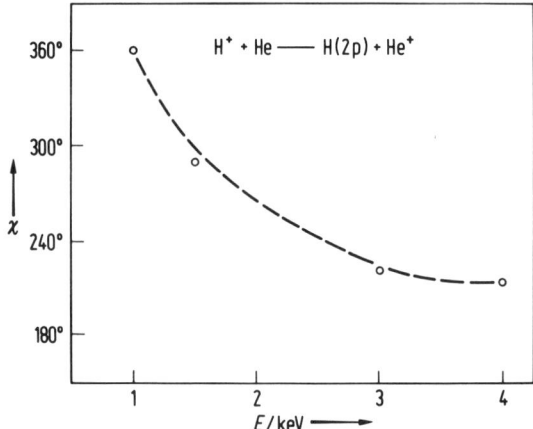

Fig.4. Relative phase χ between the magnetic sublevel amplitude f_o and f_1 as a function of the energy of the incoming photons for the same process as in Fig.3 and an impact parameter of b = ≈ 0.7 a.

measurements which can be analyzed by the above (λ,χ)-parameters is charge exchange excitation in collisions between protons (p) and atoms, i.e.,

$$P + A \rightarrow H(2P) + A^+$$
$$\hookrightarrow h\nu + H(1S) \quad ;$$

the photons of the partially polarized Lyman-α radiation and the H(1S)-atoms are measured in coincidence, Figs.3 and 4 show such an example for λ- and χ-data of proton exchange excitation with helium as a target atom. The directions of the incoming protons and the detection of the hydrogen atoms in the ground state define the planar reaction plane of the above charge exchange process. It is interesting to note that the description of this charge excitation process between protons and atoms only requires the two amplitudes f_o and $f_1 = f_{\bar{1}}$ for a complete description; interactions due to the proton spin are negligible. However, scattering processes with electron and hydrogen atoms (and other one-electron atomic or ionic systems) require separated triplet and singlet amplitudes for the elastic and inelastic interactions, e.g.,

$$e(\uparrow) + H(\uparrow) \rightarrow \quad \text{triplet processes,}$$
$$e(\downarrow) + H(\downarrow) \rightarrow \quad \text{singlet processes.}$$

We mention that superelastic electron-impact de-excitation of atoms, i.e. $h\nu + e + A^*$ $\rightarrow A + e'$, studied by optical pumping is the <u>time-inverse process</u> to the inelastic electron impact excitation investigated by the electron-photon coincidence technique. Accordingly the analysis of the superelastic technique can also lead to a complete description of excitation processes of atoms[23].

Spin reactions:

Spin-reactions with projectiles and targets of any spins are of general interest but practical cases have particularly been connected with spinless or one-electron atoms so far. The complexity of spin reactions even with these atoms is considerable.

With reference to a given axis of quantization we first consider one-electron systems with parallel and antiparallel spin, e.g. atomic and ionic projectiles (A) and target particles (B), $A(\uparrow), A(\downarrow), B(\uparrow), B(\downarrow)$ including electrons and positrons $e^{\pm}(\uparrow)$,

$e^{\pm}(\downarrow)$ and photons $h\nu(\uparrow)$, $h\nu(\downarrow)$.

By taking account of Coulomb direct and Coulomb exchange interactions only (i.e. neglecting other interactions such as spin-orbit interactions) we can list the following spin reactions and introduce relevant amplitudes and differential cross sections:

$A(\uparrow) + B(\downarrow) \rightarrow A(\uparrow) + B(\downarrow)$,   Coulomb direct amplitude f, partial differential cross section $\sigma_f = |f|^2$,

$A(\uparrow) + B(\downarrow) \rightarrow A(\downarrow) + B(\uparrow)$,   Coulomb exchange amplitude g, partial differential cross section $\sigma_g = |g|^2$,

$A(\uparrow) + B(\uparrow) \rightarrow A(\uparrow) + B(\uparrow)$,   interference amplitude f-g, partial differential cross section $\sigma_{int} = |f-g|^2$.

The normal differential cross section σ and the above differential spin cross sections can be linked to "singlet" and "triplet" cross sections as follows:

$\sigma = \tfrac{1}{2}|f|^2 + \tfrac{1}{2}|g|^2 + \tfrac{1}{2}|f-g|^2 = |f|^2 + |g|^2 - \mathrm{Re}(fg^*)$, normal differential cross section

$\sigma^{\uparrow\downarrow} = |f+g|^2 = |f|^2 + |g|^2 + 2\mathrm{Re}(fg^*) = |s|^2$,   singlet cross section
$\sigma^{\uparrow\uparrow} = |f-g|^2 = |f|^2 + |g|^2 - 2\mathrm{Re}(fg^*) = |t|^2$,   triplet cross section

----

$\sigma^{\uparrow\downarrow} - \sigma^{\uparrow\uparrow} = 4\mathrm{Re}(fg^*) = 4|fg^*|\cos\chi$
$\sigma^{\uparrow\downarrow} + \sigma^{\uparrow\uparrow} + 2|f|^2 + 2|g|^2$,

by which we define the *singlet* amplitude as $s = f+g$ and the *triplet* amplitude as $t = f-g$ following the relations

$$f = \frac{s+t}{2} \quad , \quad g = \frac{s-t}{2}.$$

By setting the direct amplitude f as real the exchange amplitude g may have a phase difference compared to the direct amplitude, i.e. $g = |g|e^{i\chi}$ (Fig.5).

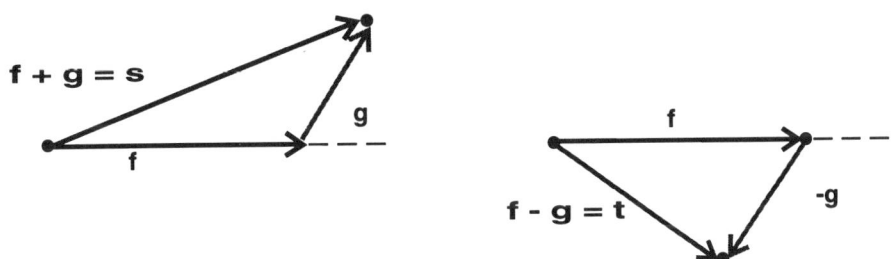

Fig.5. Geometrical representation of the amplitudes f,g,s and t by taking f as real for elastic scattering of electrons by one-electron-atoms (neglecting spin-orbit interactions).

With the above relations the differential cross section expressed in terms of singlet and triplet are

$$\sigma = \tfrac{1}{4}|s|^2 + 3/4|t|^2 = \tfrac{1}{4}|f+g|^2 + 3/4|f-g|^2 \ .$$

Full sets of $|f|$, $|g|$ and $\cos\chi$ or alternatively $|s|$, $|t|$ and $\cos\psi$ extracted from spin reactions represent a "complete" experiment (apart from knowing the signs of the phases!). Approaches, however, to such complete experiments are also valuable. e.g. measurements of $\sigma_f = |f|^2$, $\sigma_g = |g|^2$, $\sigma_s = |s|^2$, $\sigma_t = |t|^2$ or integral cross sections such as $Q_f = \iint \sigma_f d\mathbf{a}$, $Q_g = \iint \sigma_g d\mathbf{a}$, ...

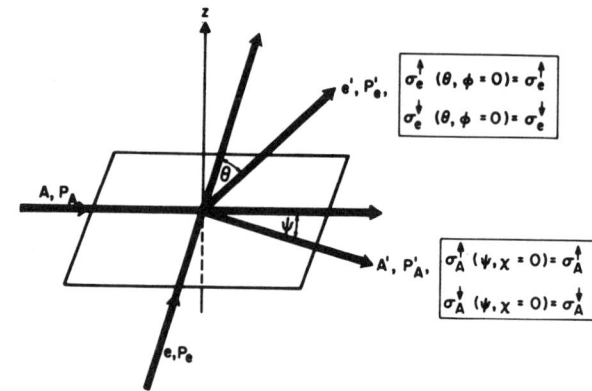

Fig.6. Scheme of the crossed-beam experiment to be analyzed with partially polarized incoming electrons e (degree of polarization P_e) and partially polarized target atoms A (degree of polarization P_A). The quantization axis is parallel to the Z axis. P'_e and P'_A represent the polarization of scattered electrons or atoms, respectively. $\sigma^\uparrow_e$ and $\sigma^\downarrow_e$ or $\sigma^\uparrow_A$ and $\sigma^\downarrow_A$ denote the partial differential cross sections for electrons or atoms with final spin up or down, respectively.

A kind of programme for complete experiments including spin-analysis of the scattering of polarized electrons by polarized atoms (one-electron atoms) can be described as follows (see Fig. 6 and Table 1): partially polarized beams of electrons are crossing a partially polarized beam of atoms.

Table 1.

Polarizations before the collision	Quantities after the collision	Information on the collision process
(1) $P_A \neq 0$, $P_e = 0$	P'_e, $\sigma(E,\theta)$	$|f|^2 = \sigma(E,\theta)\{1 - P'_e/P_A\}$
(2) $P_A \neq 0$, $P_e = 0$	P'_A, $\sigma(E,\theta)$	$|g|^2 = \sigma(E,\theta)\{1 - P'_A/P_A\}$
(3) $P_A \neq 0$, $P_e \neq 0$	$\sigma(E,\theta)$	$|f-g|^2 = \sigma + (1+P_e/P_A)(1-\sigma)$
	$I(E,\theta) = I_e^\uparrow + I_e^\downarrow$	$I(E,\theta) = \sigma(E,\theta) - P_e P_A \mathrm{Re}(f^*g)$
	$= I_A^\uparrow + I_A^\downarrow$	equivalent to $\cos\psi$

Various other cases of experimental situations indicated in the Table are connected with the analysis of spin-orbit interactions with polarizations of the electrons and atoms before and after the collisional interaction. The third column of the Table gives information on the amplitudes extracted from the spin-analysis. This kind of spin analysis summarized in the Table provides an optimum information on f, g and $\cos \chi$ of the elastic scattering of electrons by one-electron atoms (for light atoms for which spin-orbit interactions can be neglected).

Experimental programmes to such spin-analyses of the Table are rather restricted so far. However, valuable approaches towards complete experiments have already been reported; for example, a series of measurements of absolute integral exchange cross sections Q_g have been published by the NYU group[24] and the Edinburgh group[25]. Measurements of absolute differential exchange cross sections have also been made by Bederson and collaborators[24] with errors estimated to almost 30%. Ratios of elastic differential cross sections have been determined based upon the spin reaction of the first row in the Table by a JILA/Stirling cooperation[26] (Fig.7). Recent experiments[27] with polarized electrons and optically pumped sodium atoms resulted in measurements

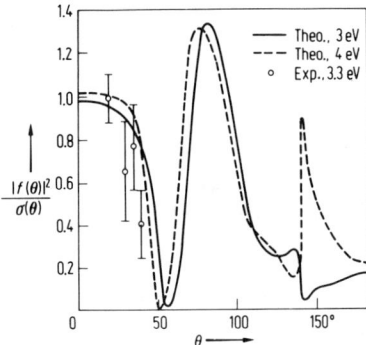

Fig.7. Experimental (Hils et al.[26]) and theoretical data (Karule and Peterkop) for $|f|^2/\sigma(\theta)$. The theoretical values of $|f|^2$ and $\sigma(\theta)$ used in this paper have been computed by Hils et al.[26] from reactance matrix elements calculated by Karule and Peterkop using a two-state close-coupling approximation. For partial waves $\ell = 0$ through $\ell = 3$ these matrix elements are published in <u>Atomic Collisions III</u>, edited by V. Ia. Veldre (Latvian Academy of Sciences, Riga, 1965) [Translation TT-66-12939 available through SLA Translation Center, John Crerar Library, Chicago]. For partial waves $\ell = 4$ through $\ell = 8$ the matrix elements were received by private communication from Karule and Peterkop.

of the ratios $|s|^2/|t|^2$ for elastic scattering and $\{|t^1|^2 + |t_{-1}|^2\}/\{|s_1|^2\}$ for inelastic excitation of the first ^{2}P-states; the indices ± 1 in the singlet and triplet amplitudes refer to the magnetic number m_ℓ of the P-states.

While information on complete sets of amplitudes of elastic and inelastic electron-atom scattering is still rather sparse a relative large body of data on *ionization asymmetries* is now available. Ionization asymmetries A are related to the number of ions produced by electron impact with the spins either parallel or anti-parallel to each other, i.e. by assuming spin conservation this results in the reactions

$$e(\uparrow) + A(\uparrow) \to A^+ + e(\uparrow) + e(\uparrow) \propto N^{\uparrow\uparrow}(A^+)$$
$$e(\uparrow) + A(\downarrow) \to A^+ + e(\uparrow) + e(\downarrow) \propto N^{\uparrow\downarrow}(A^+)$$

$N^{\uparrow\uparrow}(A^+)$ and $N^{\uparrow\downarrow}(A^+)$ are the number of ions produced of the first and second reaction, respectively. While a final complete analysis on these processes would require spin-resolved (e,2e)-experiments useful information is obtained by determining the ionization asymmetry.

$$A = \frac{N^{\uparrow\downarrow} - N^{\uparrow\uparrow}}{N^{\uparrow\downarrow} + N^{\uparrow\uparrow}}$$

Such quantities were originally first measured for atomic hydrogen (Yale University[28]) and sodium atoms (Stirling University[29]) followed by a comprehensive series of measurements on alkali atoms by Prof. Raith's group of Bielefeld University[30]. The threshold behaviour of the ionization asymmetry is of special interest. The partial waves ^{1}S, ^{3}P, ^{1}D, ^{3}F, ... of the scattered electrons can contribute at threshold resulting[31] in A = 1 for singlet waves and A = -1/3 for triplet waves. New types of (e,2e)-coincidence experiments with detection of polarized electrons will be promising as a new tool investigating ionization by electron impact.

While electrons colliding with atomic hydrogen and light alkali atoms interact by Coulomb direct and exchange reactions heavier atoms require in addition spin-orbit or even spin-spin potentials for a correct description of the scattering mechanism. In this connection I refer to the paper of Prof. Raith in these Proceedings and like to add some results on spin flip measurements in elastic polarized electron rare gas and mercury scattering by the Münster group[32]. Two spin reactions can be defined for the scattering of polarized electrons on spinless heavy atoms: As an example we refer to the scattering of spin-polarized electrons on heavy atoms with zero-spin:

(1) $e(\uparrow) + A \rightarrow A + e(\uparrow)$,   amplitude h
(2) $e(\uparrow) + A \rightarrow A + e(\downarrow)$,   amplitude k.

The first process is a "*direct*" one with the amplitude h; the second process is a "*spin-flip process*" with the amplitude k. We note that the "direct" process can coherently be superposed with an "exchange" process; however, these two processes cannot be separated from each other in this reaction because they are indistinguishably superposed. A detailed analysis of the spin components after the scattering allows to extract the moduli of the direct and spin flip amplitude and their phase difference (Fig.8). We note a pronounced interference structure in the moduli of the direct amplitude which is due to the superposition of several partial waves (as with the Ramsauer-Townsend effect!). The moduli of the spin-flip amplitude k which are caused by spin-orbit interaction are much smaller than those of the direct amplitude k; the spin-flip amplitude is primarily determined by the ($\ell=1$)-partial wave which, as a result, shows very little interference structure in the data.

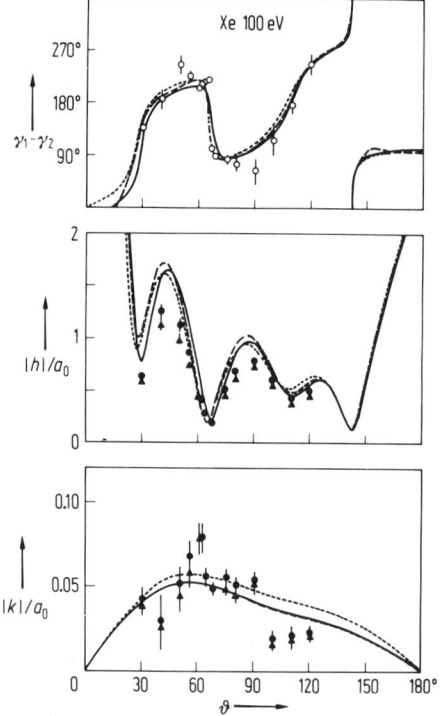

Fig.8. Moduli of the amplitudes h and k and their phase differences $\gamma_1 - \gamma_2$ for elastic scattering of polarized electrons by xenon atoms at an energy of 100 eV; experimental data points after Berger and Kessler[32]. The short and long dotted and full curves are theoretical predictions of, respectively, McEachran and Stauffer[33], Haberland et al.[34] and Awe et al.[35]. The data for $|h|$ and $|k|$ are represented in units of the Bohr radius a_0 and normalized to the measured cross section $\sigma = |h|^2 + |k|^2$; ϑ is the scattering angle.

The measurement of the complex amplitudes h and k represents a "complete experiment" in the meaning defined in the introduction of this article. "Complete", however, refers to the limits of the physical method applied which allows determining h and k but leaves out to separate the coherently superposed Coulomb direct and exchange interactions in the "direct" process as discussed above.

Obviously spin experiments and polarized photon-particle coincidences are well suited for analysis as complete experiments. Groups in Bielefeld[36] and Münster[37] are particularly involved in spin-parallel-antiparallel asymmetries in (e,2e)-collisions involving polarized electrons and polarized atoms. An Edinburgh/Stirling group is is carrying out electron-photon coincidence measurements with spin-polarized electrons and spin-polarized alkali-atoms. The analysis of such spin combined electron-photon coincidence experiments require detailed scattering geometries from which full sets of moduli of direct, exchange and their interference amplitudes $|f_m|$, $|g_m|$, $|f_m - g_m|$ or alternatively singlet and triplet amplitudes $|s_m|$ and $|t_m|$ can be obtained for the magnetic sub-states $m_t = 0$ and $m_t = \pm 1$ of the excited P-states can be obtained. Of course such spin combined electron-photon coincidence experiments for atomic hydrogen and light to medium heavy alkali atoms will certainly be possible in the near future since coincidences have already been observed recently for polarized alkali targets with unpolarized electrons as will be possible with spin polarized electrons (see paper by M.A. Chaudhry of these Proceedings).

Angular correlations of (e,2e)-processes for one-electron atoms can also be described by Coulomb direct exchange and interference reactions:

$$e_1(\uparrow) + A_2(\downarrow) \rightarrow A^+ + e_1(\uparrow) + e_2(\downarrow), \qquad f = \frac{s+t}{2}$$

$$\rightarrow A^+ + e_2(\downarrow) + e_1(\uparrow), \qquad g = \frac{s-t}{2}$$

$$e_1(\uparrow) + A_2(\uparrow) \rightarrow A^+ + e_1(\uparrow) + e_2(\uparrow), \qquad f - g = t$$

such processes for measuring the amplitudes have not yet been studied so far but relevant spin asymmetries have been measured[38], i.e. ratios $(\sigma^{\uparrow\downarrow} - \sigma^{\uparrow\uparrow})/(\sigma^{\uparrow\downarrow} + \sigma^{\uparrow\uparrow}) = (1/P_e)A = (N^{\uparrow\downarrow} - N^{\uparrow\uparrow})/(N^{\uparrow\downarrow} + N^{\uparrow\uparrow})$.

To select some further examples of electron and photon spin effects we briefly report the following examples: spin effects in *Auger electrons* as well as in *Bremsstrahlung* from polarized electron impact and in photon radiation from trapped ions and ion collision processes. While this selection appears to be arbitrary in a way, however, they demonstrate significant characteristics and examples in which spin effects are decisive for a more detailed understanding of the physics involved. These examples also supplement to a certain degree topics missing in these Proceedings.

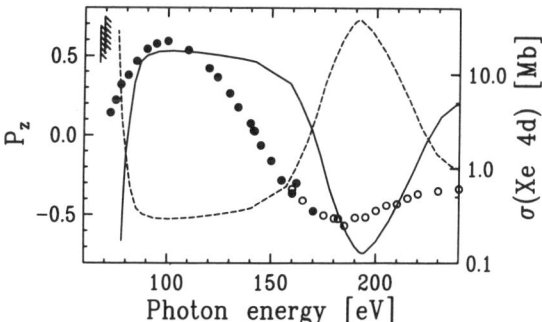

Fig.9. Spin polarization of Auger electrons produced by circularly polarized photons of various energies. The spin polarization P_z refers to the direction of the incoming circularly polarized photons. The Auger electrons are observed under 90° with reference to the incoming photons. Full curve, Xe $N_4O_{2,3}O_{2,3}$ ($^1S_0$)-Auger electrons; broken curve, $XeN_5O_{2,3}O_{2,3}$ ($^1S_0$)-Auger electrons. The points represent experimental data for the Xe 4d partial photoionization cross section: full circles, Becker et al.[39]; open circles, Linde et al.[40].

With regard to *spin polarization of Auger electrons* Lohmann et al[38] theoretically treated the Auger decay after photoionization of noble gases as a two step process. The theoretical expressions for their calculation are determined by products of the orientation (A_{10}) and alignment parameters (A_{20}) characteristic for the intermediate ion state, and relevant parameters depending on the Auger matrix transition elements. Considerable spin polarization of Auger electrons from photoionization of atoms by circularly polarized light has been predicted for Ar $L_3M_{2,3}M_{2,3}$-, Kr $M_{4,5}N_{2,3}N_{2,3}$-, Xe $N_{4,5}O_1O_{2,3}$- and $XeN_{2,3}O_{2,3}$- Auger electrons. Fig.9 shows an example. It is particularly encouraging for future experimental tests of this theory that the broad maximum of the Auger spin polarization is in the region of the maximum of partial cross section $\sigma(Xe,4d)$.

In order to investigate *spin-orbit interactions in electron-atom Bremsstrahlung* various spin-dependent experiments have been carried out. Spatial left-right asymmetries of the emission of Bremsstrahlung radiation induced by transversely polarized electrons have already been reported by various groups[41-45] previously. In such experiments, however, only the emitted Bremsstrahlung radiation has been observed disregarding the decelerated outgoing electrons. The group of W.Nakel[46] extended the art of technique by carying out electron-photon coincidence experiments with transversely polarized primary electrons inducing the Bremsstrahlung (i.e., the spin of the electrons is perpendicular to the plane defined by the detected scattered electrons and the Bremsstrahlung photons). Fig.10 shows experimental and theoretical data for the Bremsstrahlung of the photon emission asymmetry for a fixed electron scattering angle ($45°$), a fixed energy of 200 keV of the outgoing electrons and 300 keV electrons impinging on a solid gold target. While calculations in first Born approximation which results in the classical Bethe-Heitler formula[47] gives zero photon asymmetry, the Elwert-Haug[48] shows quantitatively reasonable agreement[49] with the experimental data. On the other hand the validity of the Elwert-Haug theory is limited to $Z\alpha \ll 1$ which is not fulfilled for the $Z = 79$ target and the fine structure constant α. Accordingly it would be desirable to apply relativistic theories (e.g. the one of Tseng and Pratt[50]) for calculating such spin asymmetries in Bremsstrahlung).

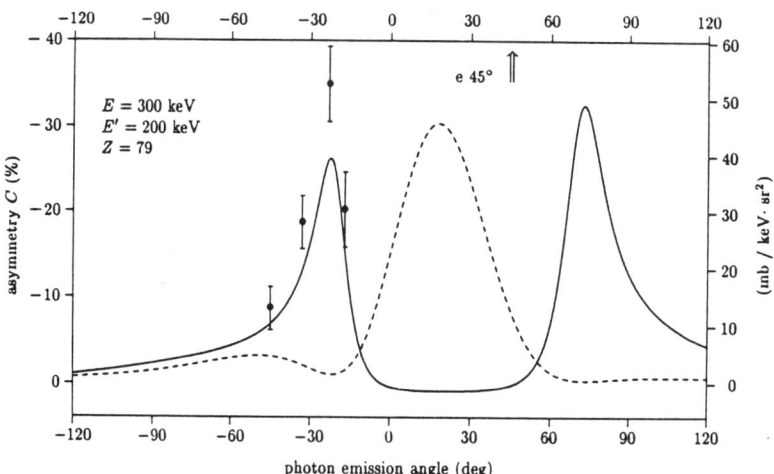

Fig.10. Photon emission asymmetry of Bremsstrahlung induced by electrons scattered at a fixed angle of $45°$ which are initially spin-polarized perpendicular to the reaction plane; the inelastically scattered electrons and the Bremsstrahlung photons are detected in coincidence. E is the energy of the incoming electrons, E' the energy of the outgoing electrons, Z = 79 of the gold target. The solid line is the theoretical result of Haug[49], the dotted line that of Elwert and Haug[48] for the triply differential cross section with unpolarized primary electrons. The experimental data points are those of Nakel's group[46].

Before I quote further selected examples of spin-polarized electron/photon physics let me summarize amplitude descriptions in electron and positron scattering by toms (Table 2). While Coulomb exchange interaction only occurs in electron-atom collisions positronium formation only takes place in positron-atom collisions. Spin-orbit interactions will be dealt with in Prof. Raith's talk which is based upon theories of Burke and Mitchell[51], Farago[52] and Khalid and Kleinpoppen[53].

Table 2: Amplitudes in Electron and Positron Scattering by Atoms

	Electron Scattering:	*Positron Scattering*
Coulomb direct interaction	f, f_m^3	f, f_m^3
Coulomb exchange interaction	g, g_m^3	———
including spin-orbit interaction	h, k^{32} $a_1-a_6)^{51,52}$, $f_1 f_2, g, h_1, h_2, k, m^{53}$	h, k^{32} $a_1-a_6{}^{51,52}$, $f_1, f_2, g, h_1, h_2 k, m^{53)}$

Without spin-orbit interaction the amplitudes for Coulomb direct and exchange processes or singlet and triplet processes describe the physics involved; in a simple way, however, tensor quantities describing spin correlations, depolarizations and polarization transfers are required for taking into account spin-orbit interactions [i.e. terms proportional to $\underline{s}\ell$, with the effective spin $\underline{s}$ and orbital quantum number ℓ in the collisional process; it might, however, be physically more appropriate to relate an amplitude description with reference to the components $s_x \ell_x$, $s_y \ell_x$ and $s_z \ell_z$ of the $(s.\ell)$ interaction of the outgoing electron!.

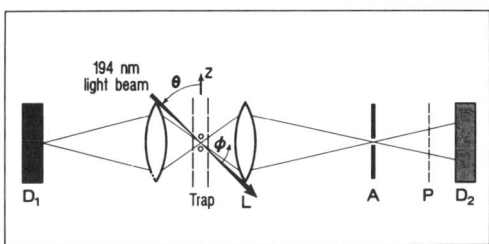

Fig.11. Scheme of the experiment for detection of Young's interference with light scattered from two ^{198}Hg$^+$ ions in a linear Paul trap (represented by two small circles near the centre of the trap). The scattered light induced by the incoming 194nm light beam is focussed onto the detector D_2 via a collecting lens L. An optical polarizer and an aperture lens L is inserted in the detected light beam. Detector D_1 serves as monitor of the ion number in the trap.

Another striking example of polarization effects which is the effect of photon polarization on Young's interference with light scattered from two trapped ions [54]. In the experiment to be described Young's interference of light scattered from two ^{198}Hg$^+$-ions in a linear Paul trap has been observed[55]. The two ions were trapped along the axis of the linear trap (Fig.11). The ions were irradiated by linearly polarized laser light so the ^{198}Hg$^+$ - $6s^2S_{1/2}$-$6p^2P_{1/2}$ fluorescence transition at 194nm could be observed. The laser beam reduces the ions' kinetic energy by Doppler laser cooling[56]. The low temperature in the trap strongly localises the ions. The laser beam also acts as the light source for the interference experiment. The experiment resembles Young's classic two-slit interference experiment with the two slits replaced by two ions. The ^{198}Hg$^+$ ground state $6s^2S_{1/2}$ and the excited $6p^2P_{1/2}$ ions are both twofold degenerate with regard to the magnetic quantum number m_j. Scattering of linearly polarized light from the two ions results either in π- or σ-scattered light. By assuming that only *one* photon is scattered in the time interval of observation the following conclusions can be drawn. For π-polarized scattered light ($\Delta m_j = 0$ transition) the ions' final states, i.e. the m_js are the same. Accordingly it cannot be determined which of the two atoms scattered the light. Therefore quantum mechanics predicts that interference takes place in the π-light scattered from the two ions. On the other hand observation of σ-polarized scattered light implies that the final state of one atom differs from its initial states in their quantum number which enables us to distinguish the scattering atom from the spectator atom and to determine "which path" or "which atom" the photon has chosen for its transition (i.e., the one atom for the possible transition and not the other one! Consequently there is no interference in the σ-light scattered from the two ions. Fig.12 clearly demonstrates Young's interference for π-light but not for σ-light scattering.

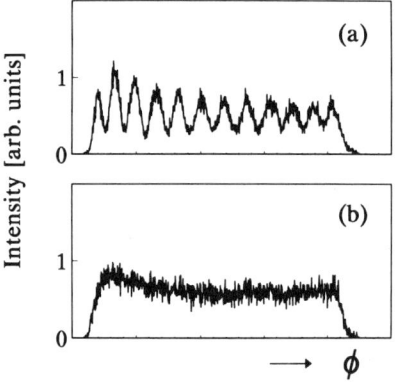

Fig.12. Polarization-sensitive detection of light scattered from two trapped ^{198}Hg$^+$ ions: (a) π-polarized scattered light, and (b) σ-polarized scattered light from the ($6s^2S_{1/2}$ - $6p^2P_{1/2}$) transition.

While spin effects in electron-atom collisions and photoionization of atoms have already reached some advanced state of investigations spin effects in heavy-particle atom collisions have less extensively been studied. Of course, there are similarities and differences between the above fields which are of interest in connection with the analysis of interactions by spin effects. Already towards the end of the 60's alkali-alkali differential spin-exchange scattering have been carried out[57] in the thermal energy range. The difficulties in the analysis of such spin-polarized atom experiments are due to the fact that it is generally impossible to extract potentials from cross sections when quantum effects are important in collision processes. Pritchard et al.[57] examined approximation models which apply to exchange processes in general and analyzed single-potential scattering with reference to spin exchange data. Relative phase shifts between triplet and singlet potentials for Li, Na, K, Rb and Cs scattering on the same and heavier alkali atoms were reported. While these data are analyzed within certain (restrictive) approximations mentioned above triplet and singlet amplitudes extracted would be free of such limitations. In principle spin-polarized alkali-atoms scattered by spin-polarized alkali atoms can be analyzed in the same way as for scattering of spin-polarized electrons by spin-polarized one-electron atoms described in Tables 1 and 2. However, as pointed out by Lutz[58], complications arise from the coupling between the motions of the electrons and nuclei in the quasi-molecular collision region at lower impact energies. The processes

$$\left.\begin{array}{l} He^+ + H \\ H^+ + He \end{array}\right\} \to H(2P) + He^+$$

have recently[59] been studied by measuring the linear polarization of the Lyman-α radiation emitted from the 2P-state of atomic hydrogen. The integral alignment $A_{20} = Q_1 - Q_0)/(Q_0 + 2Q_1)$ with the magnetic sublevel cross sections Q_m shows pronounced differences between the $He^+ + H$ and $H^+ + He$ collisions (Fig.13). By means of a semiclassical approximation it could be shown that the adiabatic potential diagrams for spin-singlet and spin-triplet systems of the above collisions show different features. It appears from the theoretical interpretation that these differences are of basic importance for the understanding of the collision dynamics of excitation and electron capture processes. However, as in the above collision between spin-polarized alkali atoms the experimental data are interpreted in terms of spin effects based upon theoretical

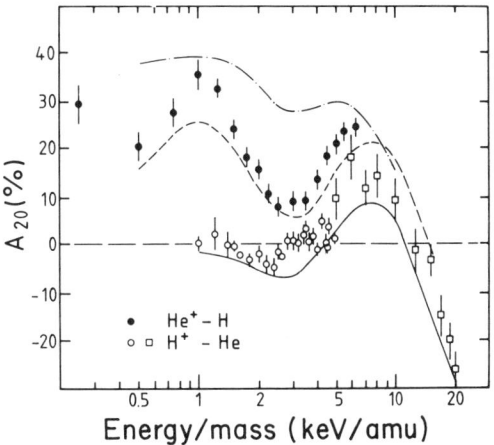

Fig.13. Integral alignment A^{20} for H(2P) production in $H^+ + H$ collisions vs. incident energy. Experimental results of Hippler et al.[59]($\bullet$) for He^+ + collision and for $He^+ + He$ collisions (O, ref.60; □ ref.61) are compared with calculations of Hippler et al.[59] (dashed and solid lines) and of Errra et al[62] (dash-dotted line for $He^+ - H$).

argumentations. We suggest that future amplitude analyses of the above excitation of H(2P) should be based upon a collision system scattering of electrons with one-electron atoms, i.e.

$$He^+(\uparrow) + H(\downarrow) \to + H(2P,\downarrow) + He^+(\uparrow), \quad \text{amplitude} \quad f = \frac{s+t}{2}$$
$$\to H(2P,\uparrow) + He^+(\downarrow), \quad \text{amplitude} \quad g = \frac{s-t}{2}$$
$$He^+(\uparrow) + H(\uparrow) \to H(2P,\uparrow) + He^+(\uparrow), \quad \text{amplitude} \quad f - g = t;$$

again the amplitudes f,g,s and t refer to Coulomb direct and exchange interactions or, alternatively, singlet and triplet interactions. The other spin collision process can be described by the following amplitudes:

$$H^+ + He(\uparrow\downarrow) \to H(2P,\uparrow) + He^+(\downarrow), \quad \text{amplitude} \quad h$$
$$\to H(2P,\downarrow) + He^+(\uparrow), \quad \text{amplitude} \quad k.$$

Because of symmetry arguments with regard to the spins the amplitudes h and k should be identical which could be tested by showing experimentally that the circular polarization of the Lyman-α radiation vanishes.

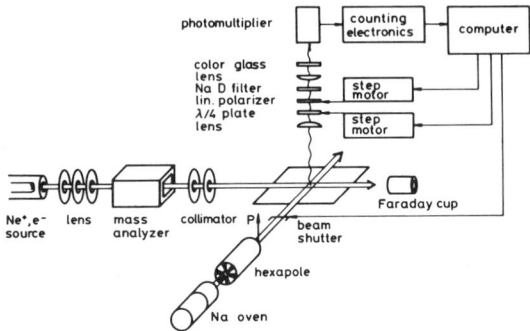

Fig.14. Experimental settings for a crossed-beam apparatus with partially polarized sodium atoms crossed by a rare gas ion beam[63],[65]. The D-line radiation is observed perpendicular to the reaction plane in the direction of the spin polarization of the sodium atoms.

Another type of spin experiment with spin polarized sodium atoms in collisions with rare gas ions has been reported by Bielefeld groups[63-65]. Again the spin effects resulting from such collisions can be described by direct, exchange and interference interactions if the rare gas ions are considered as one-hole systems (excluding He$^+$ which is a one-electron ion). The interesting observation in such experiments so far is the observation of circular polarization of the radiation of sodium D-lines with partially polarized sodium atoms in the ground state excited by rare gas ions (Fig.14). Due to the asymmetric population of the spin states of sodium ground state so will the upper excited P states of sodium have an asymmetric spin population which results in asymmetric emission of σ-light shown in the circular polarization of the D-line radiation (Fig.15).

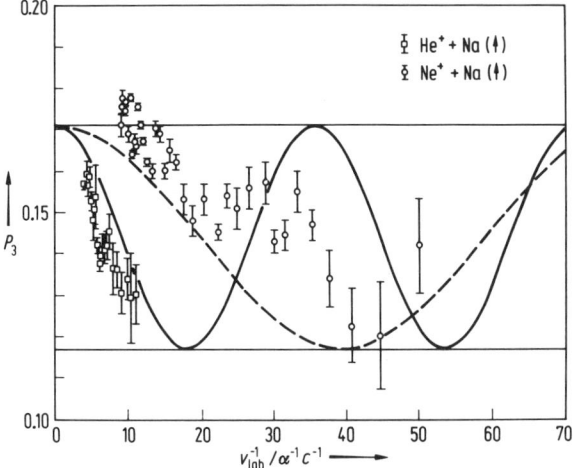

Fig.15. Experimental and theoretical data of the circular polarization of the sodium D-line radiation as a function of the inverse velocity of He$^+$- and Ne$^+$-ions. The upper horizontal line corresponds to the limit of Coulomb direct interaction (spin conservation) while the lower horizontal line refers to Coulomb exchange interaction with spin change α is Sommerfeld's fine structure constant. The theoretical curves have been calculated from a simple model of phase shifts between singlet and triplet scattering under the assumption of a motion in straight lines of the colliding nuclei for small impact parameters; dotted curve for Ne$^+$, full curve for He$^+$.

The experimental data vary between the case of a Coulomb direct process (spin not changed) and maximum depolarization of the circularly polarized emission for the Coulomb exchange process.

While progress has been made with spin effects in ion-atom collisions a full analysis of amplitudes is still far away from complete. Efforts in the future, however, are promising in particular with one-electron ions as projectiles such as Sr$^+$-ions for which about 30% spin-polarization has been obtained at an intensity of 10μA in the 5S-ground state[66].

Interactions of Polarized Electrons and/or Polarized Photons with Surfaces and Solid State Targets.

The concept of spin effects in electron and photon interactions with condensed matter is related to that of free atoms and molecules but is, in many aspects, different and more complicated. The Workshops deal with examples and show how important spin effects have been applied for analyzing the physics of electron/photon interactions with matter and their structures. While my few references given as examples are personal and selective it will demonstrate some differences in the physical processes compared with targets of free atoms/molecules and solids.

Photoemission of electrons from atoms and solids induced by circularly polarized photons links physics of polarized photons with that of polarized electrons. Fano[67] suggested the concept of producing polarized electrons by photoionization with circularly polarized photons. It was first confirmed experimentally for free cesium atoms[68] and for a Cs metal layer[69]; however, the underlying spin-orbit interaction of the Fano effect has, however, been detected prior to that in a photoionization experiment using spin-polarized heavy alkali atoms and circularly polarized light[70]. Angle and spin-resolved photo-electron spectroscopy has resulted in many applications of photoionization of atoms, molecules and solids[71]. Another type of experimental technique has been involved with photoionization of polarized atoms[72] which leads to "*complete*" descriptions without spin-orbit effects[73].

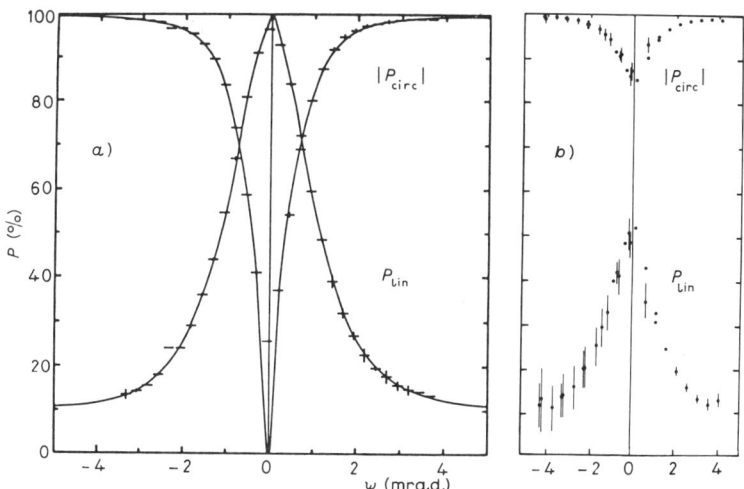

Fig.16. Circular and linear polarizations, P_{lin} and P_{circ}, as a function of the vertical angle of ψ(± 0.1 mrad) for the synchrotron radiation (a) with wavelength 50nm and (b) for ψ(± 5 mrad) at wavelength 100nm[75-78].

Some recent examples of *photoemission of electrons from solids and adsorbates* show distinct differences compared to photoionization of free atoms and molecules; they also demonstrate how atomic effects can be evident in photoionization of condensed matter. Amongst many possible examples I refer to spin-resolved photoemission of electrons of rare-gas atoms and adsorbates induced by circularly polarized light. Since most atoms and molecules have ionization energies in the vacuum ultraviolet region conventional methods for producing circularly polarized light break down. However, synchrotron radiation is linearly polarized in the plane of the storage ring but elliptically polarized with a high degree of right (left) handed circular polarization when observed above (below) the plane. Fig.16 gives an example for the degrees of circular and linear polarization of 50nm and 100nm synchrotron radiation measured behind a monochromator[74-75]. Typical intensities available which depend on various parameters of the set up are 5.10^{11} synchrotron photons s^{-1} with a circular polarization of 93%.

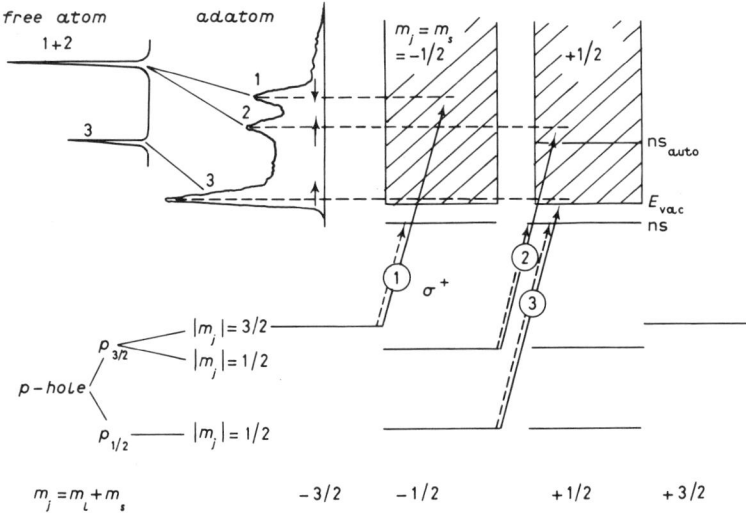

Fig.17. Energy diagram for photoionization of rare-gas atoms and rare-gas adsorbates with circularly polarized light (selection rule $\Delta m_j = +1$ for σ^+-transition). The three transitions 1-3 refer to the three peaks in the photoelectron spectrum of rare-gas adsorbates[77] while the peaks 1 and 2 coincide for free atoms [78]. The arrows with dotted lines refer to atomic transitions (after U.Heinzmann[79]).

Photoionization of rare-gas atoms induced by radiation of about 15 eV results in a p-hole with a fine structure splitting due to the possible ground state configurations $^2P_{3/2,1/2}$ of the produced ion. Accordingly two separate peaks can be detected by appropriate photoelectron spectroscopy (Fig.17, upper corner on the left). The photo-electrons of peak 3 emitted in forward or backward direction with reference to the incoming beam of the synchrotron radiation have been found to be completely spin-polarized[74,76]. This is due to the fact that no orbital angular momenta are allowed for photoelectrons in these two directions (i.e. $m_\ell \neq 0$ vanish so that only $m_s = +1/2$ electrons in the continuum state for right-handed circularly polarized ($\Delta m_j = 1$) are possible in the $P_{1/2}$ - wave, and, alternatively, for left-handed circularly polarized light with the selection rule $\Delta m_j = -1$ only $m_s = -1/2$ photoelectrons are allowed. Transitions from the $P_{3/2} |m_j| = 3/2$ and $|m_j| = +1/2$ states are associated with photoelectrons of opposite but complete spin polarization. Due to surface potential effects this symmetry with regard to spin effects of photoelectrons is broken for the case of adsorbates which results in a further splitting of the energy levels as illustrated in Fig.17. Fig.18 presents photoelectron spectra of Kr and Xe on Pt(111) as a full monolayer coverage of the adsorbate (lower part of the figure). Partial spectra for spin-up and spin-down

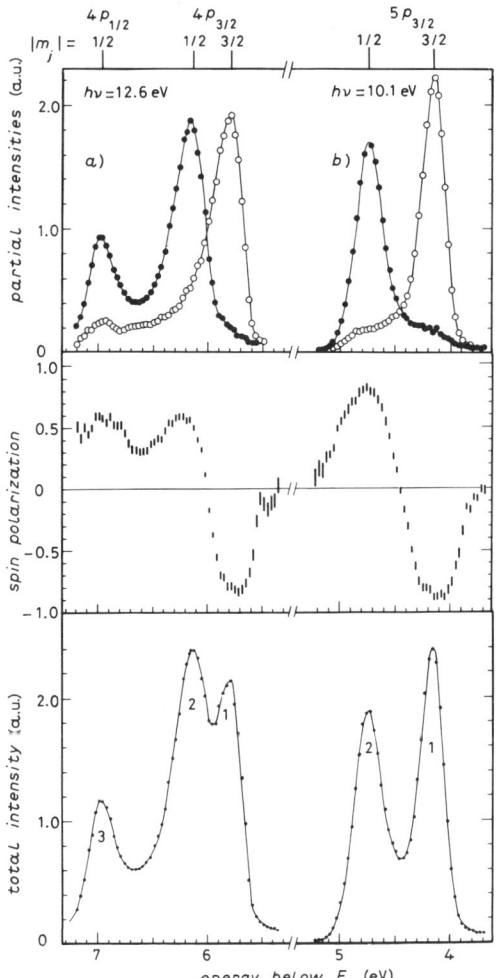

Fig.18. Spin-resolved photoelectron spectra Upper part: monolayer of Kr and Xe on platinum Pt(111); a) for Kr with photoenergy hv = 12.6 eV and b) for Xe and hv = 10.1 eV; full points for electron spin up, open points for electron spin down. Middle part: electron spin polarization. Lower part: relative total intensities of the photoelectron spectra (spin-independent). The energy of the abcissa refers to the electron energy below the threshold energy (after Heinzmann[79]).

polarizations are shown in the upper part of Fig.18. The peaks 1,2 and 3 can be characterized by the atomic hole configurations as given on top of Fig.18 and schematically introduced in Fig.17. The sequence of the m_j quantum numbers is interpreted such that 1-2 splitting is caused by lateral adatom interactions or crystal-field splitting. This interpretation is supported by the further experimental result that the 1-2 splitting vanishes in the dilute adsorption systems of xenon atoms on Pt(111) where the distance between the neighbours of the dilute system is much larger[80].

The phenomena of spin-polarized electrons from photoionization of atoms, adsorbates and nonmagnetic solids is rather general than exceptional. Cross-comparisons between results of photoionization of atoms and adsorbates are particularly useful for quantum mechanical understanding towards a complete way.

CONCLUSIONS

There are, of course, many more electron and photon spin effects in the selected areas of our Workshops which could or should be mentioned either in the talks or in my own one. However, the aim of our Workshops was not to be comprehensive but selective in the way that we look forward to exciting new and promising developments in the near future. If colleagues learn from each other and are stimulated to new ideas in these fields of research we should be satisfied with the outcome of our meetings and will carry on with exchange of knowledge and experience.

Considering the limitations of the programmes of the Workshops I cannot resist to mention some of the topics not reported. A tentative list of such topics is:
Anisotropic magneto-crystalline energies of multilayers have been calculated with the relativistic spin-polarized band theory[81].
Effects of intense lasers on bound-electron spin[82].
Differences in "spin-up" and "spin-down" angular intensities of photoelectrons emitted from inner core levels of atoms in a magnetic crystal gives a spin hologram[83].
Resonance enhancement of relativistic effects (i.e. spin effects) for scattering of very slow electrons by heavy atoms[84].
Light scattering of heavy atoms[85].
Electron capture at relativistic energies[86].
The dynamics of spin density waves[87].
Spin-down of radio pulsars: the slowing down of the pulsar rotation is connected with electron and positrons circulating in the pulsar magnetosphere[88].

REFERENCES

1. U.Fano, Revs.Mod.Phys., 29:74 (1957).
2. B.Bederson, Comments At.Mol.Phys. 1,41 and 2:65 (1969).
3. H.Kleinpoppen, Phys.Rev. A3:2015 (1971).
4. M.C. Standage and H.Kleinpoppen, Phys.Rev.Lett. 36:577 (1976).
5. N.Anderson, I.V.Hertel and H.Kleinpoppen, J.Phys.B. 17:L901-908 (1984).
6. U.Fano and J.H.Macek, Rev.Mod.Physics 45:553 (1973).
7. N.O.Anderson, J.W.Gallagher and I.V.Hertel, Physics Reports 165:1 (1988).
8. J.Bohn and U.Fano, Phys.Rev.A. 41:5953 (1990).
9. X.C. Pan and A.Chakratorty, Phys.Rev.A. 41:5962 (1990).
10. W.Hanle, Zeitschr. f. Physik. 30:93 (1924).
11. G.Csanak and D.C.Cartwright, XVth ICPEAC, Brighton (1987) (H.B.Gilbody, W.R.Newell, F.H.Read and A.C.H.Smith, eds.), and in Proceed.Int.Symp. on Correlation and Polarization in Electronic and Atomic Collisions, Belfast (1987), (A.Crowe and M.R.H.Rudge, eds.), World Scientific, p31.
12. M.Born and E.Wolf, Principles of Optics, Pergamon Press, Oxford.
13. J.W. McConkey, P.J.M.van der Burgt, J.J.Corr and P.Plessis, Proceed.Int.Symp. on Correlation and Polarization in Electronic and Atomic Collisions (P.A.Neill, K.H.Kelley, eds.) NIST Special Publication 789:115 (1990).
14. J.A.Slevin and S.Chwirot, J.Phys.B 23:165 (1990).
15. A.Crowe and M.R.H.Rudge, Comments At.Mol.Phys. 22:147 (1988).
16. J.Kessler, Advances At.Mol. and Optic.Phys. 27:81 (1991).
17. H.W.R. MacGillivray and M.C.Standage, Comments At.Mol.Phys. 26:179 (1990).
18. J.Slevin, Rep.Progr.Phys.47:461 (1984).
19. H.Kleinpoppen in Bergmann/Schaefer, Lebruch der Experimentalphysik, Vol.4:1, de Gruyter, Berlin (W.Raith, ed.) (1992).
20. R.Hippler, J.Phys.B 26:1 (1993).
    H.O.Lutz, in "Correlations and Polarization in Electronic and Atomic Collisions and (e,2e) Reactions" (eds. P.J.O. Teubner, E.Weigold), Institute of Physics. Conf.Series 122;335 (1992).

21. R.Hippler, M.Faust, R.Wolf, H.Kleinpoppen, H.O.Lutz., Phys.Rev. 36A:4644 (1987).
22. J.Majek, C.Wang, Phys.Rev.A 34:1787 (1986).
23. I.V.Hertel and W.Stoll, J.Phys.B 7:570 (1974).
24. R.E.Collin, B.Bederson, M. Goldstein, Phys.Rev. A 3:1976 (1971).
25. D.M.Campbell, H.M.Brash, P.S.Farago, Proceed.Roy.Soc.
26. D.Hils, M.V.McCusker, H.Kleinpoppen, S.J.Smith, Phys.Rev.Lett. 29:398 (1972).
27. J.J.McClelland, M.H.Kelley, R.J.Celotta, Phys.Rev.A. 40:2321 (1998), and M.H.Kelley, J.J. McClelland, S.R.Lorentz, R.E.Scholten and R.J.Celotta in Correlations and Polarizations in Electronic and Atomic Collisions and (e,2e) Reactions, eds. P.J.O.Teubner and E.Weigold, IOP Conference Series 122:23 (Bristol, 1992).
28. M.J.Alguard, V.W. Hughes, M.S.Lubell, P.F.Wainwright, Phys.Rev.Lett. 39:334 (1977).
29. D.Hils, Paris Satellite Meeting of 10th ICPEAC, unpublished (1977) and D.Hils, H.Kleinpoppen, J.Phys.B 11:L283.
30. W.Raith, in Fundamental Processes of Atomic Dynamics, eds. J.S. Briggs, H.Kleinpoppen, H.O. Lutz, Plenum Press, NATO ASI Series, Vol.181:429.
31. A.D. Stauffer, Phys.Lett. 91A:114 (1982) and C.H.Green and A.R.P.Rau, J.Phys.B 16:99 (1983).
32. O.Berger, J.Kessler, J.Phys.B 19:3539 (1986).
33. R.P.McEachran, A.D.Stauffer, J.Phys.B 19:3523 (1986).
34. R.Haberland, L.Fritschel, J.Noffke, J.Phys.A 33:2305 (1986).
35. B.Awe, F.Kemper, F.Rosicky, R.J.Feder, J.Phys.B 16:603 (1983).
36. G.Baum, W.Blask, P.Freienstein, L.Frost, S.Hesse, W.Raith, Verhandl. Deutsch. Physik.Ges. 7:677 (1991).
37. G.F. Hanne, "Correlations and POlarization in Electronic and Atomic Collisions and (e,2e) Reactions", IPC Series No.122, eds. P.J.O. Teubner, E.Weigold ,p15 (1992); S.Jones, D.H.Madison, G.F. Hanne, to be published.
38. B.Lohmann, U.Hergenhahn, N.M.Kabachnik, J.Phys. B 26:3327 (1993).
39. U.Becker, D.Szostak, H.G. Kerkhoff, M.Kupsch, B.Langer, R.Wehlitz, A.Yagishita, T.Hayaishi, Phys.Rev.A 39:3902 (1989).
40. D.W.Lindle, T.A.Ferrett, P.A.Heimann, D.A.Shirley, Phys.Rev.A 37:3808 (1988).
41. K.Güthner, Zeitschr. f. Phys.182:278 (1965).
42. Penczynski, H.L.Wehner, Zeitschr. f.Phys. 237:75 (1970).
43. H.Aehlig, Zeitschr. f.Phys. A 294:291 (1980).
44. H.R.Schaefer, W. von Drachenfels, W.Paul, Zeitschr. f.Phys. A 305:213 (1982).
45. E.Mergel, Zeitschr. f.Phys. D 17:271 (1990).
46. E.Mergel, E.Th.Prinz, C.D. Schröter, W.Nakel, "Correlations and Polarization in Electronic and Atomic Collisions and (e,2e) Reactions", IP Conference Series No.122 (eds. P.J.O.Teubner, E.Weingold) 231 (1992).
47. H.Bethe, Heitler, Proc.Roy.Soc. 146A:83 (1934).
48. G.Elwert, E.Haug, Phys.Rev. 183:90 (1969).
49. E.Haug, private communication (1975).
50. H.K.Tseng, R.H.Pratt, Phys.Rev. A 3:100 (1971).
51. P.G.Burke, J.Mitchel, J.Phys.B 7:214 (1974).
52. P.S. Farago, J.Phys.B 7:L28 (1974) and "Electron and Photon Interactions with Atoms" Plenum Press, eds. H.Kleinpoppen and M.R.C.McDowell, 235 (1976).
53. S.M.Khalid, H.Kleinpoppen, Phys.Rev. 27:236 (1983).
54. U.Eichmann, J.C.Bergquist, J.J.Bollinger, J.M.Gilligan, W.M.Itano, D.C.Wineland, Phys.Rev.-Lett. 70:2359 (1993).
55. M.G.Raizen, J.M.Gilligan, J.C.Bergquist, W.M.Itano, D.C.Wineland, Phys.Rev.A 45:6493 (1992).
56. W.M.Itano, D.C.Wineland, Phys.Rev. A 25:35 (1982).
57. D.E.Pritchard, G.M.Carter, F.Y.Chu, D.Klappner, Phys.Rev.A 2:1922 (1970), and D.E.Pritchard, F.Y.Chu, Phys.Rev. A 2:1932 (1970).
58. H.O.Lutz, "Correlations and Polarization in Electronic and Atomic Collisions and (e,2e) Reactions", IOP Conference No.122:335 (1992).
59. R.Hippler, H.Madeheim, H.O.Lutz, M.Kimura, N.F.Lane, Phys.Rev.A 40, 3446 (1989).
60. R.Hippler, M.Faust, R.Wolf, H.Kleinpoppen, H.O.Lutz, Phys.Rev. A 31:1399 (1985); 36:4644 (1987).
61. P.J.O.Teubner, W.E. Kanpilla, W.L.Fite, Phys.Rev. A 2:1763 (1970).
62. L.F.Errea, L.Mendez, A.Riera, Zeitschr. f. Phys., to be published.

63. W.Jitschin, S.Osimitsch, D.Reihl, D.M.Mueller, H.Kleinpoppen, H.O.Lutz., Phys.Rev. A 34, 3684 (1986).
64. S.Osimitsch, W.Jitschin, H.Reihl, H.Kleinpoppen, H.O.Lutz, O.Mo, A.Riera, Phys.Rev.A 40:2958 (1089).
65. S.Osimitsch, PhD thesis, Bielefeld, 1989.
66. H.Reihl, PhD thesis, Bielefeld (1994).
67. U.Fano, Phys.Rev. 178:31 (1969).
68. U.Heinzmann, J.Kessler, J.Lorenz, Phys.Rev.Lett. 25:1325 (1970).
69. U.Heinzmann, J.Kessler, B.Ohnemur, Phys.Rev.Lett. 27:1696 (1971).
70. M.S.Lubell, W.Raith, Phys.Rev.Lett. 23:211 (1969) and G.Baum, M.S.Lubell, W.Raith, Bull.Am.Phys.Soc. 14:950 (1969).
71. U.Heinzmann, "Fundamental Processes in Atomic Collision Physics", Plenum Publishing, N.Y. p.269 (1985).
72. H.Klar, H.Kleinpoppen, J.Phys.B 15:933 (1982).
73. A.Siegel, J.Ganz, W.Bußert, H.Hotop, J.Phys.B 15:2945 (1983).
74. Ch. Heckenkamp, F.Schäfers, G.Schönhense, U.Heinzmann, Phys.Rev.Lett. 52:421 (1984).
75. F.Schäfers, W.Peatman, A.Eyers, Ch.Heckenkamp, G.Schönhense, U.Heinzmann, Rev.Sci.Instr. 57:1032 (1986).
76. U.Heinzmann, "Fundamental Processes of Atomic Collision Physics" edit. by H.Kleinpoppen, J.S.Briggs, H.O.Lutz (Plenum Publ., N.Y.) p.269 (1985).
77. G.Schönhense, A.Eyers, U.Friess, F.Schäfers, U.Heinzmann, Phys.Rev.Lett. 54:547 (1985).
78. U.Heinzmann, G.Schönhense, J.Kessler, Phys.Rev.Lett. 42:1603 (1978).
79. U.Heinzmann, Proceed. Int.School of Physics "Enrico Fermi, Course CVIII, eds. M.Campagna, R.Rosei, North-Holland, p.469 (1990).
80. B.Vogt, B.Kessler, N.Müller, B.Schmiedeskamp, G.Schönhense, U.Heinzmann, Phys.Rev.Lett. 67:1315 (1991).
81. Y.Guo, W.M.Temmerman, H.Ebert, J.Magnet. and Magnetic Mat. 104-107:1772 (1992).
82. P.S. Kristic, M.H.Mittleman, Phys.Rev.A 45:514 (1992).
83. E.M.E Timmermaus, G.T.Trammell, J.P.Hannon, Phys.Rev. 72:832 (1994).
84. V.A.Dzua, V.V.Flambaum, O.P.Sushkou, Phys.Rev.A 44:4224 (1991).
85. M.Ya Agre, L.P.Rapoport, JETP, 77:382 (1993).
86. B.L. Moiseiwitsch, Adv.At.Mol. and Opt.Phys. 26:51 (1990).
87. G.Grüner, Revs.Mod.Phys. 66:1 (1994).
88. V.S.Beskin, Contemporary Physics, 34:131 (1993).

MEASUREMENT OF EXCHANGE AND SPIN-ORBIT INTERACTION EFFECTS IN ELASTIC ELECTRON-CESIUM SCATTERING

W. Raith, G. Baum, P. Baum, L. Grau,
B. Leuer, R. Niemeyer and M. Tondera

Universität Bielefeld
Fakultät für Physik
D-33501 Bielefeld

INTRODUCTION

Exchange of identical particles - here the exchange of the free electron with an atomic electron - is a quantum phenomenon. The spins play the role of markers: Elastic scattering "with exchange" can be distinguished from scattering "without exchange" if the two electrons have antiparallel spins. The spin dependence of exchange results from the different treatment of indistiguishable and distinguishable processes in quantum mechanics. In the elastic scattering of polarized electrons from polarized one-electron atoms, the differential cross section depends on the relative orientation of the polarization vectors for the electrons and the atoms P_e and P_a, respectively; the polarization P_a refers to the valence-electron polarization, the nuclear polarization is irrelevant. The spin dependence of the cross section can be described by

$$\sigma = \sigma_o[1 - A^{ex} P_e \cdot P_a] , \qquad (1)$$

where σ_o is the cross section for unpolarized particles and A^{ex} is the **exchange asymmetry**, an interesting parameter of scattering theory. Both, σ_o and A^{ex}, are functions of the electron energy E and the scattering angle θ. The minus sign in Eq.1 corresponds to the accepted definition of A^{ex}, chosen to give positive values for the "total" impact-ionization asymmetry, that is, the asymmetry averaged over all emission angles and energy partitions. - Exchange effects in electron-atom scattering are most pronounced at low energies.

Spin-orbit interaction is a classically understandable interaction between the magnetic moment of the scattered electron, oriented perpendicular to the scattering plane, and the inhomogenious motional magnetic field in the rest frame of the scattered electron. Due to this magnetic interaction, the scattering angle which would result from potential scattering alone is slightly increased or decreased by $\Delta\theta$. Because the differential cross section is anisotropic, a left/right cross-section difference

$$\sigma(\pm\theta) = \sigma_o \pm \Delta\sigma \quad \text{with} \quad \Delta\sigma \approx (d\sigma_o/d\theta) \Delta\theta \qquad (2)$$

results from it. If electrons with polarization $\mathbf{P}_e$ are elastically scattered from unpolarized atoms, the cross section is given by

$$\sigma = \sigma_o[1 + A^{s\text{-}o} \mathbf{P}_e \cdot \mathbf{n}] , \qquad (3)$$

where $A^{s\text{-}o}$ is the **spin-orbit asymmetry** and $\mathbf{n}$ is a unit vector normal to the scattering plane, pointing in the direction $\mathbf{k} \times \mathbf{k}'$ ($\mathbf{k}$ and $\mathbf{k}'$ are the electron momenta before and after the scattering, respectively). The definition of $A^{s\text{-}o}$ corresponds to that used in the Mott scattering literature where this parameter is usually called the *Sherman function* S.

Spin-orbit interaction is a relativistic effect and, therefore, strongest for high-Z target atoms. The spin dependence of the *high-energy Mott scattering* becomes unmeasurably small below, say, 10 keV. At lower energies, however, the de Broglie wavelength of the scattered electron becomes comparable to the size of the potential well of the target atom and, therefore, diffraction minima in $\sigma_o(\theta)$ result. The spin-orbit asymmetry, which is proportional to the angular derivative of the *relative* cross section according to

$$A^{s\text{-}o} = \Delta\sigma/\sigma_o \approx [(d\sigma_o/d\theta)/\sigma_o] \cdot \Delta\theta , \qquad (4)$$

changes sign at the cross section minimum and reaches large positive/negative values in the vicinity of the minimum where $d\sigma_o/d\theta$ is finite and σ_o is small. Consequently, in the so-called *low-energy Mott scattering*, pronounced spin effects with strong angular dependence can be observed even at energies below 100 eV.

In 1974 Burke and Mitchell[1] made a general theoretical analysis of relativistic electron scattering from one-electron atoms. They showed that this process involves more than the simple addition of exchange and spin-orbit interaction. The full theory depends on six complex scattering amplitudes; that corresponds to 11 independent parameters, e.g. 6 amplitude moduli and 5 phase differences. Therefore, a "complete scattering experiment" involving the experimental determination of all 11 parameters is practically impossible.

In order to compare theory and experiment in a crucial but economical way, one should find out which parameter optimally combines theoretical significance and measurement feasibility and then determine this parameter with the highest achievable accuracy. The discussion about the consequences of Burke and Mitchell's analysis for experimental atomic physics was led by P.S Farago.[2] He drew attention to the fact that the full theory allows interference of exchange and spin-orbit interaction. This interference should produce a left/right **interference asymmetry** A^{int} of the cross section if unpolarized electrons are scattered on polarized atoms with $\mathbf{P}_a$ perpendicular to the scattering plane according to

$$\sigma = \sigma_o[1 + A^{int} \mathbf{P}_a \cdot \mathbf{n}] . \qquad (5)$$

This is an effect which neither exchange nor spin-orbit coupling could provide *alone*. [Note that we consider a single scattering process, not double-scattering processes in which exchange produces a polarization in the first scattering and then spin-orbit interaction produces a cross-section asymmetry in the second scattering, or vice versa.]

THEORETICAL DISCUSSION

The 6 complex amplitudes of Burke and Mitchell's analysis[1] do not allow a straightforward comparison with our observables. The meaning of the amplitudes can be inferred from their connections with $\mathbf{P}_e$ and $\mathbf{P}_a$ in the scattering matrix (Table 1).

Table 1. The meaning of Burke and Mitchell's scattering amplitudes

Amplitude	Connection and Significance
a_1	no connection with $\mathbf{P}_e$ and $\mathbf{P}_a$; mainly **potential scattering**
a_2	coefficient of $\mathbf{P}_a \cdot \mathbf{n}$; a **new interaction** (interference?)
a_3	coefficient of $\mathbf{P}_e \cdot \mathbf{n}$; **spin-orbit interaction** (= Mott scattering)
a_4	coefficient of $(\mathbf{P}_e \cdot \mathbf{n})(\mathbf{P}_a \cdot \mathbf{n})$; **exchange interaction** only; e-e magnetic-dipole scattering is negligible at low energies
a_5, a_6	similar to a_4 but referring to directions <u>in</u> the plane of scattering

The cross section for unpolarized particles is the sum of the squares of all amplitudes:

$$\sigma_o = \sum_{i=1,\ldots 6} a_i a_i^* . \tag{6}$$

For relating our asymmetries to these amplitudes, we give here only the leading terms, omitting correction terms which are presumably small.

$$\sigma_o A^{s\text{-}o} = 2\,\mathrm{Re}\,(a_1 a_3^* + \ldots) , \tag{7}$$
$$-\sigma_o A^{ex} = 2\,\mathrm{Re}\,(a_1 a_4^* + \ldots) , \tag{8}$$
$$\sigma_o A^{int} = 2\,\mathrm{Re}\,(a_1 a_2^* + \ldots) . \tag{9}$$

The Eq.s 7-9 show that the potential-scattering amplitude a_1 is the leading factor in the formulas for all three asymmetries. The amplitude which has the pertinent connection with $\mathbf{P}_e$ and/or $\mathbf{P}_a$ enters only as a second factor. Since the amplitude a_2 for the new interference asymmetry could be very small, one should look at the full expression of Eq. 9:

$$\sigma_o A^{int} = 2\,\mathrm{Re}\,(a_1 a_2^* + a_3 a_4^*) . \tag{9a}$$

If the first term of Eq. 9a were indeed very small, A^{int} would be dominated by the second term which is the *product of the spin-orbit and the exchange amplitude*. This dependence justifies Farago's choice of calling the new effect *interference*.

EXPERIMENTAL CONSIDERATIONS

In 1974 this intruiging interference effect appeared to be at the borderline of experimental feasibility. In a first theoretical estimate for elastic e-Cs scattering D.W. Walker[3] came to the conclusion that the interference effect might become observable below 15 eV in the vicinity of a diffraction minmum of the differential cross section. Twenty years later, after technical advances have led to increases of the achievable Cs-atom polarization from 0.1 to nearly unity, we see the first glimpse of the effect Peter Farago envisioned.

For the search of the interference effect one should select experimental conditions under which the two interfering spin-dependent effects are both of measurable size at the same energy. Klewer et al.[4] measured A^{s-o} values for cesium at 13 - 25 eV. The exchange asymmetry A^{ex} had not yet been measured for cesium; for lithium Baum et al.[5] measured A^{ex} up to 30 eV. Thus we concluded that the energy region of overlap, in which both spin effects have measurable size, is about 10 to 20 eV. Gehenn and Reichert[6] measured relative values of $\sigma_o(\theta)$ for the elastic e-Cs scattering from 20 eV down to 0.8 eV; for all energies studied their data show at least one deep diffraction minimum in the studied angular range of 32 to 143°.

Of all the conceivable polarization experiments on electron-atom scattering involving polarized incoming particles and/or polarization analysis of the outgoing particles, the most feasible experiment is the measurement of cross-section asymmetries with both incoming particles polarized (Table 2). Therefore, the e^1Cs^1 scattering experiment is most suitable for obtaining accurate data to allow a crucial test of relativistic electron-atom scattering theory including exchange.

Table 2. Figures of merit of polarizers and analyzers for electron and atom beams

	ELECTRONS	ATOMS
Polarizer figure of merit $P^2 \, I_{pol}/I_{unpol}$	strained GaAs $\approx 10^{-1}$	optical pumping or sextupoles + rf transitions $10^{-1} - 10^{-2}$
Analyzer figure of merit $A^2 \, I_{signal}/I_{incident}$	Mott scattering (20 to 100 keV) $10^{-4} - 10^{-5}$	magnetic state selection $\approx 10^{-1}$
problem	signal attenuation of scattered beam	separation of scattered and unscattered atoms
	Both analyzers are well suited for monitoring the primary beam polarization.	

Since 1974, when Farago assessed the feasibility of A^{int} measurements, the experimental techniques have improved, in particular, highly polarized electron beams now have currents almost as high as unpolarized electron beams. Therefore, it was not necessary for us to start an experimental investigation of A^{int} with <u>un</u>polarized electrons. We decided to use polarized atoms *and* polarized electrons for a thorough study of exchange and spin-orbit interaction with the additional goal of observing the interference effect if possible.

The study of spin-orbit interaction requires an orientation of the electron polarization perpendicular to the scattering plane. The study of exchange requires (anti)parallel electron and atom polarization. To meet both requirements we chose to have $\mathbf{P}_e$ and $\mathbf{P}_a$ perpendicular to the scattering plane. The differential cross section then depends on the two beam polarizations in the following way:

$$\sigma = \sigma_0[1 - A^{ex}\mathbf{P}_e\cdot\mathbf{P}_a + A^{s\text{-}o}\mathbf{P}_e\cdot\mathbf{n} + A^{int}\mathbf{P}_a\cdot\mathbf{n}]. \quad (10)$$

Since our apparatus allows only measurements at scattering angles on one side of the primary beam, we cannot reverse the vector $\mathbf{n}$. But we can easily reverse both beam polarizations. The different combinations of beam-polarization directions lead to four different cross sections symbolized by two arrows of which the first represents the direction of $\mathbf{P}_e$, the second, of $\mathbf{P}_a$:

$$\begin{aligned}
\sigma(\uparrow\uparrow) &= \sigma_0[1 - P_eP_a A^{ex} + P_e A^{s\text{-}o} + P_a A^{int}], \\
\sigma(\downarrow\uparrow) &= \sigma_0[1 + P_eP_a A^{ex} - P_e A^{s\text{-}o} + P_a A^{int}], \\
\sigma(\uparrow\downarrow) &= \sigma_0[1 + P_eP_a A^{ex} + P_e A^{s\text{-}o} - P_a A^{int}], \\
\sigma(\downarrow\downarrow) &= \sigma_0[1 - P_eP_a A^{ex} - P_e A^{s\text{-}o} - P_a A^{int}].
\end{aligned} \quad (11)$$

By taking suitable combinations of two of these cross sections, two of the three asymmetries cancel out. The measured values are not absolute cross sections but rather counting rates for constant incident-beam currents. Therefore, our four relative cross-section measurements provide three independent cross-section ratios which are the three asymmetries summarized in Table 3.

Table 3. Scheme of data evaluation

Parameter X	Signal Sums I_j (X > 0)	I_k (X < 0)	$\dfrac{I_j - I_k}{I_j + I_k}$
$\mathbf{P}_e\cdot\mathbf{P}_a$	↓↑ + ↑↓	↑↑ + ↓↓	$P_eP_a A^{ex}$
$\mathbf{P}_e\cdot\mathbf{n}$	↑↑ + ↑↓	↓↑ + ↓↓	$P_e A^{s\text{-}o}$
$\mathbf{P}_a\cdot\mathbf{n}$	↑↑ + ↓↑	↓↓ + ↑↓	$P_a A^{int}$

EXPERIMENTAL ARRANGEMENT

In previous publications we already described in detail the atomic beam system[7] as well as the polarized electron beam for studies with high energy resolution.[8] The experimental lay-out is shown schematically in figure 1; figure 2 gives a scale drawing of the vacuum system.

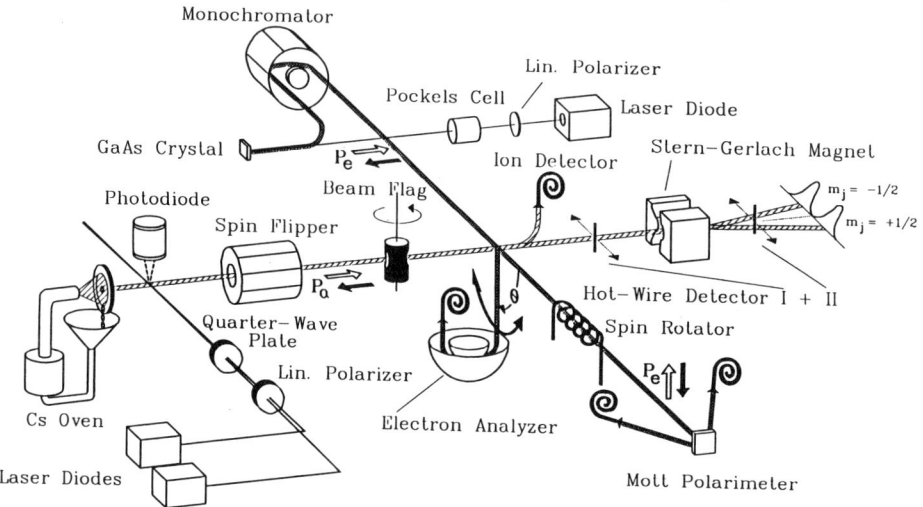

Figure 1. Experimental arrangement (not to scale)

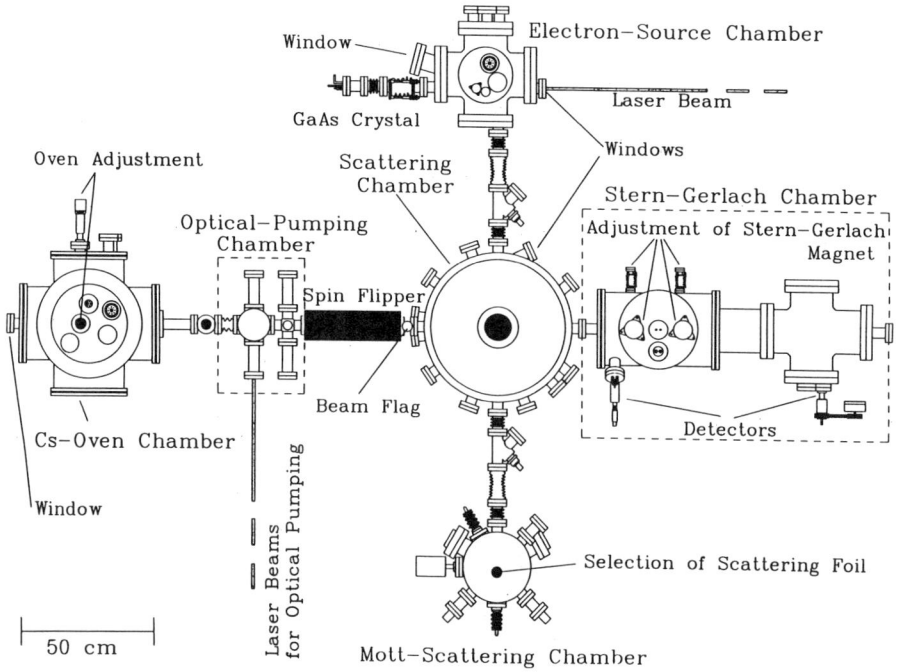

Figure 2. Top view of the vacuum system

In the earlier studies "normal" GaAs photocathodes were employed for which the principal limit of polarisation is 0.5 and only about 0.35 is actually obtained. In this experiment the polarized electron source is a "strained" GaAs photocathode[9] for which the principal limit of polarization is 1.0 and about 0.7 is obtained. The strained GaAs layer was prepared for negative electron affinity in our standard way by cesiating and oxidizing. It was illuminated by circularly polarized laser light of $\lambda = 837$ nm. The electron polarization is monitored by 20 keV Mott scattering from a thorium foil. The electron energy width, not yet reduced by the monochromator shown in figure 1, is about 0.3 eV. The detector for the scattered electrons includes energy analysis with a resolution of about 1 eV; it covered the angular range of 35 - 145° with an angular width of ±3°.

The cesium beam emerged from an oven which recirculates about 99 % of the emitted cesium. Via a vacuum lock a cesium ampoule can be brought in, cracked under vacuum and its content destilled into the oven.

As indicated in the level diagram of figure 3, the Cs beam was polarized by optical pumping out of both hfs levels of the ground state employing two laser diodes. We used wavelength-selected GaAs laser diodes. The tuning was accomplished by temperature variation, course tuning by means of a Peltier element and fine tuning by means of the injection

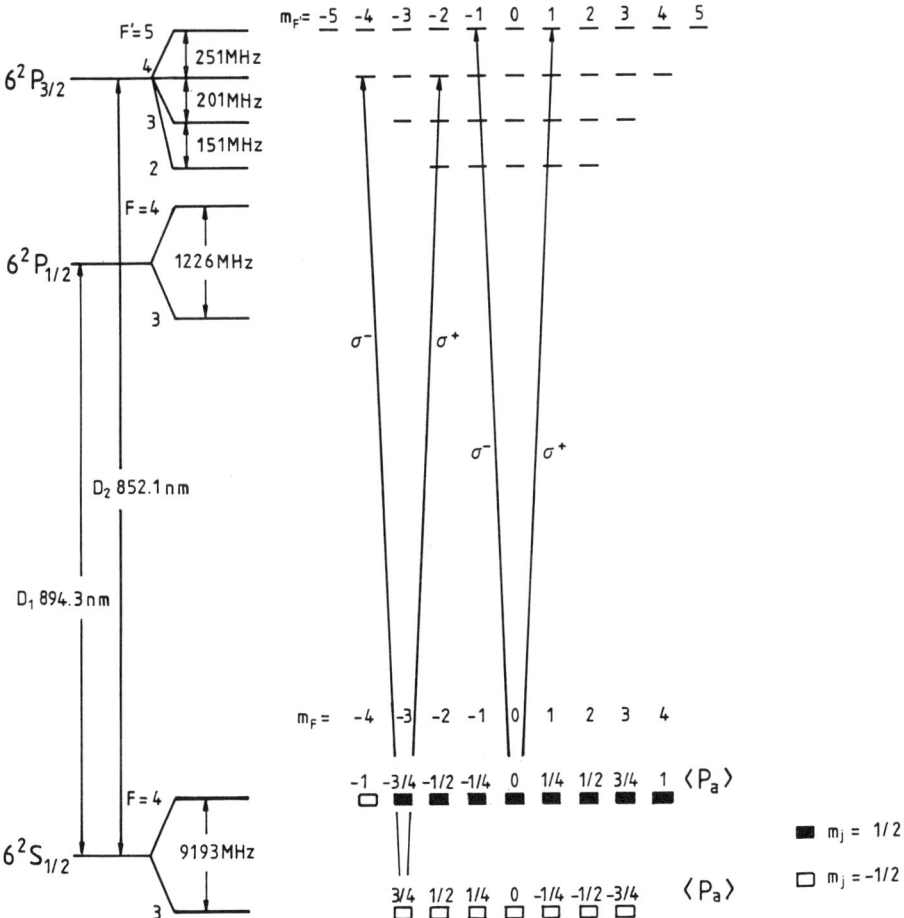

Figure 3. Term diagram of cesium showing the ground state and the first excited state as well as the transitions used for polarizing the atoms by two-laser optical pumping.

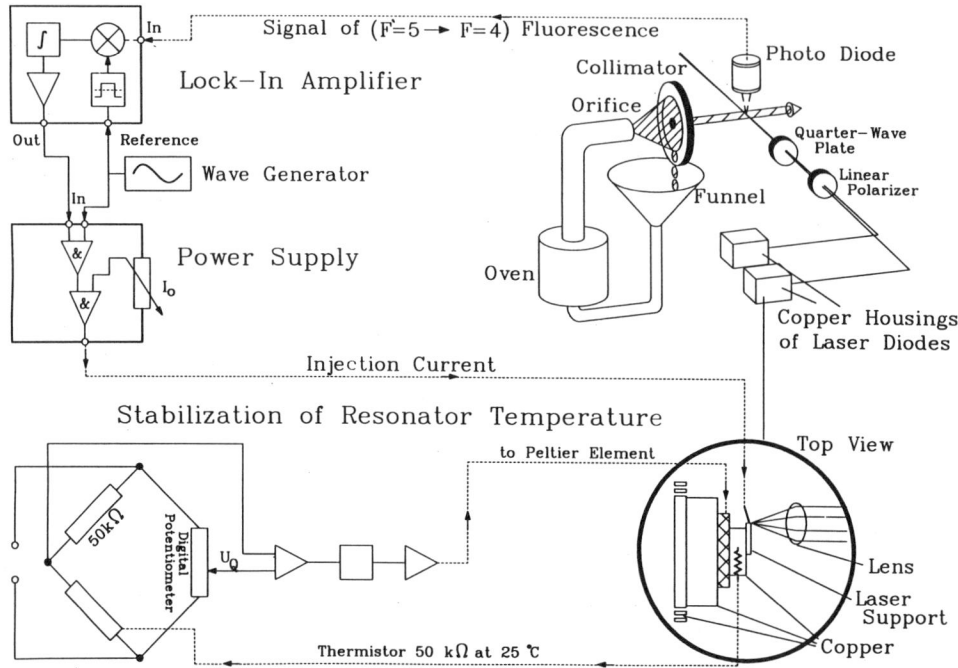

Figure 4. Schematics showing in the upper right-hand corner the pumping arrangement with detection of the fluorescence radiation. The insert (lower right) shows the laser diode mounting with the Peltier element. Its temperature is stabilized by means of a thermistor (lower left). Fine tunig of the temperature and wavelength stabilization is accomplished with electronic regulation of the injection current (upper left).

current. The wavelengths of both lasers were stabilized by modulating the injection current which caused a wavelength modulation of the light beam, measuring the intensity of the resonance fluorescence radiation and employing a feed-back circuit for regulating the injection current (figure 4).

We obtain an atomic (= valence electron) polarization of 90% for an atom density of $5 \cdot 10^9$ cm^{-3} in the interaction region. At higher densities radiation trapping impairs the optical pumping process.

We determine the atom polarization by evaluating the beam profiles produced by magnetic state separation in a Stern-Gerlach magnet. The beam profiles for σ^+ and σ^- pumping (figure 5, top) consist of two peaks which represent the beam populations according to the high-field quantum numbers $m_s = \pm \frac{1}{2}$. The low-field electronic polarization is inferred in the follwoing way: The top left of figure 5 shows the Stern-Gerlach profile of a beam pumped by σ^+ light of both lasers. The fact that the left peak is negligibly small compared with the right peak proves that the lower hfs level (F=3) of the ground state has been (almost) completely emptied by the pumping process and that the population of the (F=4, m_F=-4) Zeeman state is also extremely small. With this information the large "white" peak of the Stern-Gerlach profile of figure 5, obtained with σ^- pumping light, can be interpreted as containing only atoms in the (F=4, m_F=-4) Zeeman state for which the polarization is -1, whereas the smaller "black" peak contains atoms in the 7 Zeeman states (F=4, m_F=-3 ... +3) having low-field polarizations between -0.75 and +0.75. The actual low-field polarization achieved by the optical pumping corresponds to a Zeeman-state population distribution between a "best case" and a "worst case" for which an upper and a lower limit of P_a can be evaluated (figure 5, bottom). The upper limit is based on the assumption that all the atoms of the right-side peak belong to the (F=4, m_F=-3) Zeeman state, the lower limit is based on the assumption

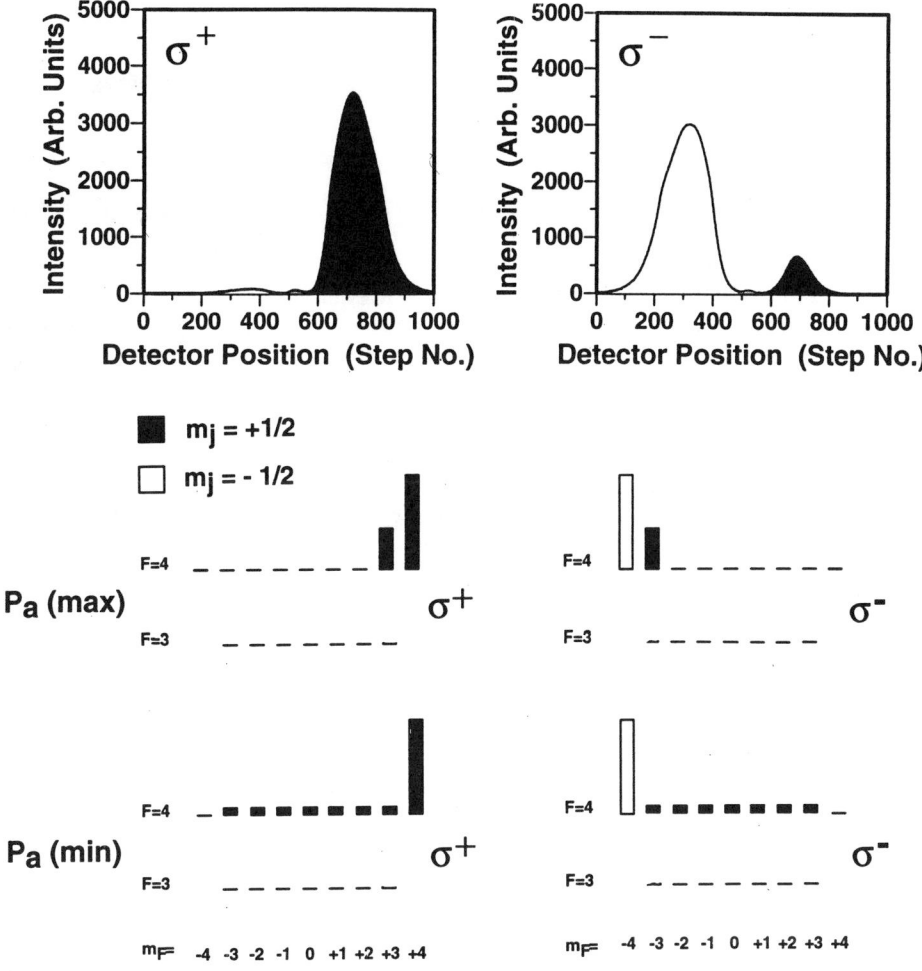

Figure 5. Top: Beam profiles after passage through Stern-Gerlach magnet for σ^+ and σ^- pumping, respectively. In the left diagram the black peak contains atoms in (F=4, m_F=-3 ... +4) states, the white peak contains atoms in (F=4, m_F=-4) and (F=3, m_F = -3 ... +3) states. In the right diagram the white peak contains only atoms in the (F=4, m_F=-4) state because the contribution from the (F=3, m_F = -3 ... +3) states is negligible; the black peak contains atoms in the (F=4, m_F=-3 ... +4) states with an unknown distribution. - Bottom: Distributions for getting an upper and a lower limit of the low-field polarization.

that those atoms are equally distributed over the 7 Zeeman states. For $|\mathbf{P}_a| \to 1$ the difference between upper and lower limit goes to zero.

Both beam polarizations $\mathbf{P}_e$ and $\mathbf{P}_a$ can be reversed by changing the circularity of the laser light used for illuminating the GaAs photocathode and for pumping the atomic beam, respectively. For reversing $\mathbf{P}_e$ in this way, we employed a carefully adjusted Pockels cell; for reversing $\mathbf{P}_a$ slowly, we rotated the quarter-wave plate, for reversing it quickly we changed the mode of a spin flipper.[10] We observed no false asymmetries caused by intensity changes associated with polarization reversals.

The data presented here were obtained with the follwing beam polarizations: P_e = 0.65 ± 0.03 and P_a = 0.89 ± 0.035. The systematic errors due to the beam-polarization uncertainties, common to all data points, amount to ±6% for A^{ex}, ±4.5% for A^{s-o}, and ±4% for A^{int}.

EXPERIMENTAL RESULTS

We began with measurements at 20 eV and continued with measurements at 13.5 eV and at 7 eV. With decreasing energy the required time increases; the data obtained at 7 eV took 100 hours of running time.

Our data of σ_o (in arbitrary units), A^{s-o}, A^{ex} and A^{int} for the three different energies are shown in figures 6-8. All the asymmetry data points are shown with statistical error bars corresponding to one standard deviation. We also show comparable σ_o data of Gehenn and Reichert[6] (normalized to ours) as well as A^{s-o} data of Klewer et al.;[4] at 20 eV the agreement is good. The asymmetry A^{ex}, for which no data exist in the literature, is negative at all angles except for a small range near the cross-section minimum. Note the enlarged scale for the diagram A^{int}; within the error margins the A^{int} measurements of figure 6 are consistent with zero.

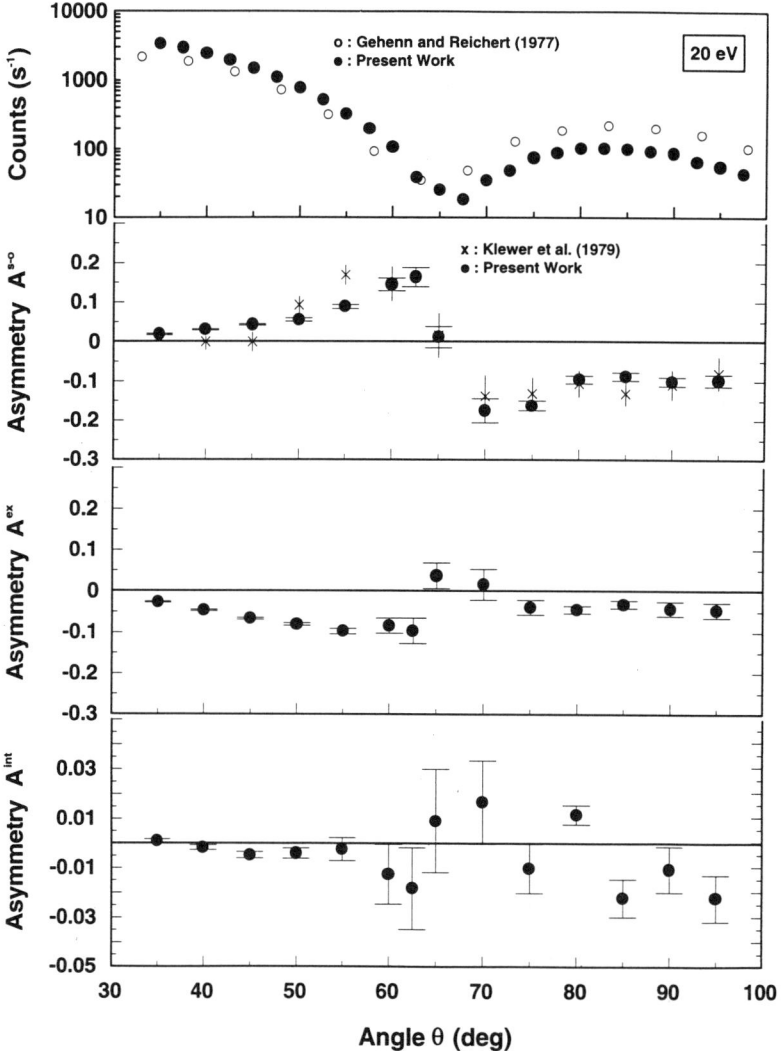

Figure 6. Measurements at 20 eV. From top to bottom: σ_o (relative), A^{s-o}, A^{ex} and A^{int}. Also shown are the σ_o data of Gehenn and Reichert[6] and the A^{s-o} of Klewer et al.[4]

The subsequent measurements at 13.5 eV (figure 7) and 7 eV (figure 8) yielded relative cross sections of which the minima are less pronounced than those in Gehenn and Reichert's data. Our A^{s-o} values at 13.5 eV differ significantly from the results of Klewer et al. The values of $|A^{ex}|$ at 13.5 eV and 7 eV are *larger* and the values of $|A^{s-o}|$ *smaller* than at 20 eV, in accordance with expectation about the energy dependence of the two effects.

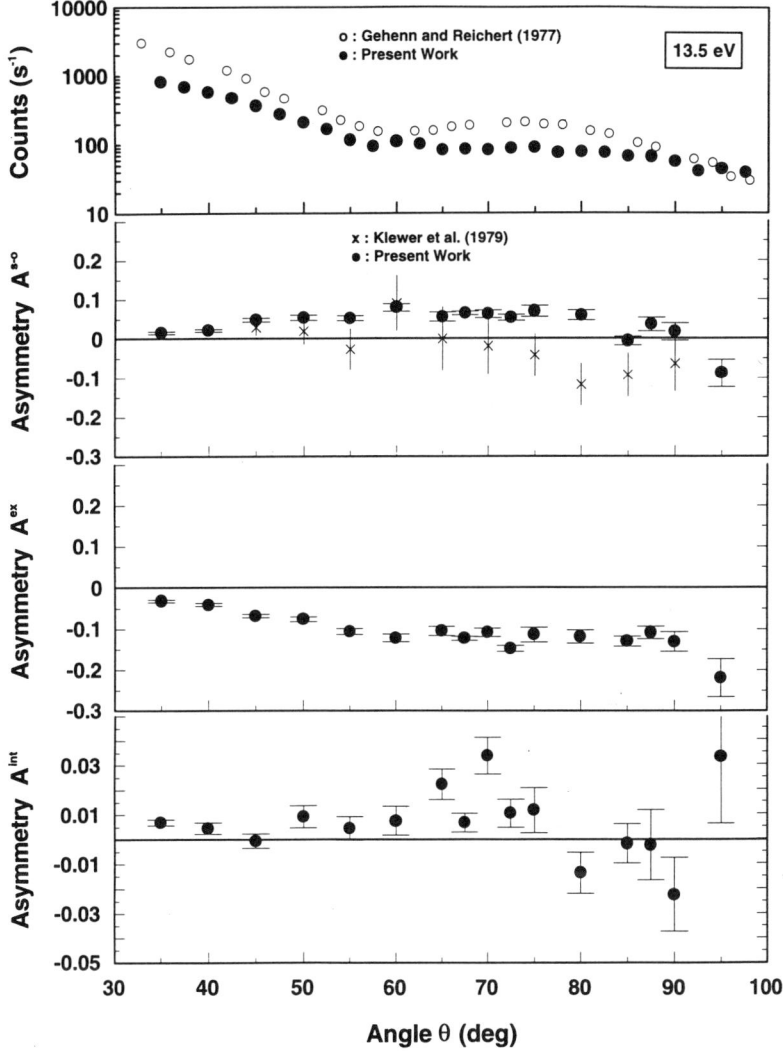

Figure 7. Measurements at 13.5 eV. From top to bottom: σ_o (relative), A^{s-o}, A^{ex} and A^{int}. Also shown are the σ_o data of Gehenn and Reichert[6] and the A^{s-o} of Klewer et al.[4]

At 13.5 eV the A^{int} data were not significantly different from zero. At 7 eV, however, *we obtained a small but definitely non-zero result for A^{int}*. The interference asymmetry is positive at small angles and negative at large angles and goes through zero near 110°, like A^{s-o}. While $A^{s-o}(\theta)$ has pronounced extrema, $A^{int}(\theta)$ has a small flat plateau of about +0.02 between 35 and 100°.

Walker[5] gives entirely different theoretical estimates of A^{int} in this energy range: For 5 eV his $A^{int}(\theta)$ curve has a narrow maximum of about 0.3 at 120° and is practically zero elsewhere; for 13.6 eV the maximum is about 0.8 and lies at 70°.

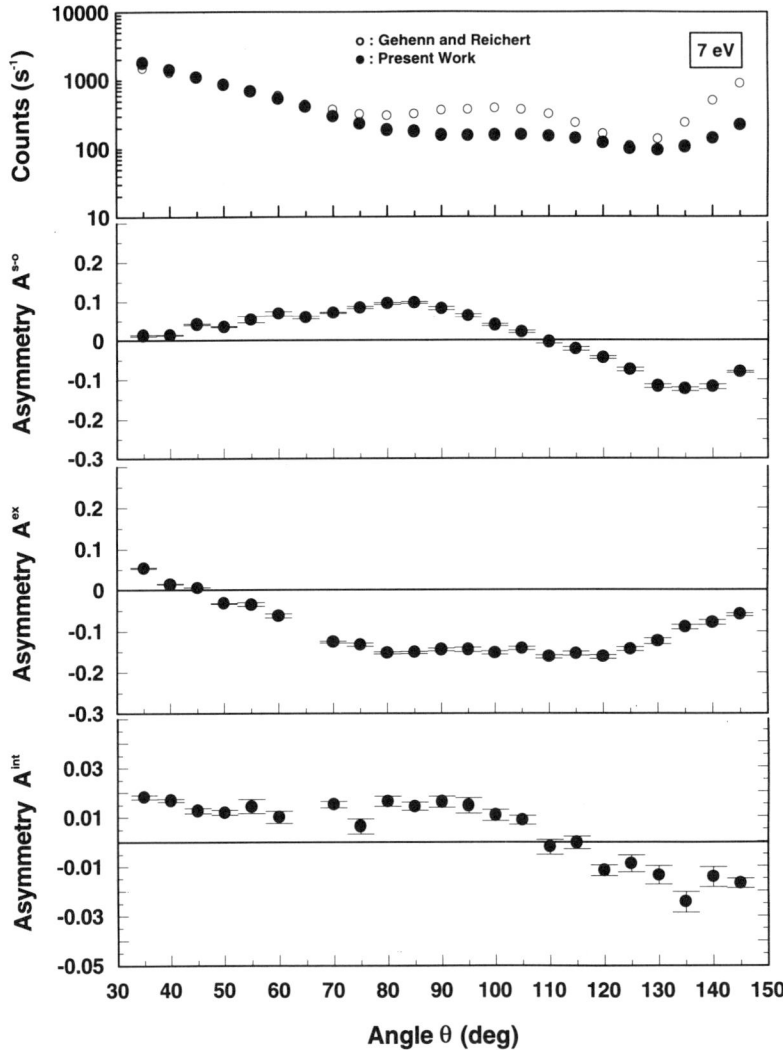

Figure 8. Measurements at 7 eV. From top to bottom: σ_o (relative), A^{s-o}, A^{ex} and A^{int}. Also shown are the σ_o data of Gehenn and Reichert.[6]

CONCLUSION

We plan to extend our measurement to much lower electron energies. It will require careful tuning, further reduction of background, use of a high-resolution electron monochromator and longer data-taking.

At energies of 0.8 - 2.0 eV measurements can be compared with extensive calculations.[11-15] These available close-coupling calculations are only valid for energies distinctly below the first ionization energy of 3.9 eV. Two theoretical approaches have been used for describing the low-energy e-Cs scattering, an 8-state Breit-Pauli approximation (Scott et al.[11]) and a 5-state Dirac approximation (Thumm and Norcross,[13] Thumm et al.[14]). They predict pronounced structures for all three asymmetries.[15] Both calculations agree satisfactorily with the experimental cross-section data. They give significantly different predictions, however, for the spin-dependent cross-section asymmetries at low energies, in particular for $A^{int}(\theta)$.

This work has been supported by the Deutsche Forschungsgemeinschaft within the research program of Sonderforschungsbereich 216.

REFERENCES

1. P.G. Burke and J.F.B. Mitchell, Spin polarization in the elastic scattering of electrons by one-electron atoms, *J.Phys.B* **7**, 214 (1974).
2. P.S. Farago, On the detection of spin-orbit interaction in the elastic scattering of electrons from one-electron atoms, *J.Phys.B* **7**, L28 (1974).
3. D.W. Walker, On the asymmetry in single scattering of electrons from one-electron atoms, *J.Phys.B* **7**, L489 (1974).
4. M. Klewer, M.J.M. Beerlage, and M.J. Van der Wiel, Polarisation and angular distribution of electrons elastically scattered from caesium atoms at 13-25 eV, *J.Phys.B* **12**, L525 (1979).
5. G. Baum, M. Moede, W. Raith, U. Sillmen, Measurement of Spin Dependence in Low-Energy Elastic Scattering of Electrons from Lithium atoms, *Phys.Rev.Lett.* **57**, 1855 (1986).
6. W. Gehenn and E. Reichert, Scattering of electrons by Cs atoms at low energies, *J.Phys.B* **10**, 3105 (1977).
7. G. Baum, B. Granitza, S. Hesse, B. Leuer, W. Raith, K. Rott, M. Tondera, and B. Witthuhn, An optically pumped, highly polarized cesium beam for the study of spin-dependent electron scattering, *Z.Phys.D*, **22**, 431 (1991).
8. G. Baum, B. Granitza, L. Grau, B. Leuer, W. Raith, K. Rott, M. Tondera, and B. Witthuhn, Spin asymmetry in electron impact ionization of caesium, *J.Phys.B* **26**, 331 (1993).
9. M. Chatwell, J. Clendenin, T. Maruyama, and D. Schultz, eds., "Photocathods for Polarized Electron Sources", SLAC-432, Stanford Linear Accelerator Center, Stanford, January 1994.
10. W. Schröder and G. Baum, A spin flipper for reversal of polarisation in a thermal atomic beam, *J.Phys.E* **16**, 52 (1983).
11. N.S. Scott, K. Bartschat, P.G. Burke, W.B. Eissner, and O. Nagy, Low-energy scattering of electrons by caesium atoms, *J.Phys.B* **17**, L191 (1984); N.S. Scott, K. Bartschat, P.G. Burke, O. Nagy, and W.B. Eissner, Low-energy scattering of electrons by caesium atoms: II, *J.Phys.B* **17**, 3755 (1984).
12. K. Bartschat, Spin-orbit and interference asymmetries in elastic scattering of electrons from heavy one-electron atoms, *J. Phys. B* **23**, 2341 (1990).
13. U. Thumm and D.W. Norcross, Relativistic R-matrix calculations for electron-alkali-metal-atom scattering: Cs as a test case, *Phys.Rev.A* **45**, 6349 (1992).
14. U. Thumm, K. Bartschat, and D.W. Norcross, Relativistic effects in spin-polarization parameters for low-energy electron-Cs scattering, *J.Phys.B* **26**, 1587 (1993).
15. K. Bartschat, Low-energy electron scattering from caesium atoms - comparison of a semirelativistic Breit-Pauli and a fullrelativistic Dirac treatment, *J.Phys.B* **26**, 3595 (1993).

THEORY OF ELECTRON SPIN EFFECTS IN ELECTRON-ATOM SCATTERING

H.R.J.Walters

Department of Applied Mathematics and Theoretical Physics
The Queen's University of Belfast
Belfast BT7 1NN
United Kingdom

1. INTRODUCTION

It is impossible in the space available here to review the whole area of spin effects in electron-atom scattering. For that the reader is referred to the excellent works of Kessler[1,2]. Rather I have selected some topics to illustrate the subject and to highlight some interesting results and problems.

The article begins in section 2 with a definition of notation and a statement of the non-relativistic spinless Hamiltonian used to describe electron collisions with the lighter atoms. There follows in section 3 a discussion of theoretical methods based on the close-coupling formalism. This starts with the basic eigenstate close-coupling approximation and leads on to the most recent developments using pseudostates. Spin asymmetry in electron scattering by atomic hydrogen is considered in section 4 and the need for more experimental work on this most fundamental of systems is stressed. Section 5 demonstrates how modern pseudostate close-coupling methods are needed to explain the kind of fine detail that spin polarized experiments are capable of revealing, the example used here is electron scattering by Na. A long outstanding problem between theory and experiment concerning the electron-photon coincidence parameters for 2p excitation of atomic hydrogen is reconsidered in section 6. If it is theory rather than experiment which is at fault in this case, then the evidence suggests that the flaw lies with the theoretical treatment of exchange, which is something that could be probed by spin polarized experiments. Spin effects in electron impact ionization are discussed in section 7. Firstly, it is shown how pseudostate close-coupling can successfully reproduce the cross section and spin asymmetry for ionization of H(1s), a problem which has long defied theorists. Next, insights into the Wannier threshold law are provided by an analysis of spin asymmetry data. The final part of this section looks at spin effects in (e, 2e) measurements which give the most complete picture of the ionization process. Section 8 introduces the relativistic Dirac Hamiltonian and its semi-relativistic Breit-Pauli approximation. The relativistic Hamiltonian is required for the study of electron collisions with heavy atoms, such as Hg, where the electrons move with relativistic speeds near the nucleus. This, in Breit-Pauli language, brings in effects such as spin-orbit coupling, mass-correction and Darwin terms. A case study is presented of electron scattering by Cs for which there exist recent calculations using both the Dirac and Breit-Pauli formalisms. Conclusions are presented in section 9.

Throughout atomic units (au) in which $\hbar = m_e = e = 1$ are used. The symbol a_0 denotes the Bohr radius.

2. NOTATION AND SPINLESS HAMILTONIAN

Consider an electron incident with momentum k_0 upon an N electron atom. The non-relativistic spinless Hamiltonian for this system is

$$H = K_0 + H_A + V \tag{1}$$

where

$$K_0 \equiv -\frac{1}{2}\nabla_0^2 \tag{2}$$

$$H_A \equiv \sum_{i=1}^{N}\left(-\frac{1}{2}\nabla_i^2 - \frac{Z}{r_i}\right) + \sum_{\substack{i,j=1 \\ i<j}}^{N}\frac{1}{|r_i - r_j|} \tag{3}$$

$$V \equiv -\frac{Z}{r_0} + \sum_{i=1}^{N}\frac{1}{|r_0 - r_i|} \tag{4}$$

Here r_i ($i = 1$ to N) (r_0) is the position vector of the i th atomic electron (incident electron) relative to the nucleus, K_0 is the kinetic energy operator for the incident electron, H_A is the atomic Hamiltonian, and V is the Coulomb interaction between the incident electron and the atom. Because of the completeness of the atomic states the collisional wave function for the system $\Psi(x_0, x_1, ..., x_N)$ may be expanded as

$$\Psi = A\left\{\sum_{n=0}^{\infty} F_n(x_0)\psi_n(x_1, ..., x_N) + \int F_\kappa(x_0)\psi_\kappa(x_1, ..., x_N)d\kappa\right\} \tag{5}$$

In (5) A is the antisymmetrization operator, $x_i \equiv (r_i, s_i)$ stands for the space, r_i, and spin, s_i, coordinates of the i th electron ($i = 0$ to N), the sum on n is over all bound atomic states ψ_n and the integral on κ is over all atomic continua, i.e., all possible ionized states of the atom¶. In the non-relativistic energy domain spin effects arise from the antisymmetrization in (5). Unless otherwise stated, it is assumed until section 8 that the systems considered can be adequately described by the spinless Hamiltonian (1).

3. THEORETICAL METHODS

Over the years a large range of approximations of varying quality and region of applicability have been developed for electron-atom scattering. Here I would like to highlight a few of the methods most pertinent to the subsequent discussions of this

---

¶ The symbol κ is used only in a formal way here. A precise notation for the ionized continua would, in general, be more complicated.

article.

3.1 Eigenstate Close-Coupling

In this approximation[3] only a finite number of important bound states are retained in (5), e.g.,

$$\Psi \approx A \sum_{n=0}^{M} F_n(\mathbf{x}_0)\psi_n(\mathbf{x}_1, ..., \mathbf{x}_N) \qquad (6)$$

This type of approximation is most viable for electron scattering off ground-state alkali metals which tends to be dominated by the ground-state and the first excited P-state, for example, in the case of Na by the Na(3^2P) and Na(3^2S) states.

3.2 Eigenstate Close-Coupling with Short-Range Correlation Terms

At low energies, where at most the channels $n = 0$ to M are open, the approximation (6) may be substantially improved by appending a set of short-range correlation terms $\chi_\nu(\mathbf{x}_0, \mathbf{x}_1, ..., \mathbf{x}_N)$ to represent the remaining closed channels in (5)[4,5], i.e.,

$$\Psi \approx A \sum_{n=0}^{M} F_n(\mathbf{x}_0)\psi_n(\mathbf{x}_1, ..., \mathbf{x}_N) + \sum_{\nu=0}^{\mu} a_\nu \chi_\nu(\mathbf{x}_0, \mathbf{x}_1, ..., \mathbf{x}_N) \qquad (7)$$

The (appropriately antisymmetrized) functions χ_ν give a bound-state-like representation of the closed channels. The functions F_n and the constants a_ν are determined by substituting (7) into the Schrödinger equation and projecting with the ψ_n and the χ_ν. If the approximation (7) is used at energies where channels with $n > M$ are open, catastrophic pseudostructures coming from the short-range terms appear in the calculated cross sections[4].

3.3 Pseudostate Close-Coupling — Intermediate Energy R-Matrix Theory

The really difficult theoretical problem arises at energies where many channels are open and where the expansion (5) is not dominated by a few bound eigenstates. This is the so-called "intermediate energy" regime. Here we need some feasible way of representing the infinity of channels in (5), especially the continuum channels. Very much in tune with modern developments is the idea of pseudostate close-coupling. In this scheme, a finite number of eigenstates of "interest" is retained in (5), while the infinity of remaining states is approximated by a finite number of discrete "pseudostates" $\overline{\psi}_n$ ($n = M + 1$ to P)

$$\Psi \approx A \left\{ \sum_{n=0}^{M} F_n(\mathbf{x}_0)\psi_n(\mathbf{x}_1, ..., \mathbf{x}_N) + \sum_{n=(M+1)}^{P} \overline{F}_n(\mathbf{x}_0)\overline{\psi}_n(\mathbf{x}_1, ..., \mathbf{x}_N) \right\} \qquad (8)$$

Of course, how the approximation (8) performs depends on the choice of pseudostates and their number. Various criteria for selecting pseudostates have been advanced[6]. Typically pseudostates are chosen so that, together with the retained eigenstates, they diagonalize the atomic Hamiltonian H_A:

$$\langle \psi_n | H_A | \psi_m \rangle = \epsilon_n \delta_{nm} \qquad \langle \overline{\psi}_n | H_A | \overline{\psi}_m \rangle = \overline{\epsilon}_n \delta_{nm}$$
$$\langle \psi_n | H_A | \overline{\psi}_m \rangle = 0 \qquad (9)$$

where ϵ_n is the energy of the eigenstate ψ_n and I shall call $\bar{\epsilon}_n$ the "energy" of the pseudostate $\bar{\psi}_n$. We may view a pseudostate $\bar{\psi}_n$ as a "clump" of eigenstates with an energy distribution[7]¶

$$f_n(\epsilon) = |\langle\psi_\epsilon|\bar{\psi}_n\rangle|^2 \tag{10}$$

about the value $\epsilon = \bar{\epsilon}_n$. In (10) ψ_ϵ stands for a (suitably normalised) eigenstate with energy ϵ and with the same symmetry (i.e., total orbital angular momentum (L, L_z) and total spin (S, S_z)) as $\bar{\psi}_n$. From this viewpoint a good choice of pseudostates, for a particular state symmetry, is one in which the $\bar{\epsilon}_n$, together with the energies ϵ_n of the retained eigenstates, span with sufficient density the eigenstate spectrum up to some finite value of energy beyond which the eigenstates in (5) do not make any substantial contribution to the transitions of interest. In addition we must, of course, include a sufficient number of state symmetries in (8).

However, nothing is without a price, and the price paid for approximating the eigenstate spectrum with pseudostates is the appearance of unphysical pseudostructure in the calculated cross sections[8-12]. Since the pseudostates enter the formalism in the same way as bound eigenstates, see (8), they have, like the latter, discrete thresholds with associated resonance structure. It is now accepted that one way to deal with this unphysical structure is to energy average the scattering amplitude (note, not the cross section) through it[10-13]. With the advent of greater computing capability it has become possible to do calculations with a large number of pseudostates (see, for example, references 14 to 16). Increasing the number of pseudostates enables a finer division of the eigenstate spectrum into "clumps" and hence a closer approximation to the exact expansion (5). It is therefore not surprising to find that the pseudostructure diminishes as the number of pseudostates is increased, becoming relatively innocuous at large but now manageable basis sizes[14,17].

It has been customary to construct pseudostates from orbitals whose radial part is expanded in a Slater basis, i.e., in terms of functions of the form

$$r^n e^{-\alpha r} \tag{11}$$

where n is a positive integer and α a constant (see, for example, reference 18). More recently a Laguerre basis has been used with much success[15,17,19], i.e.,

$$\left(\frac{\lambda(k-1)!}{(2l+1+k)!}\right)^{\frac{1}{2}} (\lambda r)^{l+1} L_{k-1}^{2l+2}(\lambda r) e^{-\lambda r/2}$$
$$k = 1, 2, 3, ... \tag{12}$$

where l is the angular momentum of the orbital, $L_{k-1}^{2l+2}(\lambda r)$ is a Laguerre polynomial[20] and λ is a parameter. The Laguerre basis (12) has two significant advantages over the Slater one (11). Firstly, it is more systematic so that convergence can be easily studied. Secondly, the basis functions (12) (for fixed λ and varying k) are mutually orthogonal so that adding a new function to the set is guaranteed to add something new to the expansion; by contrast the Slater functions are not mutually orthogonal and linear dependence problems can arise.

The common feature of the bases (11) and (12) is that, like eigenstates, they decay exponentially. There is, however, a third type of basis which is in use in the context of the new Intermediate Energy R-Matrix approximation (IERM)[14,16,21-33]. Here the basis functions, which are generated numerically, are "true" continuum functions, i.e., not exponentially damped, but are chopped to zero beyond some finite distance $r = a$. While the computations that have so far been performed in the IERM approximation

---

¶ This is actually the modulus squared of the definition given in reference 7

may be considered merely as large coupled pseudostate calculations with such a basis, the IERM approach is more general. By using propagation techniques[21,27] it should be possible to extend the IERM system wave function to very large distances in two of the electron coordinates simultaneously, thereby enabling a detailed description of the interaction between two electrons out to a large distance and hence a **realistic** representation of the ionization channels in (5). This is an exciting prospect for the future.

4. SPIN ASYMMETRY FOR ATOMIC HYDROGEN

Consider an electron with momentum k_0 and spin state χ_0 incident upon a hydrogen atom in the state $\psi_0 \chi_0^H$ where ψ_0 and χ_0^H are respectively the space and spin parts of the atomic wave function. Let the electron be scattered with momentum k_f and spin χ_f and let the atom state be changed to $\psi_f \chi_f^H$. The scattering amplitude for this process is

$$f_{f0} = -\frac{1}{2\pi} \left\langle e^{i k_f \cdot r_0} \chi_f(s_0) \psi_f(r_1) \chi_f^H(s_1) \left| \left(-\frac{1}{r_0} + \frac{1}{|r_0 - r_1|} \right) \right| \Psi \right\rangle \tag{13}$$

where Ψ is the correctly antisymmetrized system wave function. Assuming that the Hamiltonian is spinless, see (1), Ψ may be written

$$\Psi = \phi(r_0, r_1) \chi_0(s_0) \chi_0^H(s_1) - \phi(r_1, r_0) \chi_0(s_1) \chi_0^H(s_0) \tag{14}$$

where

$$\phi(r_0, r_1) \xrightarrow{r_0 \to \infty} e^{i k_0 \cdot r_0} \psi_0(r_1) + \text{scattered waves}$$
$$\phi(r_0, r_1) \xrightarrow{r_1 \to \infty} \text{scattered waves only} \tag{15}$$

The amplitude (13) can be expressed in terms of direct and exchange amplitudes f and g defined as

$$f = -\frac{1}{2\pi} \left\langle e^{i k_f \cdot r_0} \psi_f(r_1) \left| \left(-\frac{1}{r_0} + \frac{1}{|r_0 - r_1|} \right) \right| \phi(r_0, r_1) \right\rangle \tag{16}$$

$$g = -\frac{1}{2\pi} \left\langle e^{i k_f \cdot r_1} \psi_f(r_0) \left| \left(-\frac{1}{r_1} + \frac{1}{|r_0 - r_1|} \right) \right| \phi(r_0, r_1) \right\rangle \tag{17}$$

or in terms of singlet and triplet amplitudes f^S and f^T

$$f^S = f + g \qquad f^T = f - g \tag{18}$$

which describe the scattering in states of definite total spin. In general to fully determine f_{f0} a knowledge of $|f|$, $|g|$ and the phase γ of f relative to g is needed, the overall phase of f_{f0} being arbitrary.

For scattering of unpolarized electrons by unpolarized hydrogen atoms the differential cross section is given by

$$\sigma_u = \frac{k_f}{k_0} \left\{ \frac{3}{4} |f - g| + \frac{1}{4} |f + g| \right\} \tag{19}$$

where the components

$$\sigma^S = \frac{k_f}{k_0} |f + g|^2 \tag{20}$$

$$\sigma^T = \frac{k_f}{k_0} |f - g|^2 \tag{21}$$

of (19) are the cross sections for pure singlet(S) and triplet(T) scattering.

Consider now the case where a spin polarized electron is scattered by a polarized hydrogen atom but the spins in the final state are not analysed. If $\sigma(\uparrow\uparrow)$ ($\sigma(\uparrow\downarrow)$) is the cross section for a collision in which the initial spins are parallel (antiparallel) then

$$\sigma(\uparrow\uparrow) = \frac{k_f}{k_0}|f-g|^2 = \sigma^T \tag{22}$$

$$\sigma(\uparrow\downarrow) = \frac{k_f}{k_0}(|f|^2 + |g|^2) = \frac{1}{2}(\sigma^S + \sigma^T) \tag{23}$$

The spin asymmetry parameter, A, is defined by

$$A = \frac{\sigma(\uparrow\downarrow) - \sigma(\uparrow\uparrow)}{\sigma(\uparrow\downarrow) + \sigma(\uparrow\uparrow)} = \frac{1-r}{1+3r} \tag{24}$$

where $r = \sigma^T/\sigma^S$. A positive value for A implies that singlet scattering dominates triplet scattering, a negative value the reverse. From an experimental viewpoint only relative cross sections, rather than the more difficult absolute measurements, are needed to determine A. Although A has been defined in terms of differential cross sections σ the same definition (24) applies for integral cross sections. Measurements of the asymmetry parameter A have been made for $90°$ elastic scattering[34] and for ionization[34-36].

4.1 Spin Asymmetry for e⁻ + H(1s) Elastic Scattering

Van Wyngaarden and Walters[37] have calculated the spin asymmetry parameter for e⁻ + H(1s) elastic scattering in the energy range 0.136 to 300eV from the best theoretical information available to them at the time. From 0.136 to 8.7eV their results should be highly reliable, being based upon very accurate variational calculations. Figure 1 compares their calculations and two simpler approximations, namely, 1s-2s-2p eigenstate close-coupling and static exchange, with the only available experimental data for elastic scattering[34], at $90°$. While the experimental and theoretical trends are obviously in agreement, the error bars on the measurements, which correspond to one standard deviation (68% confidence limits), are too large to draw any detailed conclusions. Superficially, the experimental data seem to be in better agreement with the simpler approximations. Clearly more refined measurements and measurements over a larger energy and angular range would be highly desirable.

4.2 Spin Asymmetry for e⁻ + H(1s) Inelastic Scattering

Van Wyngaarden and Walters[38] have also calculated the spin asymmetry parameter for 2s, 2p and 2s + 2p excitation at energies from 13.6 to 300eV. Figure 2 shows their results for 2s excitation in the energy range 13.6 to 17.7eV. Much structure is observed. Figure 3 illustrates the interlacing of the individual singlet and triplet cross sections, σ^S and σ^T, which gives rise to this structure at 17.7eV ($k_0^2 = 1.30$au). Note that the pronounced behaviour in A appears where σ^S and σ^T are relatively small. In this respect the asymmetry is misleading in that it can over-dramatize a trivial situation. Much better would be a direct knowledge of σ^S and σ^T but this would require absolute measurements.

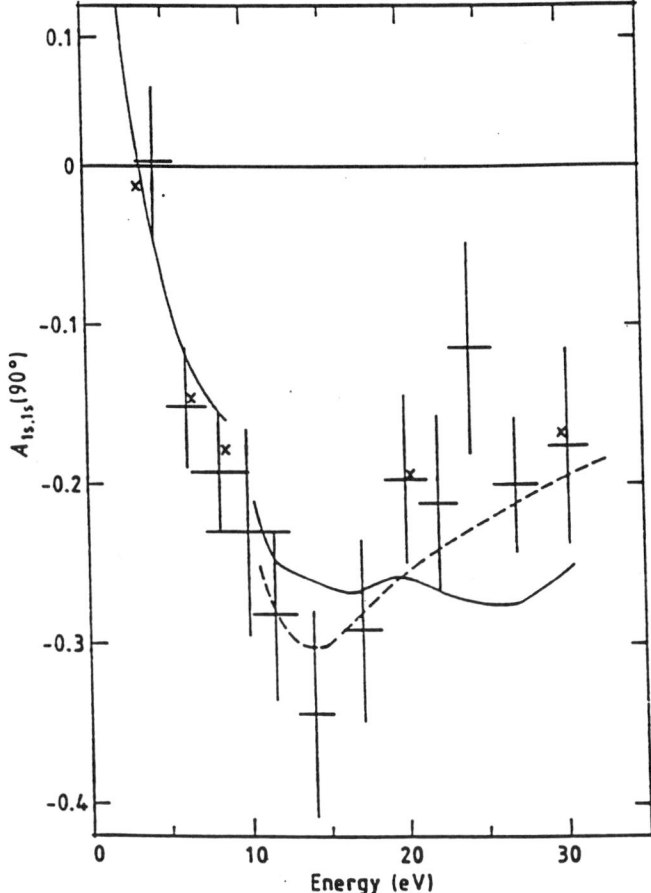

Figure 1. Spin asymmetry parameter $A_{1s,1s}(90°)$ for $e^- + H(1s)$ elastic scattering at $90°$. Solid curve, theoretical calculation of van Wyngaarden and Walters[37]; dashed curve, 1s-2s-2p eigenstate close-coupling approximation; ×, static-exchange approximation. Experimental data are from Fletcher et al[34].

5. SPIN ASYMMETRY AND $L_\perp$ FOR ELECTRON-SODIUM SCATTERING

Some recent experiments on electron-Na scattering using spin polarized electrons and atoms[39-41] demonstrate how useful spin polarized measurements are to advancing our theoretical understanding of electron-atom collisions. In these experiments the spin asymmetries in $e^- + Na(3^2S)$ elastic scattering and $Na(3^2S) \longrightarrow Na(3^2P)$ excitation have been determined as well as the angular momentum transfer to the 3^2P state perpendicular to the collision plane, $L_\perp$, for both singlet and triplet scattering.

As far as valence excitations are concerned the alkali metals behave like one-electron atoms and the formalism used for atomic hydrogen (section 4) may be taken over. In particular, the spin asymmetry is again defined by (24) while $L_\perp^{S,T}$ is given by

$$L_\perp^{S,T} = \frac{|f_+^{S,T}|^2 - |f_-^{S,T}|^2}{|f_+^{S,T}|^2 + |f_-^{S,T}|^2} \tag{25}$$

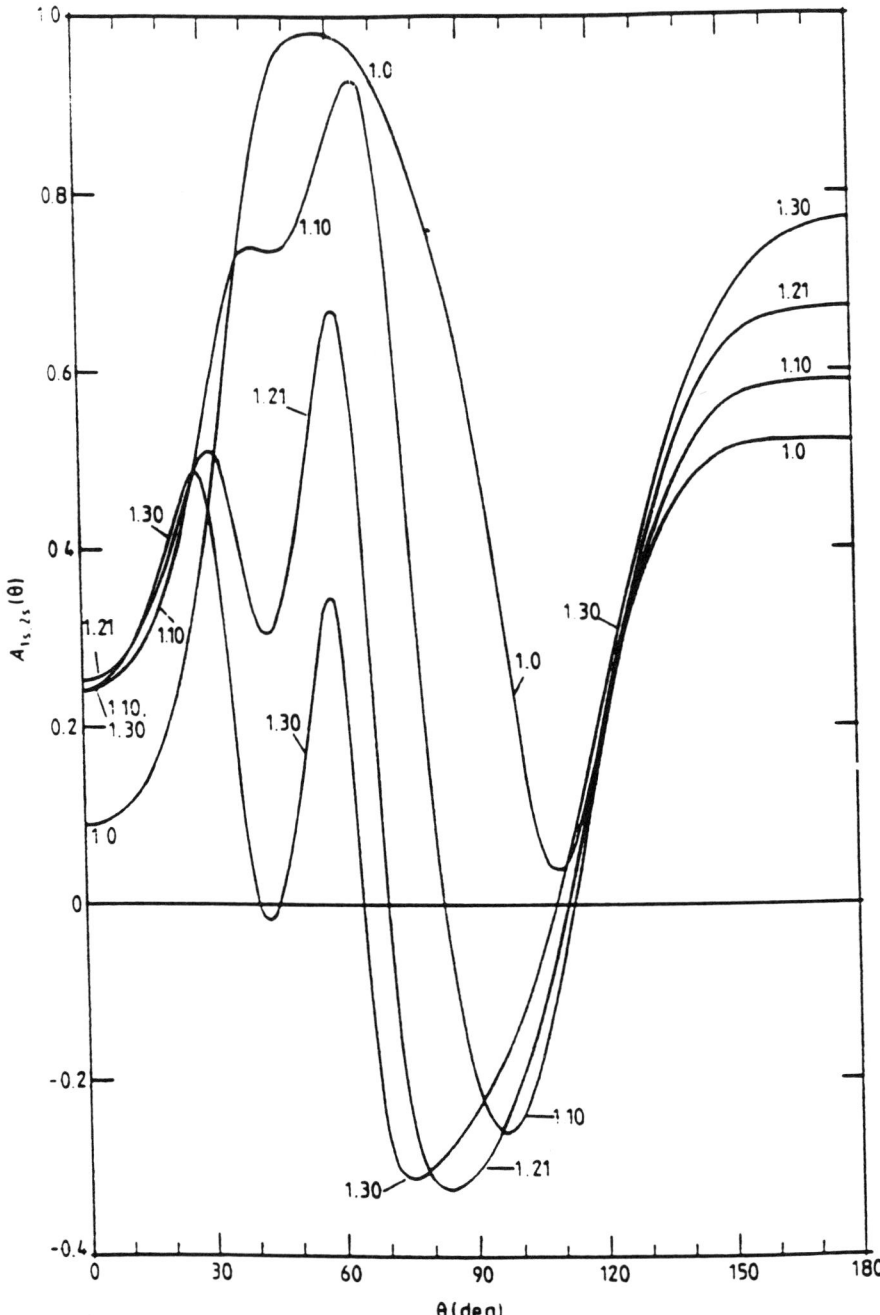

Figure 2. Theoretical results of van Wyngaarden and Walters[38] for the spin asymmetry $A_{1s,2s}$ for 2s excitation of atomic hydrogen at $k_0^2 = 1.0$, 1.10, 1.21 and 1.30 au (i.e, 13.6, 15.0, 16.5 and 17.7eV). The values of k_0^2 are indicated on the curves.

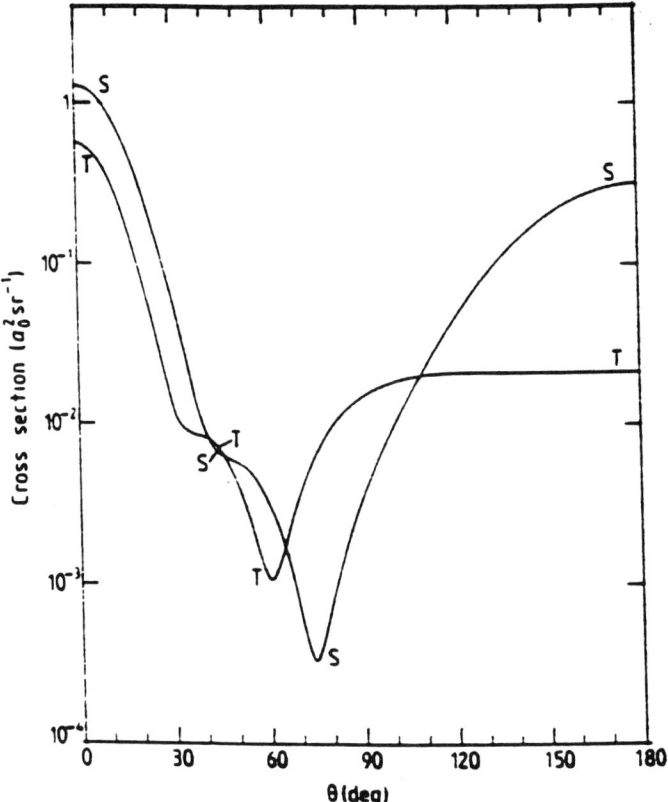

Figure 3. Theoretical results of van Wyngaarden and Walters[38] for the singlet(S) and triplet(T) differential cross sections for 2s excitation of atomic hydrogen at $k_0^2 = 1.30$ au (17.7eV).

where $f_+^{S,T}$ ($f_-^{S,T}$) is the singlet/triplet scattering amplitude for exciting the 3^2P magnetic substate with M = +1 (M = −1), the quantization axis being perpendicular to the scattering plane. Results for scattering at 20eV are exhibited in figure 4. This figure shows excellent agreement between experiment and a recent large pseudostate close-coupling calculation by Bray[42] but poor accord on the spin asymmetry parameters between experiment and an eigenstate close-coupling calculation taken to convergence in the bound eigenstates[43]. Clearly, therefore, the improvement given by the pseudostate approximation derives from its representation of couplings to the continuum states of the atom. As remarked in section 3.1, electron collisions with ground state alkalis present perhaps the best scenario for eigenstate close-coupling methods and indeed such methods yield good answers for the grosser features such as unpolarized cross sections. Figure 4 demonstrates the power of spin polarized measurements to probe much more deeply into the dynamics.

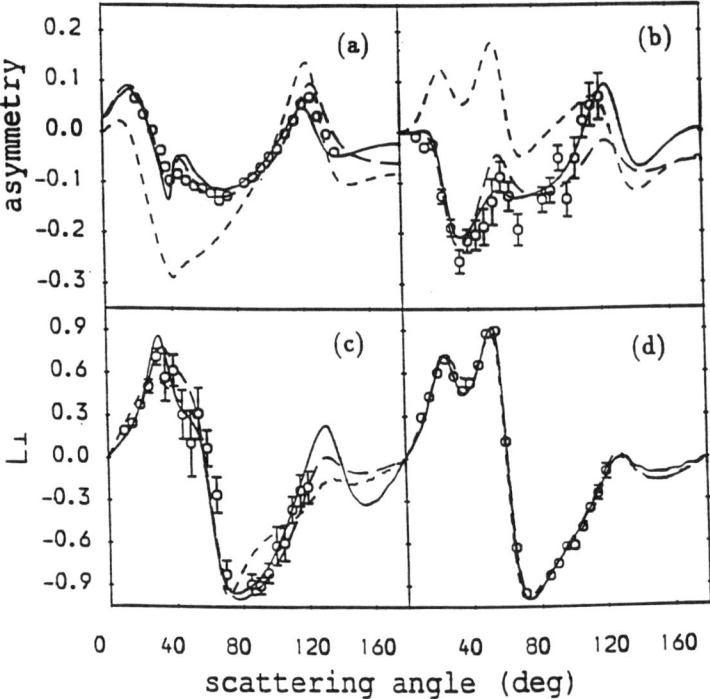

Figure 4. Electron scattering by ground state Na at 20eV : (a) spin asymmetry parameter for elastic scattering; (b) spin asymmetry parameter for 3^2P excitation; (c) $L_\perp^S$ for 3^2P excitation; (d) $L_\perp^T$ for 3^2P excitation. Solid curve, pseudostate close-coupling[42]; short-dash curve, eigenstate close-coupling[43]; long-dash curve, coupled-channel optical model[43]; o, experimental data of Kelley et al[41].

6. THE ELECTRON-PHOTON CORRELATION PROBLEM FOR 2P EXCITATION OF ATOMIC HYDROGEN

In 1986 van Wyngaarden and Walters[44-46] made extensive pseudostate close-coupling calculations of electron scattering by H(1s) in the energy range 54.4 to 350eV. Comparison with experimental data revealed some significant discrepancies. In particular there was a serious problem with regard to the electron-photon correlation parameters λ, R and I (see reference 44 for definitions) for (spin unpolarized) 2p excitation at 54.4eV. These results are shown in figures 5(a)-(c). Here it is seen that the coupled-pseudostate theory fails to predict the depth of the large angle dip in the measured λ parameter, gives a positive value for R at large angles in contrast to the negative result from experiment, and does not do well in comparison with the experimental values for the I parameter. The sign difference between theory and experiment for R is particularly disturbing since it is a qualitative rather than a quantitative discrepancy. and it is clear from figure 4 of reference 47 that a negative R parameter is definitely required to explain the phase of the observed sinusoidal coincidence count rate curve at 100^0. Subsequent calculations by Bray and Stelbovics[15] using a larger pseudostate basis and by Scholz et al[29] in the Intermediate Energy R-Matrix approximation (IERM) have served only to confirm the results of van Wyngaarden and Walters. The unanimity of sophisticated theoretical approximations in predicting a positive R value at large angles is striking. There is also a substantial body of experimental data on λ, R and I at 35eV[48]. Here again both IERM and the work of Bray and Stelbovics fail to predict the large angle dip in the λ parameter and, mysteriously, are not in agreement with the R parameter at small angles where one expects the theory to be at its best.

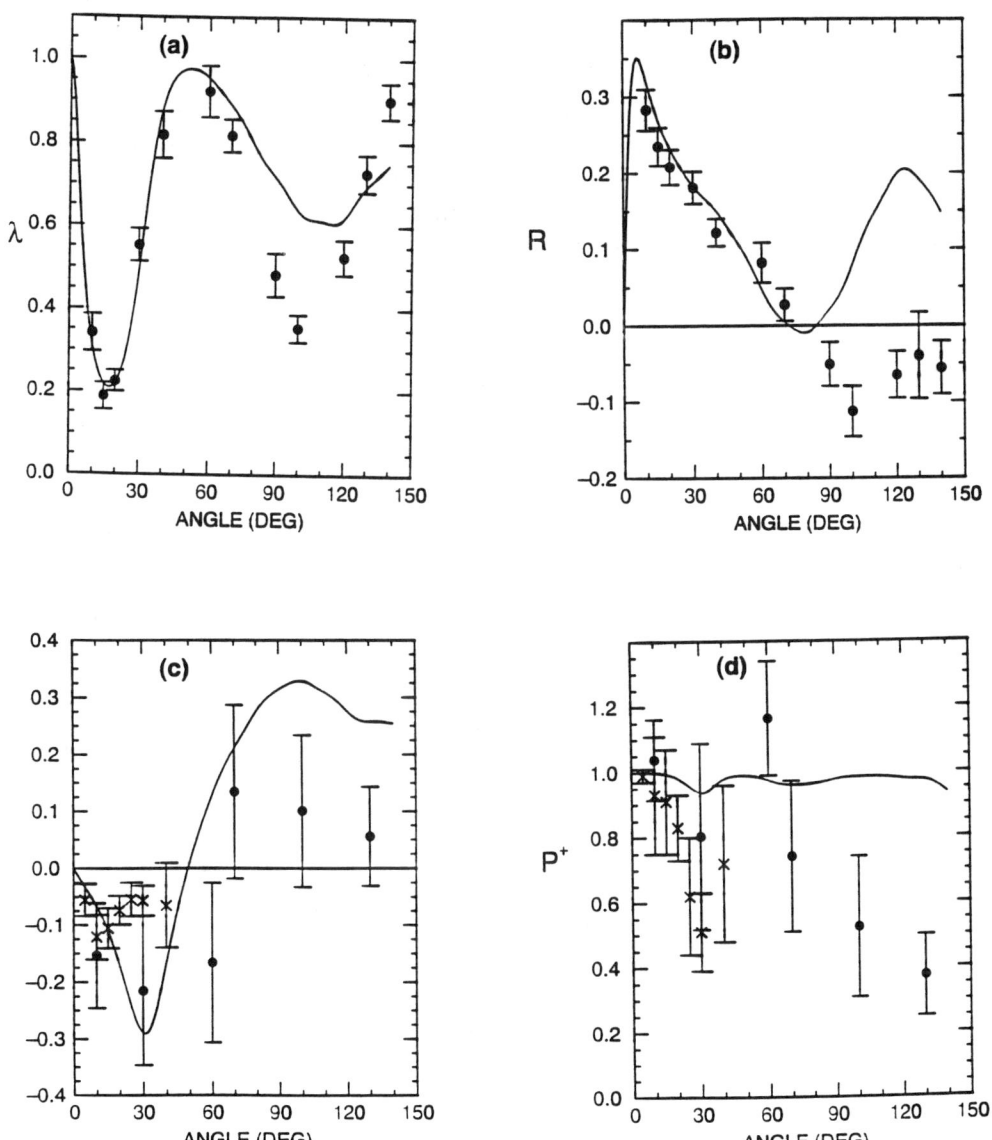

Figure 5. Angular correlation parameters λ, R, I and P[+] for 1s → 2p excitation of atomic hydrogen at 54.4eV : solid curve, pseudostate close-coupling approximation[44]; •, experimental data of Williams[47,50]; ×, experimental data of Nic Chormaic et al[51]. Error bars correspond to two standard deviations.

Andersen et al[49] have introduced a new set of parameters to replace λ, R and I. These new parameters are more closely related to the physics of the problem. In figure 5(d) one such parameter, the degree of polarization P^+, is shown for the 1s → 2p transition at 54.4eV. The significance of P^+ is that it takes the value unity if exchange scattering is zero. Figure 5(d) compares the coupled-pseudostate calculation of van Wyngaarden and Walters[44] with experimental results for P^+ deduced from the measurements of Williams[47,50] and from the more recent work of Nic Chormaic et al[51]. Whereas the theoretical curve remains close to unity over the angular range shown ($0°$ to $140°$), suggesting that exchange effects are not very important, the experimental data, especially the new data of Nic Chormaic et al[51] in the smaller angular range $5°$ to $40°$, can deviate markedly from the value unity, implying the opposite. Obviously an experiment with spin polarized particles would be extremely useful in resolving this problem. Indeed, there is a great need for a detailed set of experiments, even with unpolarized particles, at a sequence of energies up to at least 54.4eV to enable a study to be made of how the discrepancies between theory and experiment at 35 and 54.4eV develop and change with energy. Such a body of data would be very valuable.

7. IONIZATION

7.1 Ionization of H(1s)

Recently[52] a notable success has been scored by the coupled-pseudostate approximation in describing the total cross section and corresponding spin asymmetry parameter (24) for electron impact ionization of H(1s) at energies above the threshold region. This cross section had defied accurate theoretical treatment for a long period of time. The ionization cross section has been extracted from the coupled-pseudostate calculations by employing the ansatz

$$\sigma_{ion} = \sigma_{tot} - \sum_{n=0}^{M} \sigma_n - \sum_{\substack{n=(M+1) \\ \bar{\epsilon}_n < 0}}^{P} \bar{\sigma}_n \left\{ \sum_{\substack{m=0 \\ \epsilon_m < 0}}^{\infty} f_n(\epsilon_m) \right\} \qquad (26)$$

In this ansatz the ionization cross section is calculated by removing from the total cross section, σ_{tot}, that part coming from bound state excitation. The first sum over n in (26) deletes the bound eigenstate cross sections, σ_n, the second removes the cross sections $\bar{\sigma}_n$ for exciting pseudostates $\bar{\psi}_n$ with negative energy weighted by the fraction of each $\bar{\psi}_n$ overlapping the bound eigenstate spectrum, this fraction is obtained from the energy distribution function $f_n(\epsilon)$ of (10). The total cross section, σ_{tot}, is just the sum of all eigenstate, σ_n, and pseudostate, $\bar{\sigma}_n$, cross sections in the approximation. The results are shown in figure 6. The agreement with experiment for both the ionization cross section and the spin asymmetry parameter seen in this figure provides very substantial support for the coupled-pseudostate approach to electron-atom scattering.

7.2 Threshold Ionization

Ever since the pioneering work of Wannier[54] using classical mechanics the threshold behaviour of the cross section for electron impact ionization of atoms and ions has been a subject of much interest (see, for example, references 55 to 64). In the Wannier picture the threshold configuration of the final state consists of the two outgoing electrons moving in opposite directions away from the ion core and with equal speed. The

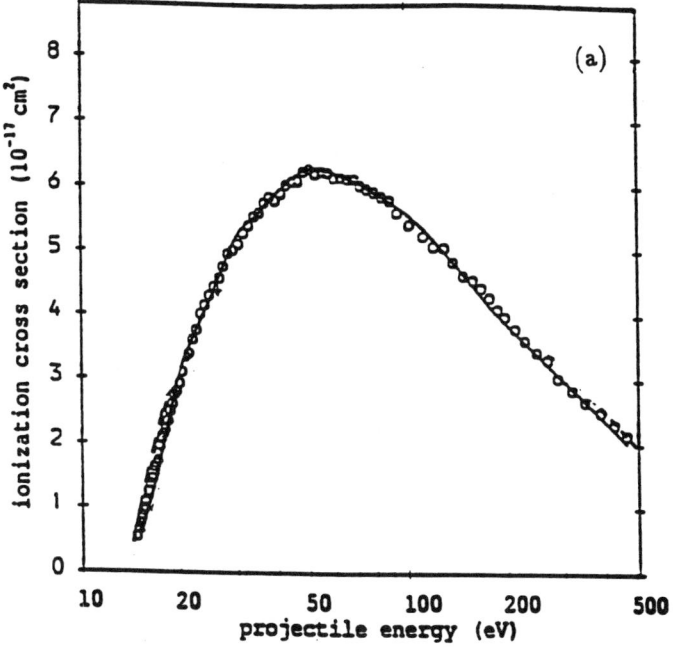

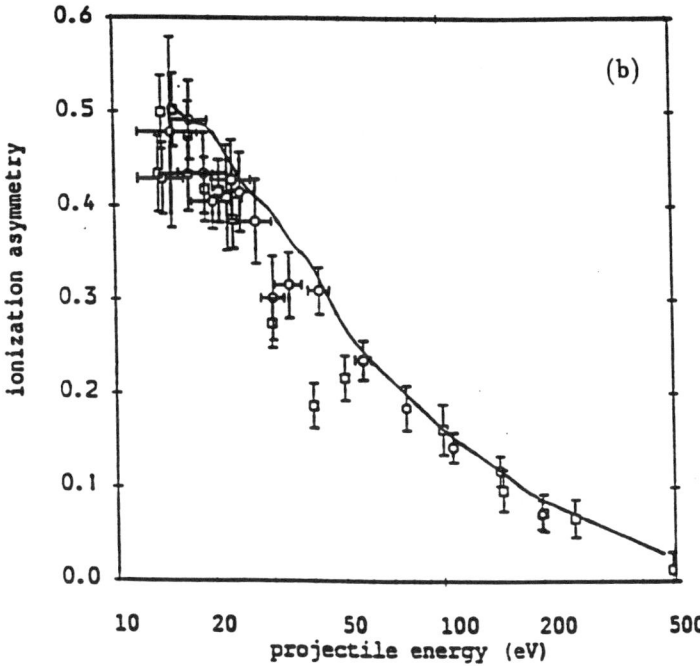

Figure 6. Cross section (a) and spin asymmetry parameter (b) for electron impact ionization of H(1s). Solid curve is the coupled-pseudostate calculation of reference 52. Experimental data for the total ionization cross section are from Shah et al[53], for the asymmetry parameter from Fletcher et al[34] and Crowe et al[36].

Wannier threshold law for the ionization cross section, σ_{ion} is

$$\sigma_{ion} = aE^\eta + bE^{3\eta} \tag{27}$$

where

$$\eta \equiv \frac{1}{4}\left[\left(\frac{100Z' - 9}{4Z' - 1}\right)^{1/2} - 1\right] \tag{28}$$

a and b are constants, Z' is the residual ion charge and E is the residual energy (i.e., the sum of the kinetic energies of the two electrons). For $Z' = 1$ η takes on the value 1.127. The power η depends only upon the correlation between the two electrons as they move outwards and not upon what happens near the ion, i.e., not on the details of the ionization mechanism. An important question is "over what energy range E is (27) valid ?" In an experiment on ionization of H(1s) McGowan and Clarke[65] found that their data were consistent with the threshold law (27) up to about $E = 0.4$eV.

The experiment of McGowan and Clarke was performed using unpolarized electrons and atoms. More recently Guo et al[35] have made measurements of the spin asymmetry parameter in the threshold region. Writing the threshold law for the singlet and triplet cross sections as

$$\sigma_S = a_S E^\eta + b_S E^{3\eta} \qquad \sigma_T = a_T E^\eta + b_T E^{3\eta} \tag{29}$$

the spin asymmetry (24) becomes

$$A = \frac{(a_S - a_T) + (b_S - b_T)E^{2\eta}}{(a_S + 3a_T) + (b_S + 3b_T)E^{2\eta}} \tag{30}$$

leading to

$$\frac{dA}{dE} = \frac{8\eta}{(1 + 3\sigma_T/\sigma_S)^2} \frac{a_S b_S E^{2\eta - 1}}{(a_S + b_S E^{2\eta})^2} \left\{\frac{a_T}{a_S} - \frac{b_T}{b_S}\right\} \tag{31}$$

Exactly at threshold $dA/dE = 0$ but at finite E it is non-zero, its sign being given by the bracket in (31). In an analysis of the data of Guo et al[35], Guo and Lubell[66] find that A can be fitted with a (constant) positive slope dA/dE of (4.4 ± 1.5) eV^{-1} in the energy regime where the unpolarized results of McGowan and Clarke fit the Wannier law. From plausibility arguments Guo and Lubell estimate that a_S, a_T, b_S and b_T of (29) satisfy the inequalities $a_T < a_S, b_T > b_S$ which would imply that dA/dE of (31) should be negative, contrary to the experiment. The implication then is that the Wannier law, if it holds at all, must hold over a much smaller energy region than the unpolarized experiment of McGowan and Clarke suggests, a result not inconsistent with some theoretical work of Kazansky and Ostrovsky[63].

Non-Wannier behaviour may be admitted into the formulae (29) by replacing the constants $a_{S,T}$ and $b_{S,T}$ by functions of energy, $f_{S,T}(E)$ and $g_{S,T}(E)$ say. Assuming that $f_{S,T}(E)$ and $g_{S,T}(E)$ are analytic about $E = 0$, and so can be expanded in a power series in E, Guo and Lubell[66] and Lubell[67] have made an analysis of the spin asymmetry data for ionization of H, metastable He, Li, Na, K and Cs. In each case, except perhaps for Li, they find that the threshold value of dA/dE is significantly different from zero, contrary to the pure Wannier result (31). They also find that non-Wannier behaviour becomes visible in the individual singlet and/or triplet cross sections at substantially lower energies than for the spin-averaged cross section. In the case of H(1s) this energy is as low as 0.06eV and 0.03eV for the singlet and triplet cross sections respectively, which contrasts with the 0.4eV of the unpolarized cross section of McGowan and Clarke[65]. However, the assumption that $f_{S,T}(E)$ and $g_{S,T}(E)$ are analytic about $E = 0$, on which the details of the above analysis are based, must be called into question. In some recent theoretical work[68] terms involving $\sqrt{E}$ and $E \ln E$, which are not analytic at $E = 0$, have appeared in higher order corrections to the Wannier form (29).

Prompted by the deviations from the Wannier law seen in the spin asymmetry for electron impact ionization, Freedman et al[69] have re-examined the experimental data for two-electron photodtachment from H^- and K^- near threshold in which the two escaping electrons emerge in a pure spin singlet state. Applying more sophisticated statistical methods to the data than used previously, they find significant structure in the results, which is not compatible with the Wannier law.

Although much of the analysis descibed in this section is of a quite speculative nature, it would appear that the study of spin effects in the threshold behaviour of two-electron escape could prove very valuable indeed. More experimental studies with higher precision are to be strongly urged. In particular[67] a measurement of the ionization asymmetry for Rb, a re-measurement of the same for Li, and more two-electron photodetachment studies of negative ions are badly needed. Even considerably more accurate spin averaged ionization cross sections for valence-1 systems are very desirable.

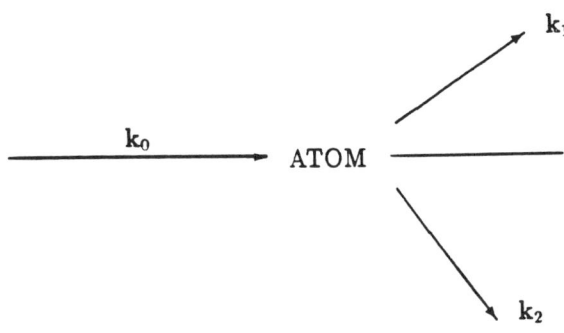

Figure 7.

The most sensitive probe of ionization is provided by (e, 2e) measurements[70]. The set-up is illustrated in figure 7. Here an incident electron with momentum (energy) k_0 (E_0) ionizes the atom and the two outgoing electrons in the final state with momenta (energies) k_1 (E_1) and k_2 (E_2) are detected in coincidence. The results of an (e, 2e) measurement are expressed in terms of the triple differential cross section (TDCS), $d^3\sigma/d\Omega_1 d\Omega_2 dE$, which is the cross section for finding the two outgoing electrons within solid angles $d\Omega_1$ and $d\Omega_2$ about the directions of k_1 and k_2 respectively and with energies in the range dE about the values E_1 and E_2.

Until very recently (e, 2e) experiments had been conducted with spin unpolarized beams and targets. In 1992 Baum et al[71] (see also references 72 and 73) reported an (e, 2e) determination of the spin asymmetry (24) for a Li target. These measurements were made in energy-sharing coplanar geometry in which the incident and outgoing electrons all move in the same plane and $E_1 = E_2$. The angle of k_1 was fixed at 45^0 to the incident direction (described in the anticlockwise sense, see figure 7) and the TDCS was measured as a function of the angle θ_2 between k_2 and the incident direction (described in the clockwise sense). The experimental results for $E_0 = 54.4 eV$ and

$35^0 \leq \theta_2 \leq 80^0$ are shown in figure 8 where they are compared with the distorted-wave Born approximation (DWBA) calculations of Zhang et al[74]. Because, for reasons of getting a sufficiently large count rate, the angular resolution of the experiment is quite big, $\pm 10^0$, it is necessary to convolute the calculations with the experimental resolution to get a valid comparison with the data. When this is done quite good agreement between theory and experiment is obtained, figure 8. More refined measurements would be highly desirable to test the theory even more rigorously.

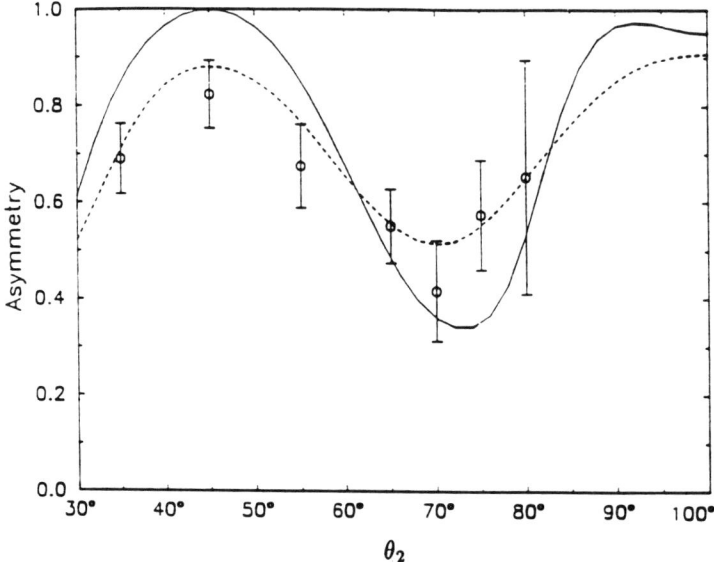

Figure 8. Spin asymmetry parameter for $e^- + \text{Li}(2^2S) \to \text{Li}^+ + e^- + e^-$ in coplanar energy-sharing geometry with $E_0 = 54.4\text{eV}$, $E_1 = E_2 = 24.5\text{eV}$, $\theta_1 = 45^0$. Experimental data are from Baum et al[71]; solid curve is the DWBA calculation of Zhang et al[74]; dashed curve is the same calculation convoluted with the experimental angular resolution.

Another interesting application of spin analysis in (e, 2e) processes has been highlighted by Hanne[75] and Jones et al[76]. It has been known for some time that spin dependent effects can be observed in excitation of unpolarized atoms by polarized electrons if the final fine structure state of the atom is resolved. This is known as the "fine structure effect"[2,77]. Here the fine structure state serves only to define a particular coherent combination of orbital and spin angular momenta of the atom, the actual collision process requiring only the spinless Hamiltonian (1) for its description. A similar situation should apply to (e, 2e). Jones et al[76] have made calculations of the TDCS for ionization of ground state Xe with incident electrons polarized with spin-up and spin-down relative to the scattering plane and where the final fine structure states $^2P_{1/2}$ and $^2P_{3/2}$ of the ion are resolved. Their results are shown in figure 9 for the case of near threshold energy-sharing, $E_1 = E_2 = 1\text{eV}$, coplanar geometry where the two outgoing electrons move in opposite directions, $\theta_{12} = 180^0$, and the cross section is given as a function of the angle θ_1 between electron 1 and the incident direction. It is clear that the spin effects are quite significant.

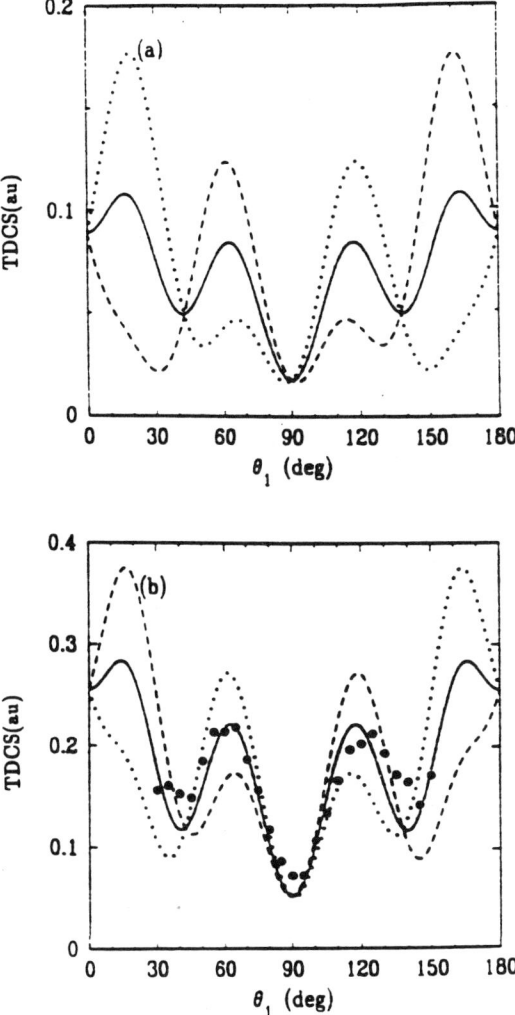

Figure 9. TDCS for electron impact ionization of Xe in coplanar energy-sharing geometry with $E_1 = E_2 = 1\text{eV}$ and $\theta_{12} = 180^0$: (a) leaving the ion in the $5p^5\ ^2P_{1/2}$ state ($E_0 = 15.44\text{eV}$); (b) leaving the ion in $5p^5\ ^2P_{3/2}$ state ($E_0 = 14.13\text{eV}$). Dashed (dotted) curve, for incident electron with spin up (down); solid curve, for unpolarized incident electron. Experimental data in (b) are the relative measurements of Rösel et al [78] for unpolarized electrons.

8. RELATIVISTIC HAMILTONIAN

As the nuclear charge Z increases relativistic effects become more important not only in the description of the target state but even for low incident energies of the scattered electron. An appropriate relativistic generalization of the spinless Hamiltonian (1) is provided by the Dirac Hamiltonian[79]

$$H^D = \sum_{i=0}^{N} \left(c\boldsymbol{\alpha}\cdot\mathbf{p}_i + (\beta - \mathbf{I})c^2 - \frac{Z}{r_i} \right) + \sum_{\substack{i,j=0 \\ i<j}}^{N} \frac{1}{|\mathbf{r}_i - \mathbf{r}_j|} \tag{32}$$

where $\mathbf{p}_i = -i\nabla_i$, $\boldsymbol{\alpha}$ and β are the usual Dirac matrices[80], $\mathbf{I}$ is the unit matrix, c is the velocity of light and wave functions are four component spinors. The Hamiltonian (32) takes account of the spin-orbit interaction of each electron (see (34) below) but not of other relativistic phenomena such as spin-spin, spin-other orbit or retardation effects.

A semi-relativistic approximation to H^D is provided by the Breit-Pauli Hamiltonian[81,82]:

$$H^{BP} = H^{NR} + H^{SO} + H^{MASS} + H^{DARWIN} \tag{33}$$

where H^{NR} is the non-relativistic spinless Hamiltonian (1) and

$$H^{SO} = \frac{1}{2}\left(\frac{1}{c}\right)^2 \sum_{i=0}^{N} \frac{1}{r_i}\frac{d\overline{V}}{dr_i}(r_i)(\mathbf{l}_i \cdot \mathbf{s}_i) \tag{34}$$

$$H^{MASS} = -\frac{1}{8}\left(\frac{1}{c}\right)^2 \sum_{i=0}^{N} \nabla_i^4 \tag{35}$$

$$H^{DARWIN} = \frac{1}{8}\left(\frac{1}{c}\right)^2 \sum_{i=0}^{N} \nabla_i^2 \overline{V}(r_i) \tag{36}$$

H^{SO} is the spin-orbit interaction, $\mathbf{l}_i$ and $\mathbf{s}_i$ being respectively the orbital and spin angular momentum operators for the i th electron, H^{MASS} and H^{DARWIN} are the so-called "mass-correction" and "Darwin" terms and $\overline{V}(r_i)$ is the average potential felt by the i th electron (this will be discussed more fully below in the context of spin-orbit coupling in Cs).

An interesting question is "how do H^{BP} and H^D compare in practice ?" As an answer to this question we are fortunate in having two recent eigenstate close-coupling calculations of electron scattering by Cs ($Z = 55$), one in the Breit-Pauli formalism, by Bartschat[83], the other using the Dirac Hamiltonian (32), by Thumm and Norcross[84-86]. Both calculations treat the Cs atom as a one-electron system with a core and use very similar, although not absolutely identical, core interaction potentials. The Breit-Pauli formulation serves to demonstrate the nature of the approximations made.

The Breit-Pauli Hamiltonian used by Bartschat takes the form

$$H^{BP} = \frac{1}{2}\sum_{i=0}^{1}\left(-\nabla_i^2 + 2V_{core}(r_i) + \left(\frac{1}{c}\right)^2 \frac{1}{r_i}\frac{d\overline{V}}{dr_i}(r_i)(\mathbf{l}_i \cdot \mathbf{s}_i)\right) + \frac{1}{|\mathbf{r}_1 - \mathbf{r}_0|} + V_{diel}(\mathbf{r}_0, \mathbf{r}_1) \tag{37}$$

In (37) V_{core} represents the interaction of the outer electron (i.e., the valence or scattered electron) with the core. V_{core} consists of an electrostatic component, V_{static}, an exchange contribution, V_{exch}, and a polarization term V_{pol}^l:

$$V_{core}(r) = V_{static}(r) + V_{exch}(r) + V_{pol}^l(r) \tag{38}$$

V_{pol}^l depends upon the orbital angular momentum l of the outer electron and is given by

$$V_{pol}^l(r) = -\frac{\alpha_d(l)}{2r^4}\left\{1 - \exp\left[-\left(\frac{r}{r_c(l)}\right)^6\right]\right\} \tag{39}$$

where the dipole polarizability of the core $\alpha_d(l)$ and the cut-off radius $r_c(l)$ are treated as adjustable parameters. The core polarization induced by one outer electron also gives rise to an interaction with the second outer electron[87], this is represented by the dielectronic potential, V_{diel}, in (37). V_{diel} is parameterized as

$$V_{diel}(\mathbf{r}_0, \mathbf{r}_1) = -\frac{\alpha_d(\mathbf{r}_0 \cdot \mathbf{r}_1)}{r_0^3 r_1^3}\left[\left(1-\exp\left\{-\left(\frac{r_0}{r_{diel}}\right)^6\right\}\right)\left(1-\exp\left\{-\left(\frac{r_1}{r_{diel}}\right)^6\right\}\right)\right]^{\frac{1}{2}} \quad (40)$$

where α_d is the (averaged over l, see(39)) dipole moment of the core and r_{diel} is an adjustable cut-off.

In earlier work by Scott et al[88,89] the average potential $\overline{V}$ in the spin-orbit term of (37) was taken to be the bare nuclear potential, i.e.,

$$\overline{V}(r) = -\frac{Z}{r} \quad (41)$$

However, this ignores the screening effect of the other electrons in the atom. To take account of screening Bartschat has considered two possibilities, an ab initio approximation with

$$\overline{V}(r) = V_{core}(r) \quad (42)$$

and a screened-charge approximation

$$\overline{V}(r) = -\frac{Z_{scr}(l)}{r} \quad (43)$$

where the screening depends upon the orbital angular momentum of the outer electron.

The adjustable parameters $\alpha_d(l)$, $r_c(l)$, r_{diel} and $Z_{scr}(l)$ have been chosen to give good energies for the Rydberg series in Cs and for the electron affinity of Cs$^-$. This fitting also allows indirectly for the mass-correction and Darwin terms (see (35) and (36)) which are not explicitly included in the model Hamiltonian (37).

The Dirac calculations of Thumm and Norcross and the Breit-Pauli calculations of Bartschat use the five Cs eigenstates 6s $^2S_{1/2}$, 6p $^2P^0_{1/2,3/2}$ and 5d $^2D_{3/2,5/2}$. In addition Bartschat has also made 8-state calculations by adding the 7s $^2S_{1/2}$ and 7p $^2P^0_{1/2,3/2}$ eigenstates to the set. Good agreement between the Dirac and Breit-Pauli work is obtained for the integrated cross sections. Agreement on differential cross sections is also good but not perfect. Where the real discrepancies occur is for spin asymmetry parameters. Figure 10 shows the spin asymmetries $S, A, A_\perp$ and $A_\parallel$ for elastic scattering by ground state Cs at 2.04eV. $S(A)$ is defined as the left-right asymmetry in the differential cross section when polarized (unpolarized) electrons are scattered by unpolarized (polarized) atoms; $A_\parallel$ ($A_\perp$) is the up-down asymmetry in the differential cross section when polarized electrons are scattered by polarized atoms with polarizations in (perpendicular to) the scattering plane. Non-relativistically (i.e., with the Hamiltonian (1)) $S = A = 0$ and $A_\parallel = A_\perp$. Figure 10 shows significant differences between the Breit-Pauli and Dirac predictions and also substantial deviations from the non-relativistic result $A = 0$ in the Dirac case.

Figure 11 shows the spin parameters S_p and S_A for the 6s $^2S_{1/2} \to$ 6p $^2P^0_{1/2,3/2}$ excitations at 2.04eV. S_p is the polarization of an initially unpolarized beam after scattering by an unpolarized atom. S_A is the left-right asymmetry in the differential cross section when polarized electrons are scattered by unpolarized atoms. Here again there are substantial differences between the Breit-Pauli and Dirac predictions. Clearly there is a question mark over the Breit-Pauli approximation in so far as spin asymmetry parameters are concerned, although it does well on cross sections. An experimental test of the predictions of figures 10 and 11 would be highly desirable.

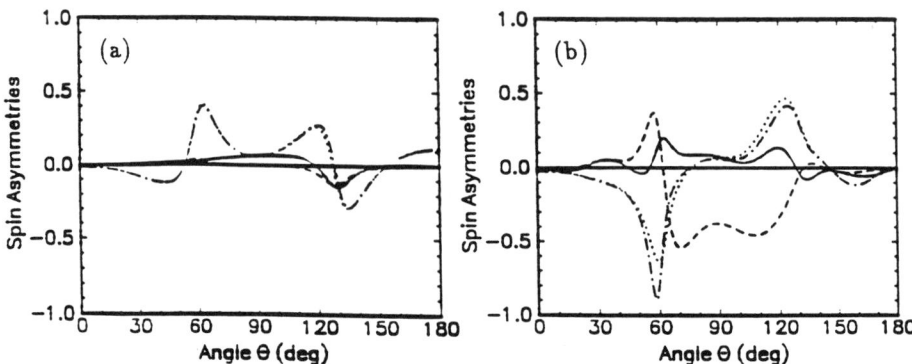

Figure 10. Spin asymmetries for e⁻ + Cs elastic scattering at 2.04eV calculated in a 5-state close-coupling approximation using (a) the Breit-Pauli Hamiltonian[83] and (b) the Dirac Hamiltonian[90] : solid curve, S; dashed curve, A; dash-dot curve, $A_\perp$; dotted curve, $A_\parallel$. See text for states and definitions of asymmetries.

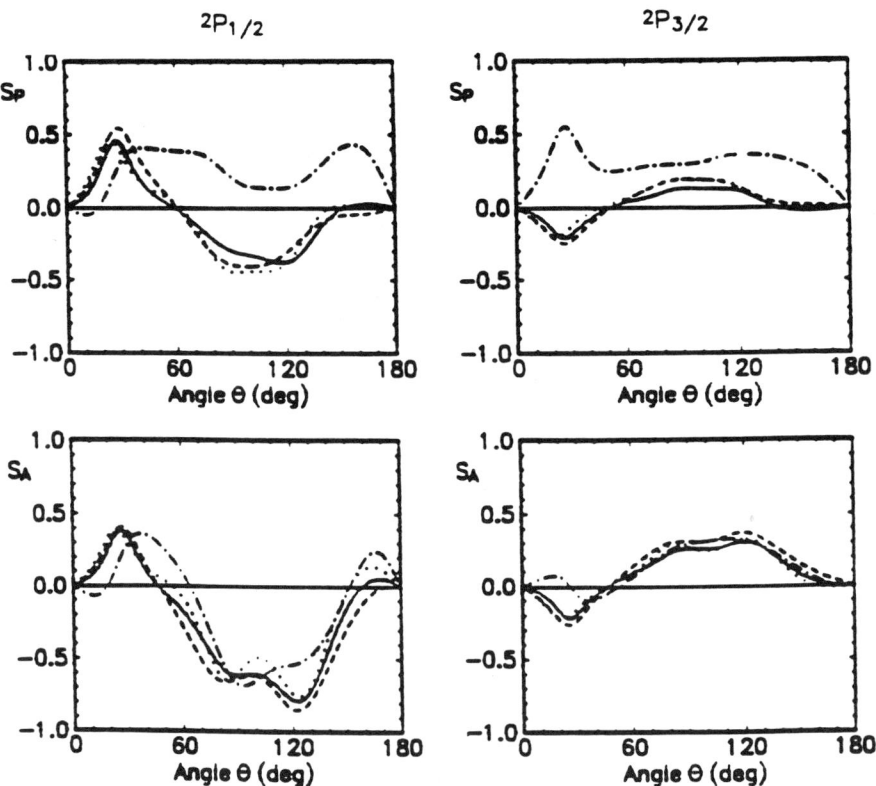

Figure 11. Spin parameters S_P and S_A (see text for definitions) for electron impact excitation of the 6s $^2S_{1/2} \to$ 6p $^2P^0_{1/2,3/2}$ transitions in Cs at an incident energy of 2.04eV : solid curve, 8-state Breit-Pauli calculation[83]; dashed curve, 5-state Breit-Pauli approximation[83]; dash-dot curve, 5-state Dirac approximation[90]; dotted curve, Breit-Pauli calculation of Scott et al[88,89].

9. CONCLUSIONS

From a theorist's viewpoint, I would strongly urge the development of experiments and refinement of techniques using spin polarized electrons and atoms. As I have tried to illustrate here, such experiments provide a very stiff test of approximations and a challenge to our theoretical understanding. The latter is well illustrated by the case of threshold ionization (section 7.2) where spin asymmetry measurements are substantially altering our views on the range of applicability of the Wannier threshold law. Experiments which combine spin analysis with a determination of an electron-photon coincidence parameter, such as the experiment on Na described in section 5, are highly useful. Indeed the addition of spin analysis to coincidence experiments of the electron-photon type (section 6) or (e, 2e) type (section 7.3), difficult though that may be in practical terms, would be very valuable. In short, it is my opinion that theory would greatly benefit from a very much broader base of experimental information derived from polarized electron/atom measurements than presently exists.

Some specific experimental projects which are suggested by this short review may be summarized as follows. The programme of spin asymmetry measurements on atomic hydrogen needs to be greatly widened. Atomic hydrogen is the most fundamental test for scattering theory since it is the simplest atom and its wave functions are known exactly. To have only a knowledge of the elastic spin asymmetry at 90^0 (figure 1), and that tantalizingly just on the verge of agreement with theory, is frustrating. The disagreement between experiment and theory on the electron-photon coincidence parameters for 2p excitation of atomic hydrogen (section 6) remains a major unsolved problem. If the experiments are correct, the present situation suggests that there is something not quite right about the treatment of exchange in the theoretical calculations. This is an ideal problem to study using spin polarized electrons/atoms, if it can be done. But even if it cannot be done, a very much greater quantity of data on this system using unpolarized electrons and hydrogen atoms, and with greater accuracy, would be very welcome indeed to theorists. Spin asymmetry measurements on threshold ionization (section 7.2) are turning into a success story but much more experimental information is required, especially on "one-electron" systems, i.e., atomic hydrogen and the alkali metals; also more studies of threshold two-electron photodetachment of negative ions are highly desirable. The promising work on spin asymmetry in (e, 2e) collisions initiated by Baum and co-workers (section 7.3) needs to be taken much further and with a smaller angular resolution; the importance of (e, 2e) measurements to the understanding of ionization cannot be understated. Last, but not least, the study of relativistic effects in electron-atom collisions should be further encouraged. In particular, theorists need to know to what extent the Breit-Pauli approximation can be trusted. The two recent calculations on Cs using the Breit-Pauli and Dirac Hamiltonians (section 8), which lead to significant differences in the prediction of some spin parameters, provide an ideal case for experimental investigation.

Finally, with regard to the status of theory, it is my feeling that pseudostate close-coupling (section 3.3), with a sufficiently large basis, is probably the most reliable and generally applicable theoretical method available to us today for the study of electron-atom collisions.

ACKNOWLEDGEMENTS

I am greatly indebted to Ann Kernoghan for help with the preparation of the diagrams and to Dr Patrick Norrington for advice on relativistic scattering. This work was also supported by the Science and Engineering Research Council.

REFERENCES

1. J.Kessler,"Polarized Electrons", Springer-Verlag, Berlin (1985).
2. J.Kessler, Adv.At.Mol.Phys. **27** 81 (1990).
3. I.Percival and M.J.Seaton, Proc.Camb.Phil.Soc. **53** 654 (1957).
4. P.G.Burke and A.J.Taylor, Proc.Phys.Soc. **88** 549 (1966).
5. A.J.Taylor and P.G.Burke, Proc.Phys.Soc. **92** 336 (1967).
6. A.E.Kingston and H.R.J.Walters, Comm.At.Mol.Phys. **11** 177 (1982).
7. H.R.J.Walters, J.Phys.B **21** 1277 (1988).
8. P.G.Burke and T.G.Webb, J.Phys.B **3** L131 (1970).
9. P.G.Burke and J.F.B.Mitchell, J.Phys.B **6** 320 (1973).
10. D.H.Oza and J.Callaway, Phys.Rev.A **27** 2840 (1983).
11. D.H.Oza, Phys.Rev.A **30** 1101 (1984).
12. J.Callaway, Phys.Rev.A **32** 775 (1985).
13. P.G.Burke, K.A.Berrington and C.V.Sukumar, J.Phys.B **14** 289 (1981).
14. T.T.Scholz, J.Phys.B **24** 2127 (1991).
15. I.Bray and A.T.Stelbovics, Phys.Rev.A **46** 6995 (1992).
16. B.R.Odgers, M.P.Scott and P.G.Burke, J.Phys.B **27** 2577 (1994).
17. I.Bray and A.T.Stelbovics, Phys.Rev.Lett. **69** 53 (1992).
18. W.C.Fon, K.A.Berrington, P.G.Burke and A.E.Kingston, J.Phys.B **14** 1041 (1981).
19. A.T.Stelbovics, J.Phys.B **22** L159 (1989).
20. I.S.Gradshteyn and I.M.Ryzhik, "Tables of Integrals, Series, and Products", Academic Press, New York (1980).
21. P.G.Burke, C.J.Noble and M.P.Scott, Proc.Roy.Soc. **A410** 289 (1987).
22. T.Scholz, P.Scott and P.G.Burke, J.Phys.B **21** L139 (1988).
23. T.T.Scholz, M.P.Scott, P.G.Burke and C.J.Noble, in "Electronic and Atomic Collisions (Invited Papers of the XV International Conference on the Physics of Electronic and Atomic Collisions)", H.B.Gilbody, W.R.Newell, F.H.Read and A.C.H.Smith, eds., North-Holland, Amsterdam, page 215 (1988).
24. M.P.Scott, T.T.Scholz, H.R.J.Walters and P.G.Burke, J.Phys.B **22** 3055 (1989).
25. K.Higgins, P.G.Burke and H.R.J.Walters, J.Phys.B **23** 1345 (1990).
26. T.T.Scholz, H.R.J.Walters and P.G.Burke, J.Phys.B **23** L467 (1990).
27. M.LeDourneuf, J.M.Launay and P.G.Burke, J.Phys.B **23** L559 (1990).
28. T.T.Scholz, PhD Thesis, The Queen's University of Belfast (1990).
29. T.T.Scholz, H.R.J.Walters, P.G.Burke and M.P.Scott, J.Phys.B **24** 2097 (1991).
30. T.T.Scholz, in "Electronic and Atomic Collisions (Proceedings of the Seventeenth International Conference on the Physics of Electronic and Atomic Collisions)", W.R.MacGillivray, I.E.McCarthy and M.C.Standage, eds., Adam Hilger, Bristol, page 181 (1992).
31. H.R.J.Walters, T.T.Scholz, M.P.SCott and P.G.Burke, in "Correlations and Polarization in Electronic and Atomic Collisions and (e, 2e) Reactions", P.J.O.Teubner and E.Weigold, eds., Institute of Physics (Conference Series Number 122), Bristol, page 59 (1992).

32. M.P.Scott and P.G.Burke, J.Phys.B **26** L191 (1993).
33. M.P.Scott, B.R.Odgers and P.G.Burke, J.Phys.B **26** L827 (1993).
34. G.D.Fletcher, M.J.Alguard, T.J.Gay, V.W.Hughes, P.F.Wainwright, M.S.Lubell and W.Raith, Phys.Rev.A **31** 2854 (1985).
35. X.Q.Guo, D.M.Crowe, M.S.Lubell, F.C.Tang, A.Vasilakis, J.Slevin and M.Eminyan, Phys.Rev.Lett. **65** 1857 (1990).
36. D.M.Crowe, X.Q.Guo, M.S.Lubell, J.Slevin and M.Eminyan, J.Phys.B **23** L325 (1990).
37. W.L.van Wyngaarden and H.R.J.Walters, J.Phys.B **19** 1817 (1986).
38. W.L.van Wyngaarden and H.R.J.Walters, J.Phys.B **19** 1827 (1986).
39. J.J.McClelland, M.H.Kelley and R.J.Celotta, Phys.Rev.A **40** 2321 (1989).
40. R.E.Scholten, S.R.Lorentz, J.J.McClelland, M.H.Kelley and R.J.Celotta, J.Phys.B **24** L653 (1991).
41. M.H.Kelley, J.J.McClelland, S.R.Lorentz, R.E.Scholten and R.J.Celotta in "Correlations and Polarization in Electronic and Atomic Collisions and (e, 2e) Reactions", P.J.O.Teubner and E.Weigold, eds., Institute of Physics (Conference Series Number 122), Bristol, page 23 (1992).
42. I.Bray, Phys.Rev.A **49** R1 (1994).
43. I.Bray and I.E.McCarthy, Phys.Rev.A **47** 317 (1993).
44. W.L.van Wyngaarden and H.R.J.Walters, J.Phys.B **19** 929 (1986).
45. W.L.van Wyngaarden and H.R.J.Walters, J.Phys.B **19** L53 (1986).
46. H.R.J.Walters, in "Electronic and Atomic Collisions (Invited Papers of the XV International Conference on the Physics of Electronic and Atomic Collisions)", H.B.Gilbody, W.R.Newell, F.H.Read and A.C.H.Smith, eds., North-Holland, Amsterdam, page 147 (1988).
47. J.F.Williams, J.Phys.B **14** 1197 (1981).
48. J.Slevin, M.Eminyan, J.M.Woolsey, G.Vassilev, H.Q.Porter, C.G.Back and S.Watkin, Phys.Rev.A **26** 1344 (1982).
49. N.Andersen, J.W.Gallagher and I.V.Hertel, Phys.Repts. **165** 1 (1988).
50. J.F.Williams, Australian J.Phys. **39** 621 (1986).
51. S.Nic Chormaic, S.Chwirot and J.Slevin, J.Phys.B **26** 139 (1993).
52. I.Bray and A.T.Stelbovics, Phys.Rev.Lett. **70** 746 (1993).
53. M.B.Shah, D.S.Elliott and H.B.Gilbody, J.Phys.B **20** 3501 (1987).
54. G.H.Wannier, Phys.Rev. **90** 817 (1953).
55. R.Peterkop, J.Phys.B **4** 513 (1971).
56. A.R.P.Rau, Phys.Rev.A **4** 207 (1971).
57. H.Klar and W.Schlecht, J.Phys.B **9** 1699 (1976).
58. C.H.Greene and A.R.P.Rau, Phys.Rev.Lett. **48** 533 (1982).
59. J.M.Feagin, J.Phys.B **17** 2433 (1984).
60. A.R.P.Rau, Comm.At.Mol.Phys. **14** 285 (1984).
61. A.R.P.Rau, Phys.Repts. **110** 369 (1984).
62. F.H.Read, in "Electron Impact Ionization", T.D.Märk and G.H.Dunn, eds., Springer-Verlag, Vienna, page 42 (1985).
63. A.K.Kazansky and V.N.Ostrovsky, J.Phys.B **25** 2121 (1992).
64. J-M.Rost, Phys.Rev.Lett. **72** 1998 (1994).
65. J.W.McGowan and E.M.Clarke, Phys.Rev. **167** 43 (1968).

66. X.Q.Guo and M.S.Lubell, J.Phys.B **26** 1221 (1993).
67. M.S.Lubell, Phys.Rev.A **47** R2450 (1993).
68. J.H.Macek and S.Yu.Ovchinnikov, Phys.Rev.A **50** 468 (1994).
69. J.R.Friedman, X.Q.Guo and M.S.Lubell, Phys.Rev.A **46** 652 (1992).
70. C.T.Whelan, H.R.J.Walters, A.Lahmam-Bennani and H.Ehrhardt (eds.), "(e, 2e) and Related Processes", Kluwer, Dordrecht, (1993).
71. G.Baum, W.Blask, P.Freienstein, L.Frost, S.Hesse, W.Raith, P.Rappolt and M.S.Streun, Phys.Rev.Lett. **69** 3037 (1992).
72. L.Frost, in "Correlations and Polarization in Electronic and Atomic Collisions and (e, 2e) Reactions", P.J.O.Teubner and E.Weigold, eds., Institute of Physics (Conference Series Number 122), Bristol, page 161 (1992).
73. L.Frost, P.Freienstein, S.Hesse, W.Blask, G.Baum and W.Raith, in "Invited Papers and Progress Reports, 3rd European Conference on (e, 2e) Collisions and Related Problems", G.Stefani, ed., Area della Ricerca di Roma, Rome, page 295 (1990).
74. X.Zhang, C.T.Whelan and H.R.J.Walters, J.Phys.B **25** L457 (1992).
75. G.F.Hanne, in "Correlations and Polarization in Electronic and Atomic Collisions and (e, 2e) Reactions", P.J.O.Teubner and E.Weigold, eds., Institute of Physics (Conference Series Number 122), Bristol, page 15 (1992).
76. S.Jones, D.H.Madison and G.F.Hanne, Phys.Rev.Lett. **72** 2554 (1994).
77. G.F.Hanne, Phys.Repts. **95** 95 (1983).
78. T.Rösel, R.Bär, K.Jung and H.Ehrhardt, in "Invited Papers and Progress Reports, 2nd European Conference on (e, 2e) Collisions and Related Problems", H.Ehrhardt, ed., Universität Kaiserslautern, Kaiserslautern, page 69 (1989).
79. I.P.Grant, Adv.Phys. **19** 747 (1970).
80. P.A.M.Dirac, "The Principles of Quantum Mechanics", Oxford University Press, Oxford (1958).
81. B.H.Bransden and C.J.Joachain, "Physics of Atoms and Molecules", Longman, London (1983).
82. H.E.Bethe and E.E.Salpeter, "Quantum Mechanics of One- and Two-Electron Atoms", Springer-Verlag, Berlin (1957).
83. K.Bartschat, J.Phys.B **26** 3595 (1993).
84. U.Thumm and D.W.Norcross, Phys.Rev.Lett. **67** 3495 (1991).
85. U.Thumm and D.W.Norcross, Phys.Rev.A **45** 6349 (1992).
86. U.Thumm and D.W.Norcross, Phys.Rev.A **47** 305 (1993).
87. C.D.H. Chisholm and U.Öpik, Proc.Phys.Soc. **83** 541 (1964).
88. N.S.Scott, K.Bartschat, P.G.Burke, W.B.Eissner and O.Nagy, J.Phys.B **17** L191 (1984).
89. N.S.Scott, K.Bartschat, P.G.Burke, O.Nagy and W.B.Eissner, J.Phys.B **17** 3775 (1984).
90. U.Thumm, K.Bartschat and D.W.Norcross, J.Phys.B **26** 1587 (1993).

CORRELATION STUDIES OF ELECTRON IMPACT EXCITATION – PAST, PRESENT, FUTURE

Albert Crowe

Department of Physics
The University
Newcastle upon Tyne
NE1 7RU
U.K.

INTRODUCTION

The type of process discussed here can be written as

$$e^-(\underline{k}_{in}) + A \rightarrow A^* + e^-(\underline{k}_{out})$$
$$\downarrow$$
$$A' + h\nu \qquad (1)$$

where $\underline{k}_{in}$ and $\underline{k}_{out}$ are the momenta of the incident and scattered electrons respectively. A' may be the initial atomic state A or some state intermediate in energy between A and the excited state A^*. A key element in correlation studies is the detection of the scattered electron with well defined momentum, ensuring that the subset of excited states actually studied are identically prepared in the collision process. This also defines a collision plane, resulting in important symmetry implications in analysis of the experiments. Detailed information on the excitation process is then retrieved from studies of the polarisation and/or angular distributions of the emitted photons in coincidence with the scattered electrons – the polarisation and angular correlation techniques, respectively. The general discussion relates to particles with a random orientation of spin but reference will be made to recent work using spin-polarised particles.

A comprehensive review and analysis of these studies until mid-1987 has been given by Andersen et al. (1988). Since then aspects of the field have been reviewed by Slevin and Chwirot (1990), Becker, Crowe and McConkey (1992) and Crowe (1993). This contribution will review advances in the field, concentrating on more recent work, some of which indicate the directions in which the studies might usefully proceed in future.

An alternative approach for the study of such processes is the time-reversed process, often referred to as superelastic scattering where A' (in practice, the ground state A) is laser excited to A*. Electrons are then superelastically scattered from A* to yield the left hand side of (1), information being obtained from the intensity of the superelastically scattered electrons as a function of the laser photon polarisation. Apart from quoting results from this method, it will not be discussed in detail. The reader is referred to a recent discussion by Mac Gillivray and Standage (1991) on the equivalence of the techniques and to Sang et al. (1994) for a detailed discussion of the current status of the method applied to sodium.

ANALYSIS OF ELECTRON-PHOTON CORRELATION EXPERIMENTS

The theoretical background and analysis of these experiments have been discussed in detail by a number of authors. Particularly important are the early works of Macek and Jaecks (1971), Fano and Macek (1973) and Blum (1981). These references provide detailed theoretical discussions of the process in terms of, for example, state multipoles and the density matrix formalism. In the present discussion only a simple description of the alignment and orientation of the excited state and their relation to measured quantities are presented.

After excitation from an initial pure state, the identically prepared excited states isolated in these experiments can be described by a linear combination of the magnetic substates i.e. by $\sum a_M(E,\theta)|LM\rangle$, where $a_M(E,\theta)$ are the excitation amplitudes associated with states with quantum numbers L, M. The correlation experiments determine, as a function of the incident electron energy E and scattering angle θ, the magnitudes and phases of the complex amplitudes a_M.

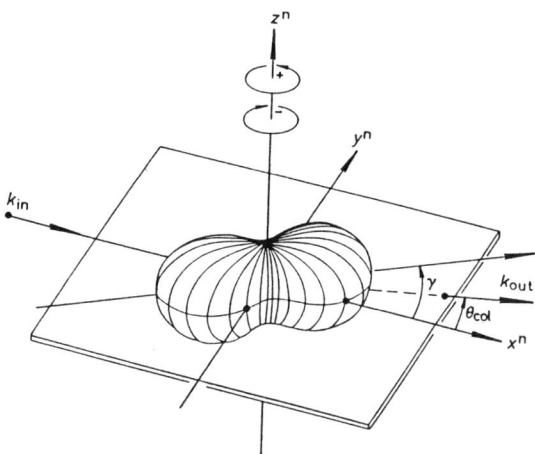

Figure 1. Schematic illustration of excited p-state charge cloud produced in a specific collision defined by $\underline{k}_{in}$ and $\underline{k}_{out}$.

Different parameterisations have been used to present this information. Many results have been presented in terms of the original (λ, χ) parameters of Eminyan et al. (1973, 1974). Here the parameterisation in terms of the shape and dynamics of the excited state charge cloud introduced by Andersen and Hertel (1986) is used. Figure 1 shows a schematic illustration of an excited p-state, $|\psi(\theta,\varphi)|^2$. The state will be described with respect to the natural co-ordinate frame (Hermann and Hertel, 1982) with the collision plane, defined by $\underline{k}_{in}$ and $\underline{k}_{out}$, being the (x, y) plane. Then, for an S–P excitation process, assuming only excitation of states with positive reflection symmetry, the excited state is given simply by

$$|P\rangle = a_{+1}(E, \theta)|1+1\rangle + a_{-1}(E, \theta)|1-1\rangle. \tag{2}$$

The amplitudes a_M are readily related to a parameter set describing the excited state (Andersen and Hertel, 1986). These are

γ, the direction of the charge cloud with respect to $\underline{k}_{in}$ (x)

P_ℓ the shape (relative width to length) of the charge cloud in the (x, y) plane

$\langle L_\perp \rangle$, the angular momentum, perpendicular to the scattering plane, of the excited state

ρ_{00}, the relative height of the charge cloud measured along the z-axis.

A detailed analysis linking these parameters with the experimental observables requires a knowledge of the states involved. For the most straightforward cases, the in-plane angular correlation method in which the angular distribution of the decay photons, measured in the scattering plane, is observed in coincidence with the scattered electron, gives the shape of the excited charge cloud in the scattering plane i.e. γ and P_ℓ. The angular correlation method gives no information on the dynamics of the system, although in the simplest cases the magnitude of $\langle L_\perp \rangle$ can be inferred. Where the charge cloud has a height along the z-axis, a second angular correlation measurement in a different geometry is required (e.g. King et al., 1985) to determine that height, ρ_{00}.

In the polarisation correlation method, the shape of the charge cloud in the scattering plane is given by the intensity distribution of the radiation emitted perpendicular to the scattering plane following transmission by a rotating linear polariser. In practice, the relative Stokes parameters P_1 and P_2 are usually measured where

$$I_z P_1 = I_z(0°) - I_z(90°) \tag{3a}$$

$$\text{and} \quad I_z P_2 = I_z(45°) - I_z(135°) \tag{3b}$$

where I_z is the total photon intensity emitted perpendicular to the scattering plane and $I_z(\alpha°)$ is the intensity transmitted by a linear polariser relative to the x^n (electron beam) direction. Then

$$P_\ell = (P_1^2 + P_2^2)^{1/2} \tag{4a}$$

$$\text{and} \quad \gamma = \tfrac{1}{2}\arg(P_1 + iP_2). \tag{4b}$$

A measurement of the circular polarisation P_3, where

$$I_z P_3 = I_z(RHC) - I_z(LHC) \tag{5}$$

and (RHC), (LHC) refer to the handedness of the radiation, gives a direct measure of $L_\perp$, i.e.

$$L_\perp = -P_3. \tag{6}$$

Determination of ρ_{oo} requires a measurement of the linear polarisation of the radiation emitted in another direction. The most common measurement is made in the y^n direction, giving

$$I_y P_4 = I_y(0°) - I_y(90°). \tag{7}$$

The relationships (4a) and (6) apply only in the simplest cases. However, a major strength of this parameterisation is the ease with which it can be generally applied. A detailed discussion of the extension to more complex situations involving, for example, states of higher angular momentum and multiplicity, including fine and hyperfine depolarisation of the observed radiation has been given by Andersen et al. (1988) and is outlined where appropriate below.

STUDY OF SPECIFIC SYSTEMS

The n^1P States of Helium

The ground states of helium and the alkaline earth atoms Be, Mg, Ca, Sr, Ba have a common s^2 outer shell structure and correlation studies of excitation to their $(np)^{1,3}P$ states can be simply analysed to give a complete description of the processes. This is particularly true for the lightest system, helium, where the states involved are LS coupled, the only spin effect is due to electron exchange and there is no depolarisation of the emitted radiation due to hyperfine interactions. For the heavier systems, spin-orbit coupling effects may play a significant role in the excitation process and hence complicate the analysis. A detailed discussion of studies on the heavier alkaline earth elements is given by Beyer elsewhere in this volume.

Excitation of the 2^1P state of helium has been the subject of most experimental study since the pioneering angular correlation work of Eminyan et al. (1973, 1974). Experimental studies have been carried out for incident electron energies from 22 eV (0.8 eV above threshold), Neill and Crowe (1984) to 500 eV (Steph and Golden, 1980), often over a wide range of scattering angles. The wavelength of the 2^1P-1^1S line (58.4 nm) makes this system unattractive for study using the polarisation correlation technique. However, strictly, study of this system was not complete until the circular polarisation measurements of Khakoo et al. (1986) were carried out. Chwirot and Slevin (1994) have recently discussed the relative merits of the angular and polarisation correlation techniques for VUV transitions.

Figure 2 illustrates the variation in shape and dynamics of the 2^1P state of helium using the data of Eminyan et al. (1974) and Hollywood et al. (1979) at an incident electron energy of 81.2 eV. The change in direction of angular momentum around 65° is inferred from Khakoo et al. (1986). It is interesting that the only recent experimental study at this energy (Mikoza et al., 1994a) agrees well with previous experimental data but concludes that even at small scattering angles no single theoretical model predicts good agreement with the experimental data.

Recent theoretical work has concentrated on use of the R-matrix method close to threshold (eg. Fon et al., 1993a).

Use of the polarisation correlation technique was pioneered by Standage and Kleinpoppen (1976) for the 3^1P-2^1S (501.6 nm) radiation emitted following excitation of the 3^1P state. Apart from the highly unfavourable branching ratio, use of this radiation is to be preferred to the undispersed VUV (3^1P-1^1S) radiation, the coincidence signal from which may include a considerable cascade contribution from higher lying excited states.

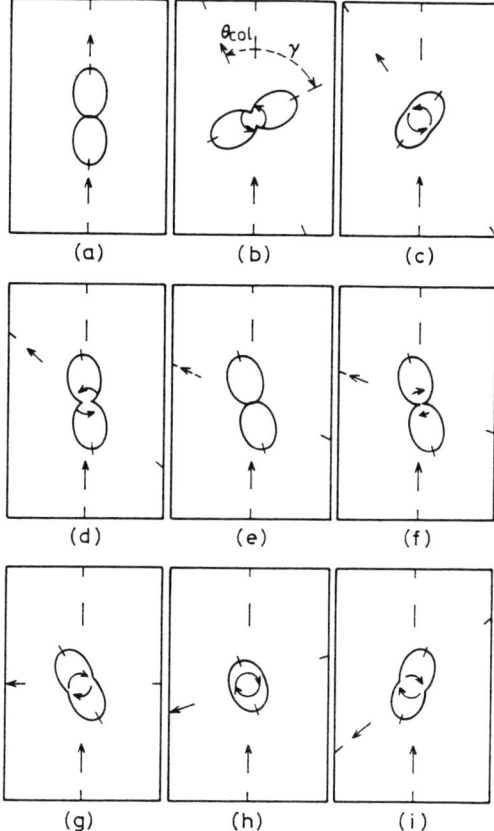

Figure 2. The shape of the He (2^1P) charge cloud in the scattering plane excited by 81.2 eV electrons for electron scattering angles θ_{col} of (a) 0°(180°) (b) 25° (c) 35° (d) 50° (e) 65° (f) 70° (g) 90° (h) 110° (i) 130°. The magnitudes and directions of the angular momentum are given by the length and direction of the circular arrows within each charge cloud. A full circle corresponds to $\hbar$.

The more recent low energy (26.5, 29.6 eV) data of Neill and Crowe (1988) and Neill et al. (1989) using both angular and polarisation correlation methods show a variation of both $P_\prime$ and $L_\perp$ with electron scattering angle qualitatively similar to those predicted by the 19-state R-matrix method of Fon et al. (1990). However the charge cloud alignment γ predicted theoretically disagrees with experiment at electron scattering angles greater than 60°. The comparisons between theory and experiment for γ, $P_\prime$ and $L_\perp$ at 29.6 eV are given in figure 3. The change of sign of $L_\perp$ from positive values at smaller scattering angles to negative values at larger scattering angles is typical of the behaviour observed for both 2^1P and 3^1P excitation over a wide range of incident electron energies. Many models have been put forward to qualitatively explain this behaviour (e.g. Andersen and Hertel, 1986). As yet no systematic study of the theoretical prediction of n independence (Csanak and Cartwright, 1986) of these results have been carried out.

Mikoza et al. (1994a) have given careful consideration to radiation trapping effects in their measurements, but their low angle results at 80 eV confirm, within statistical error, previous experimental data at that energy.

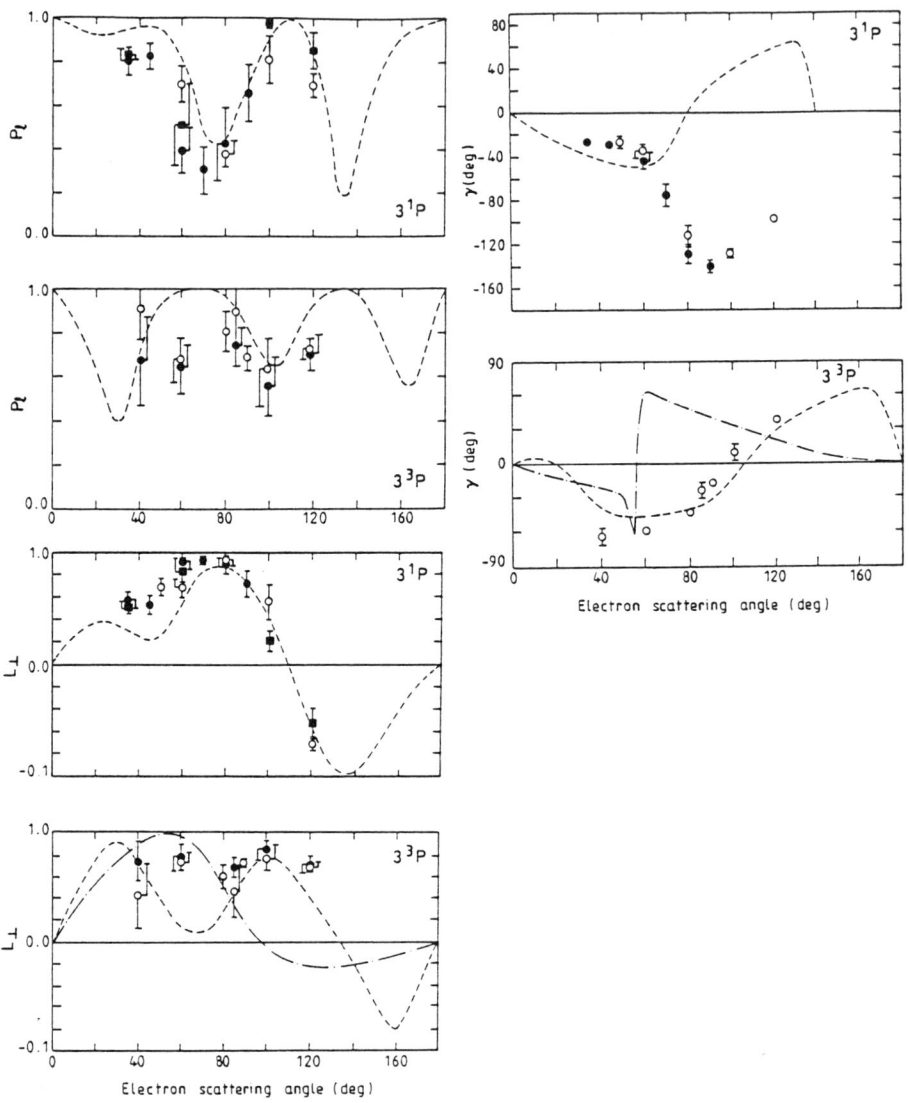

Figure 3. Parameters P_ℓ, γ and $L_\perp$ for the 3^1P and 3^3P states of helium excited by 29.6 eV electrons. *Solid circles* – polarisation correlation data of Neill et al. (1989) for 3^1P and Donnelly and Crowe (1989) for 3^3P; *open circles* – angular correlation data of Neill and Crowe (1988) for 3^1P and Donnelly et al. (1988) for 3^3P; *solid squares* – values of $|L_\perp|$ from P_ℓ using $L_\perp^2 = 1 - P_\ell^2$ and P_ℓ from P_3 for 3^1P from Neill et al. (1989); *dashed lines* – 19 state R-matrix calculation of Fon et al. (1990); *dot-dash lines* – FOMBT calculation of Cartwright and Csanak (1986).

The 3^3P States of Helium

For singlet state excitation in helium, each of the doublet scattering amplitudes a_M consist of a direct and exchange component which are experimentally indistinguishable. However, excitation of two electron atoms initially in a spin zero state to a triplet state provides an opportunity to study a pure exchange process. Experimentally, simple identification of the emitted radiation from the excited triplet state is sufficient to isolate the exchange process without the complexity of spin-polarised electron experiments.

Analysis of the correlation data must take account of the fine structure depolarisation of the observed radiation in reconstructing the shape and dynamics of the excited state. For the n^3P states, the polarisations before and after depolarisation are related by simple numerical coefficients (Andersen et al., 1988).

Because of the difficulties of observing the 2^3P-2^3S IR radiation, experimental studies have been confined to the 3^3P state. Figure 3 shows variation of the γ, P_ℓ and $L_\perp$ parameters as a function of electron scattering angle for the 3^3P state at an electron energy of 29.6 eV (Donnelly et al., 1988; Donnelly and Crowe, 1989). Here the predictions of the 19-state R-matrix calculations of Fon et al. (1990) for γ show the same qualitative behaviour as observed experimentally. While much of the P_ℓ data are consistent with this theory, there is a clear disagreement at a scattering angle of 60°. Similar large discrepancies are observed at this angle for $L_\perp$. It should be noted that values of $L_\perp$ calculated from P_3 values of Fon et al. (1990) differ significantly in this angular range from those attributed to this calculation by Donnelly and Crowe (1989). The experimental data strongly disagree with the FOMBT calculations of Cartwright and Csanak (1986). Close to threshold there is also strong disagreement between the 29-state R-matrix calculation of Fon et al. (1993b) and the experimental data of Humphrey et al. (1987). At higher energies data from Stirling (Beyer et al., 1988) and Utrecht (Beijers et al., 1987a; Batelaan et al., 1990) can be compared with the DWBA calculations of Bartschat and Madison (1988).

Figure 3 also enables comparisons to be made between 3^1P and 3^3P excitation at 29.6 eV. For each parameter marked differences are obvious – the behaviour of γ at larger scattering angles, the relatively constant values of both P_ℓ and $L_\perp$ for the 3^3P state compared with well defined variations for the 3^1P state.

The 3^1D State of Helium

In this case the excited state can be written

$$|n^1D\rangle = a_{+2}(E,\theta)|2+2\rangle + a_0(E,\theta)|2,0\rangle + a_{-2}(E,\theta)|2-2\rangle \tag{8}$$

with $a_{\pm 1} \equiv 0$ due to the negative reflection symmetry of the $|2+1\rangle$ and $|2-1\rangle$ states in the natural co-ordinate frame. The $M = \pm 2$ states lie in the scattering plane, analogous to the $M = \pm 1$, P states. However, the $M = 0$ state lies perpendicular to the scattering plane, giving 'height' to the charge cloud along the $0z$ direction. This implies that an inplane linear polarisation measurement (equation 7) is required to determine the non-zero value of ρ_{00} in this case. P_ℓ and γ can be defined in terms of the measured Stokes parameters P_1 and P_2 as for ^1P states. From Andersen et al. (1988), $L_\perp$ is given by

$$L_\perp = -2I_z P_3 = 2(a_2^2 - a_{-2}^2) = \frac{4P_3(1+P_4)}{(1-P_1)(1-P_4)-4} \tag{9}$$

and ρ_{00} by

$$\rho_{00} = a_0^2 = \frac{\frac{3}{2}(1+P_1)(1-P_4)}{4-(1-P_4)(1-P_1)} \tag{10}$$

Recently three groups have reported experimental values of the complete parameter set (γ, P_ℓ, $L_\perp$, ρ_{00}) over a wide range of electron energies and scattering angles (Table 1). By contrast, the only theoretical study to report all parameters is the multichannel eikonal calculation (DMET) of Mansky and Flannery (1991).

Table 1. Experimental measurements of the parameter set (γ, P_ℓ, $L_\perp$, ρ_{00}) for the 3^1D state of helium.

	Electron Energies (eV)	Scattering Angles (deg)
Crowe (1990)	29.6	40 – 100
Batelaan et al. (1991a)	40	31.2 – 60
	45	15 – 50
	60	15 – 31.2
Mikoza et al. (1994b)	40	10 – 40
	60	15, 25
	81.6	15
McLaughlin et al. (1994a)	26.5	40 – 120
Donnelly et al. (1994)	29.6	40 – 120
McLaughlin et al. (1994b)	40	40 – 120

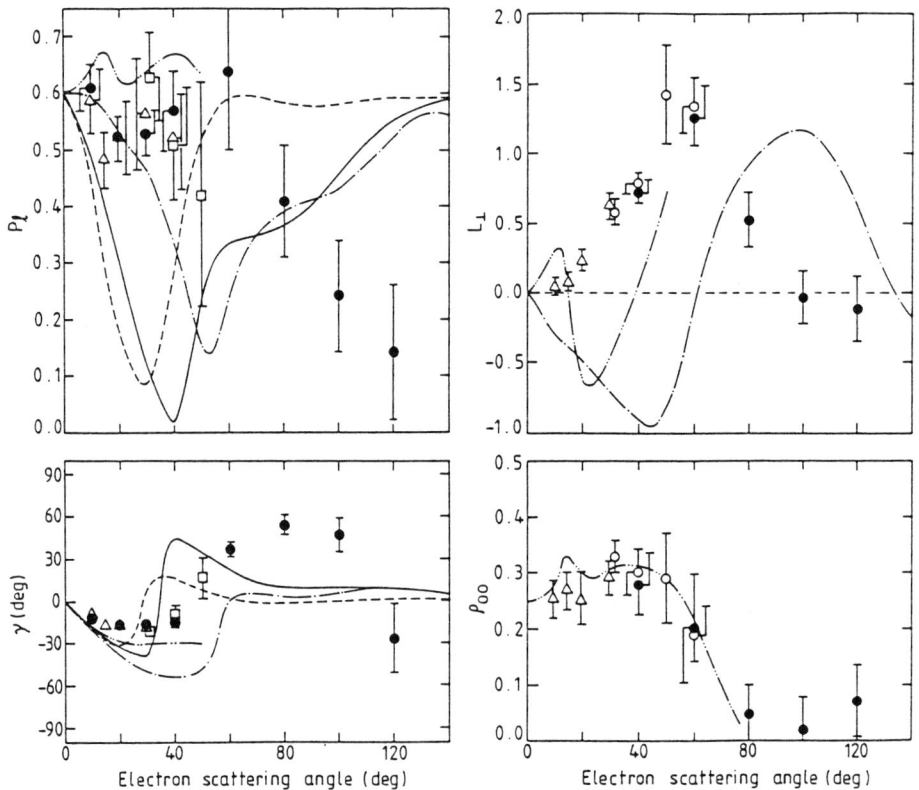

Figure 4. The parameters P_ℓ, γ, $L_\perp$ and ρ_{00} for the 3^1D state of helium at an incident electron energy of 40 eV. *Solid circles* – Donnelly and Crowe (1988a), McLaughlin et al. (1994b) and private communication; *open circles* – Batelaan et al. (1991a); *squares* – Beijers et al. (1987b); *triangles* – Mikoza et al. (1994b); *solid line* – Bartschat and Madison (1988) – DWBA-EP; *dashed line* – Bartschat and Madison (1988) – DWBA-GP; *dash-dot line* – Cartwright and Csanak (1987) – FOMBT; *dash-double dot line* – Mansky and Flannery (1991) - DMET.

Figure 4 shows the variation of the parameters for the 3^1D state at 40 eV where data are available over a wide range of electron scattering angles and comparison can be made between the experimental data from the three groups. The excellent agreement between the three independent experiments allows firm conclusions to be reached regarding the angular variation of each parameter. P_ℓ remains close to its theoretical value of 0.6 at zero degrees, out to at least 40°, γ takes on only small negative values (maximum ~ 20°) at small scattering angles and becomes positive beyond 45°, while ρ_{∞} has essentially a constant value ~ 0.25 – 0.3 out to 50°. At this energy, the behaviour of $L_\perp$ with angle has similarities with that seen for the 2^1P state of helium i.e. positive at small scattering angles becoming marginally negative at larger angles (McLaughlin et al., 1994b). However at higher energies, data from all three groups show negative $L_\perp$ at small angles in contrast to 2^1P.

The lack of any consistent agreement between theory and experiment is obvious from figure 4. However, figure 5, showing the measured circular polarisation P_3 as a function of electron scattering angle up to 60° at an energy of 40 eV is included to highlight the lack of a reliable theoretical calculation for S–D excitation processes in simple systems. Assuming that the excellent agreement between the different experiments points to reliable data, no theory describes the excitation. Furthermore the different theories show no agreement with each other. There is little evidence that these perturbative approaches show improved agreement with experiments at higher energies. No calculations have been performed appropriate to the near-threshold energies.

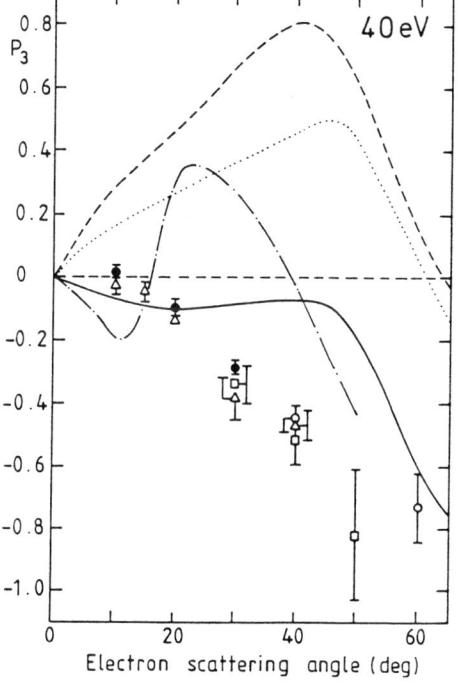

Figure 5. Variation of the Stokes parameter P_3 for the 3^1D state of helium at 40 eV.

Circles (open and closed) – Donnelly and Crowe (1988) and private communication; *squares* – Beijers et al. (1987b); *triangles* – Mikoza et al. (1994b); *solid line* – Bartschat and Madison (1988) – DWBA-EP; *dashed line* – Bartschat and Madison (1988) – DWBA-GP; *dotted line* – Cartwright and Csanak (1987) – FOMBT; *dash-dot line* – Mansky and Flannery (1991) - DMET.

It has been recognised (Andersen et al., 1983) that observation of the dipole radiation alone leaves an ambiguity in the complete description of S→L (L>1) excitation processes. The method of removal of the ambiguity of Neitzke and Andersen (1984) for heavy particle collisions using external fields is not appropriate for low energy electron collisions.

Alternative methods, involving measurement of P_3 in the vicinity of resonance states (Batelaan et al., 1991b) and triple coincidence (e, γ_1, γ_2) where γ_1 and γ_2 relate, in this case, to the 3^1D–2^1P and 2^1P–1^1S cascade transitions respectively (Wang et al., 1994a) have recently been discussed.

The 3^3D State of Helium

For the 3^3D state, the only independent complete polarisation correlation study is that of Crowe et al. (1994) for an incident electron energy of 40 eV. Figure 6 shows the variation of the excited state parameters with scattering angle at this energy. The measured Stokes parameters have been corrected for the effects of fine structure depolarisation. Unlike the ^3P fine structure correction, the measured and corrected parameters are no longer related by a simple numerical coefficient. Expressions for the corrected P_ℓ, $L_\perp$ and ρ_{oo} have been given by Crowe et al. (1994). Values of γ remain very close to zero for angles up to 60°, beyond which they change dramatically. Small differences with Batelaan et al. (1991a) for both P_ℓ and ρ_{oo} arise only from different measured P_4 values since their analysis uses P_1 and P_2 from Donnelly and Crowe (1988b). The values of $L_\perp$ decrease from a relatively small value ($|L_\perp|_{max} = 2\hbar$) at the lowest measured angle of 40°. Here the agreement between the data at angles other than 120° is largely a measure of the agreement between the measured P_3 parameters. At present there is no corresponding complete set of theoretical parameters.

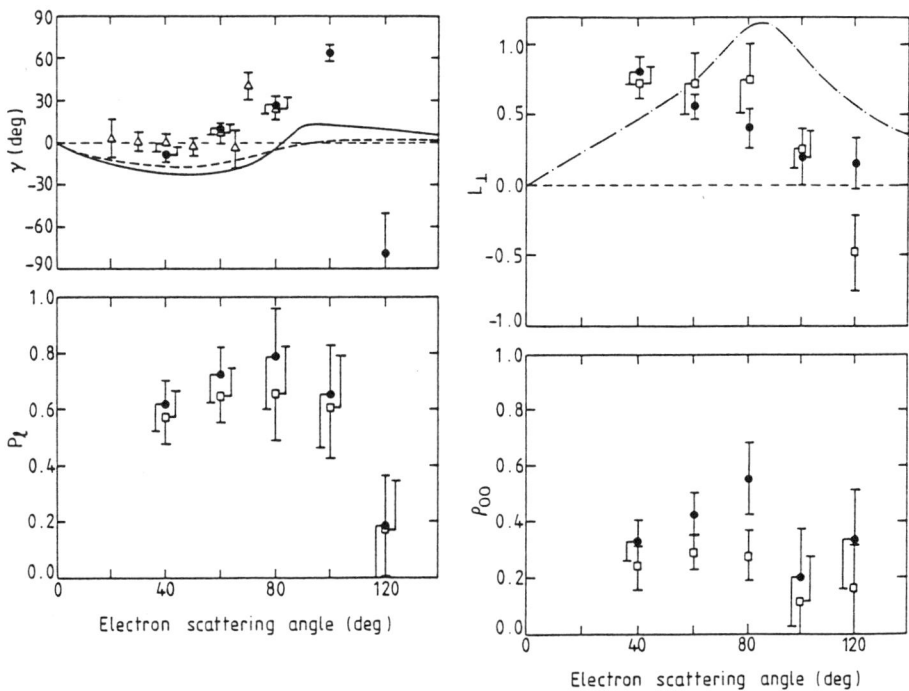

Figure 6. Variation of the P_ℓ, γ, $L_\perp$ and ρ_{oo} parameters for the 3^3D state of helium at 40 eV. *Solid circles* – Crowe et al. (1994); *squares* – Batelaan et al. (1991a); *triangles* – Beyer et al. (1989); *full line* – Bartschat and Madison (1988) – DWBA-EP; *dashed line* – Bartschat and Madison (1988) – DWBA-GP; *dash-dot line* – Csanak and Cartwright (1987) – FOMBT, at 45 eV.

The ^{2}P States of Hydrogen and Sodium

It might be expected that scattering processes involving electrons on one electron systems, particularly atomic hydrogen, would be well understood. However, serious discrepancies between the results of correlation experiments for the simple 1s–2p excitation and corresponding theoretical calculations at large scattering angles must constitute one of the greatest unsolved fundamental problems in atomic collision physics.

Unlike the excitation of singlet and triplet states in helium, correlation studies of doublet states with unpolarised electrons are incomplete. Experiments with spin polarised beams and/or spin analysis can isolate the singlet and triplet scattering amplitudes arising from the parallel and antiparallel spin combinations of the continuum and atomic electrons in the scattering process. Such experiments have yet to be performed in atomic hydrogen. A substantial advantage of the parameter set $(\gamma, P_\ell, L_\perp)$ is that it can be extended to incomplete or partially coherent excitation processes such as unpolarised electron-hydrogen scattering. Again fine structure depolarisation of the detected radiation has to be taken into account in determining these parameters from polarisation or angular correlation data.

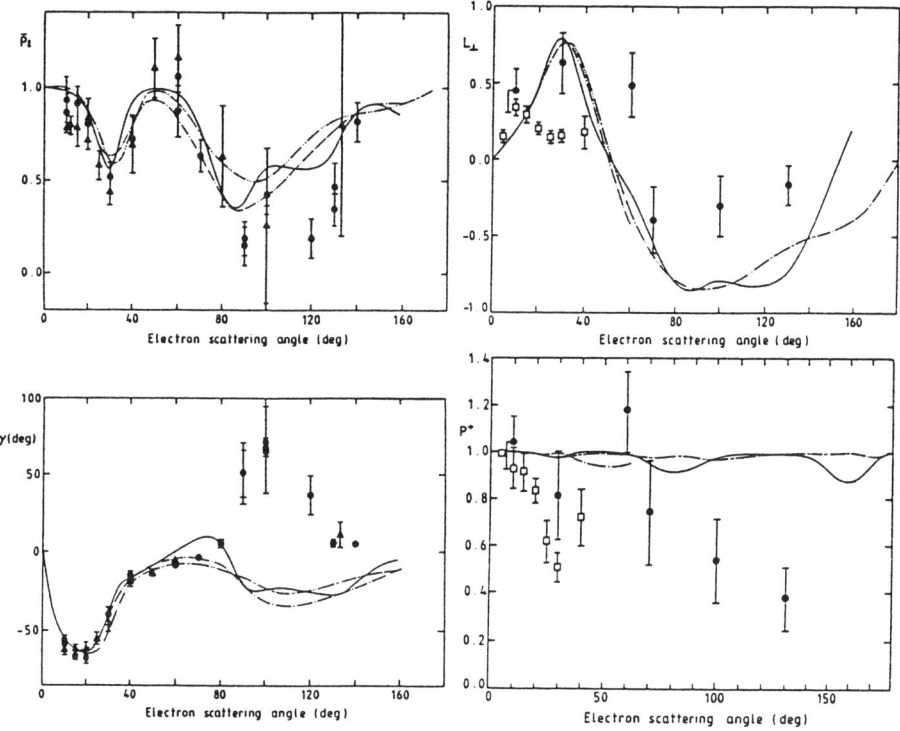

Figure 7. Variation of the parameters P_ℓ, γ, $L_\perp$ and P^+ for excitation of the 2^2P state of hydrogen at 54.4 eV. *Circles* – Williams (1981, 1986); *triangles* – Hood et al. (1979), Weigold et al. (1990); *squares* – Nic Chormaic et al. (1993); *solid line* – IERM (Scholz et al. 1991); *dash-dot line* – DWB2 (Madison et al., 1991); *dash-double dot line* – 80CC (Bray and Stelbovics, 1992).

Figure 7 shows the complete parameter set for excitation of H(2p) at an electron energy of 54.4 eV together with the degree of polarisation P^+ defined by $P^+ = \left(P_3^2 + P_\ell^2\right)^{1/2}$. The only complete parameter set is due to Williams (1981, 1986). For P_ℓ and γ there is substantial agreement with the experimental data of Hood et al. (1979)

and Weigold et al. (1980). A very large number of calculations have been performed on this system using a variety of approaches and at smaller angles most reproduce the above experimental data well. However, at large angles this is not true and at least five calculations using different theoretical approaches ranging chronologically from van Wyngaarden and Walters (1986) to Wang et al. (1994b) agree well with each other but not with experiment.

The extent of the problem is best emphasised by comparing the measured and predicted angular correlations (figure 8). It can be seen that for an electron scattering angle of 100° the measured and predicted angular correlations are completely out of phase. A similar discrepancy is noted between theory and the experiment of Slevin et al. (1982) at 35 eV. The comment, "at this time, it seems unlikely that it (the discrepancy) will be resolved by improved theoretical calculations", of Wang et al. (1994b) is a typical theoretician's view of this problem.

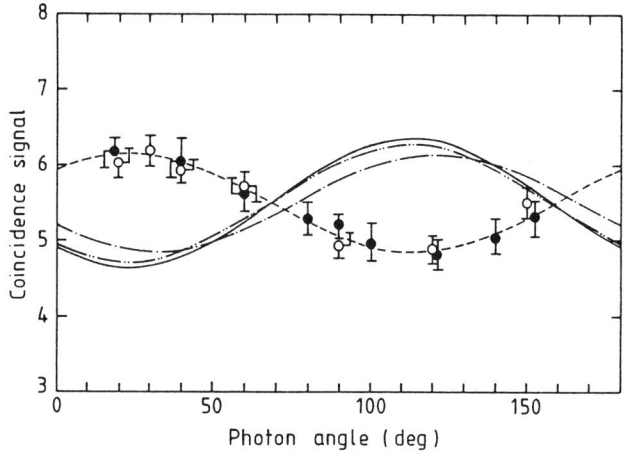

Figure 8. Experimentally measured and theoretically predicted angular correlation for H(2p) excitation at an electron scattering angle of 100° and an incident energy of 54.4 eV. *Experimental data* – Williams (1981, 1986); *dotted line* – fit to experimental data; *theoretical predictions* – notation as figure 7.

Although care must be exercised (Scholten et al., 1993), P^+ can be taken as a measure of the importance of exchange scattering. The theoretical calculations all indicate P^+ close to unity at all scattering angles, suggesting exchange is generally unimportant. This contrasts with the large angle data of Williams (1986). The more recent polarisation correlation data of Nic Chormaic et al. (1993) at smaller scattering angles is inconsistent with both previous experiment and theory. The deep minimum in P^+ around 30° appears inconsistent with the conventional view that exchange scattering should be unimportant at these scattering angles for this intermediate energy.

With regard to this problem it is interesting to discuss results for 3^2P excitation in sodium. Here considerable debate exists between different experimental groups concerning the value of P^+ at small scattering angles (5–20°) in the energy range 20–25 eV. Both above and below this energy range, there is general agreement that P^+ is essentially unity. Figure 9(a) shows the P^+ data at an electron energy of 20 eV (equivalent to 22.1 eV for correlation studies) from three different groups, all using the superelastic scattering technique. While data from the Brisbane group (Farrell et al., 1989, Sang et al., 1994) show substantially non-unity values of P^+ for scattering angles less then 20°, the Flinders group (e.g. Scholten et al., 1993) measure values which are very close to unity. Although they have large errors, the early data of Hermann et al. (1977, 1980) tend to support the Brisbane data. Various calculations suggest only small deviations of P⁺ from unity at small angles.

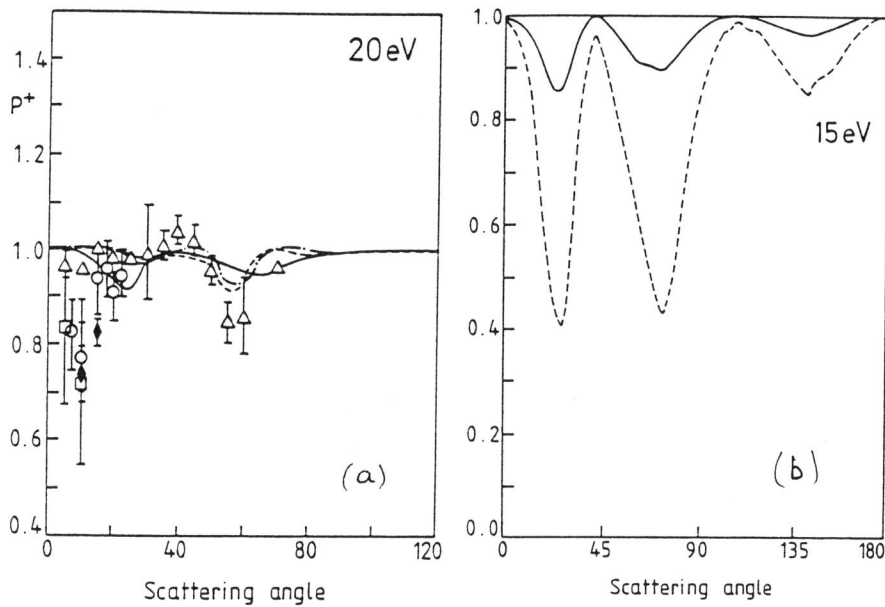

Figure 9(a). The degree of polarisation P⁺ for superelastic scattering of 20 eV electrons from the laser excited 3^2P state of sodium. *Circles* – Sang et al. (1994); *triangles* – Teubner et al. (1989); *squares* – Hertel et al., from Andersen et al. (1988); *diamonds* – Farrell et al. (1989); *full line* – DWBA2 theory of Sang et al. (1994); *dashed line* – 15-state close coupling theory of Bray and McCarthy (1993); *dot-dash line* – CCO calculation of Bray and McCarthy (1993). **(b).** Values of P⁺ (*solid line*) and S (*dotted line*) for 15 eV superelastic scattering from Na (3^2P) from DWBA2 calculations of Sang et al (1994).

Sang et al. (1994) have introduced a parameter S given by

$$S = P_1^S P_1^T + P_2^S P_2^T + P_3^S P_3^T \tag{11}$$

where superscripts S and T refer to singlet and triplet scattering respectively, as a more direct measure of exchange scattering. It is related to P⁺ through the relation

$$P^{+2} = 1 + \frac{3\sigma^S \sigma^T}{8\sigma^2}(S-1) \tag{12}$$

where σ is the spin averaged, σ^S the singlet, and σ^T the triplet differential cross sections. An indication of the increased sensitivity of S to exchange effects relative to P⁺ is shown in figure 9(b). Both parameters are calculated for the superelastic scattering of 15 eV electrons by the 3^2P state of sodium using the DWBA2 theory (Sang et al. 1994).

The use of spin polarised electrons to determine S experimentally would clearly give a highly sensitive probe of exchange effects. A series of superelastic scattering experiments of spin polarised electrons from spin polarised sodium atoms produced by pumping with circularly polarised laser light have been carried out by a NIST group, mostly for the 3s–3p transition in sodium. Figure 10 shows typical data from these experiments. At the higher energy (40 eV), there is little difference in the values of the singlet and triplet $L_\perp$ at all scattering angles, indicating that exchange is unimportant. At 4.1 eV, very different values of the singlet and triplet $L_\perp$ are observed at most angles, indicative of large exchange effects. The experimental data are well reproduced by a recent coupled channel optical calculation (Bray and McCarthy, 1993) at all energies and by a DWB2PE calculation (Madison et al., 1992) at higher energies.

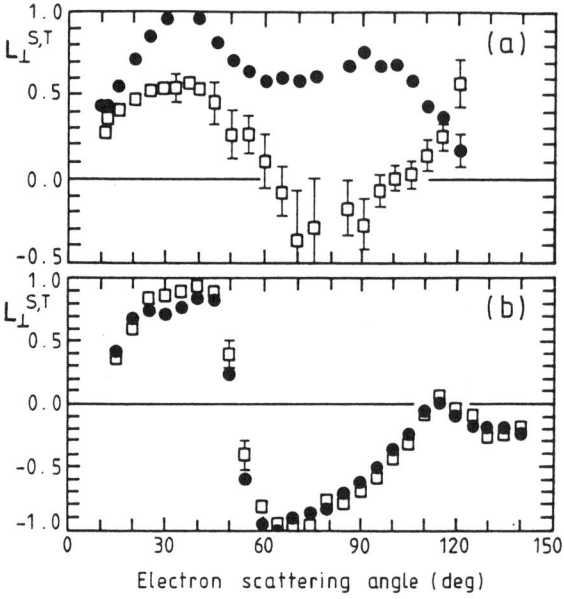

Figure 10. Values of spin resolved $L_\perp$ for the 3^2P state of sodium at (a) 4.1 eV (b) 40 eV, from Kelley et al. (1992).

Full circles – triplet scattering; *open squares* – singlet scattering.

Andersen and Bartschat (1994) have combined existing experimental and theoretical data for the Na(3s–3p) transition to show that a complete description of the excitation process can be obtained. They suggest use of $L_\perp^S$, $L_\perp^T$, γ^S, γ^T, r and Δ^{+1} as an appropriate complete parameter set, where r is the ratio of the triplet to singlet differential cross sections and Δ^{+1} is a singlet-triplet phase angle associated with the M = +1 state. The NIST experiments (Kelley et al, 1992, and references therein) determine r in addition to $L_\perp^S$ and $L_\perp^T$. Combining this information with P_ℓ and γ from unpolarised electron studies can give values of γ^S and γ^T. Andersen and Bartschat recognised that data from an experiment by Hegemann et al. (1991) in which they measured the ratio of the spin polarisations of scattered to incident electrons for scattering from unpolarised sodium atoms could be linked to the final parameter Δ^{+1}.

Heavy Atom Excitation

In all cases discussed until now it could reasonably be assumed that only excited state wave functions having positive reflection symmetry about the scattering plane could be excited. This assumption leads to considerable simplification of the analysis in these cases. For example, excited p-states have no height along the z-axis and for coherent excitation (e.g. $^{1,3}P$ states of helium) only two independent parameters and the differential cross section are required to completely describe the excitation. In general, this assumption cannot be made. For example, for a J = 1 state of a heavy atom (in the natural frame), the

$M_J = 0, \pm 1$ states may all be excited in contrast to only $M_L = \pm 1$ as noted previously for L = 1 states of light atoms.

Excitation of the $M_J = \pm 1$ states can be parameterised using γ, P_ℓ^+ and $L_\perp^+$, where the "+" refers to the component of the J = 1 oscillator density with positive reflection symmetry. The density matrix element ρ_{00} is used to describe the component with negative reflection symmetry. Physically it an be regarded as the relative height of the excited state charge cloud along the 0z axis and as a measure of spin flip transitions.

Most work in this area has been carried out for the $\left(^2P_{1/2}\right)$ ns' $[\frac{1}{2}]_1^0$ and $\left(^2P_{3/2}\right)$ ns $[\frac{1}{2}]_1^0$ states of the heavier rare gases, each consisting of varying components of pure Russell-Saunders $^1P_1$ and $^3P_1$ states, depending on the rare gas. Work in this area has been reviewed by Becker et al. (1992). Only two interesting examples of these studies are discussed here. Figure 11(a) shows values of ρ_{00} for the $4s[\frac{3}{2}]_1^0$ state of argon excited by 20 eV electrons reported by Corr et al. (1992). At this relatively low incident energy, significantly non-zero values of ρ_{00} are observed. The different theoretical approaches predict ρ_{00} values of similar magnitude to those observed experimentally, although the experimental angular variation is not reproduced. Wang et al. (1994c) have extracted the ratio of exchange to direct scattering differential cross sections from the ρ_{00} values for the $4s[\frac{3}{2}]_1^0$ state.

Figure 11(b) shows values of γ for the $5s[\frac{3}{2}]_1^0$ state of krypton excited by 30 eV electrons measured by Murray et al. (1990) using the angular correlation technique and Zheng and Becker (1993) from P_1 and P_2. A feature of these studies over a broad range of energies and scattering angles is the substantial agreement with the distorted wave Born calculations of Bartschat and Madison (1987, 1992).

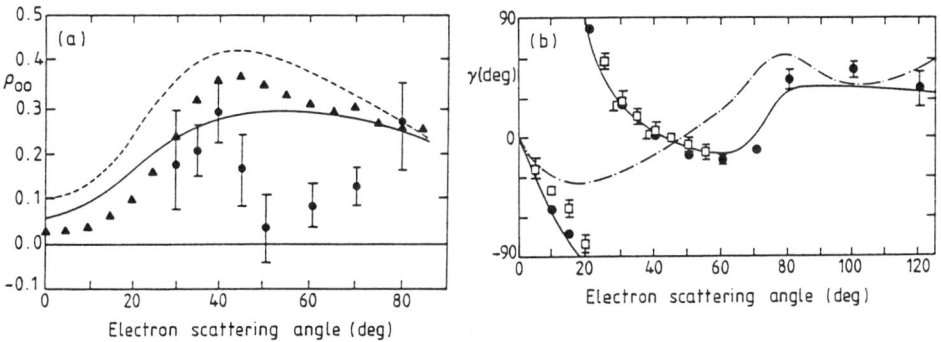

Figure 11(a). ρ_{00} for the $4s[^3/_2]_1$ state of argon excited by 20 eV electrons. *Circles* – Corr et al. (1992); *triangles* – FOMBT of da Paixao et al. (1984); *full curve* – DWBA of Bartschat and Madison (1987); *dashed curve* – RDW of Zuo et al. (1991); **(b).** γ for the $5s[^{3/2}]_1$ state of krypton at 30 eV. *Full circles* – Murray et al. (1990); *open squares* – Zheng and Becker (1993); *full curve* – DWBA of Bartschat and Madison (1987, 1992); *dash-dot curve* – FOMBT of Meneses et al. (1985).

Baerveldt et al. (1994) have reported the first values of P_1, P_2, P_3, P_4 for higher excited states of the rare gases viz. the $3p'[\frac{3}{2}]_2$, $3p[\frac{3}{2}]_2$, $3p[\frac{5}{2}]_3$ and $3p[\frac{5}{2}]_2$ states of neon. With the exception of the $3p[\frac{5}{2}]_3$ state which is a pure $^3D_3$ state, interpretation of the data using any parameter set would appear to be extremely difficult because of their complex LS compositions.

Clearly use of spin polarised beams is essential to gain a clear understanding of spin effects in heavy atom excitation. Sohn and Hanne (1992) have used the polarisation correlation method to study excitation of the 6^3P_1 state of mercury using transversely polarised electrons. A detailed analysis and parameterisation of these experiments has been given by Raeker et al. (1993). A schematic illustration of an atomic charge cloud (J = 1 or $3/2$) after excitation by spin polarised electrons is shown in figure 12(a). In simple physical terms, the charge cloud is no longer only rotated through an angle γ in the scattering plane. Its principal axis now lies out of the scattering plane and the whole charge cloud is subject to a further rotation about this out of plane axis. The direction of the principal axis is described by the angles (γ, δ, ε) of figure 12(b), the reader being referred to Raeker et al. (1993) for a detailed description of the various co-ordinate systems shown in the figure.

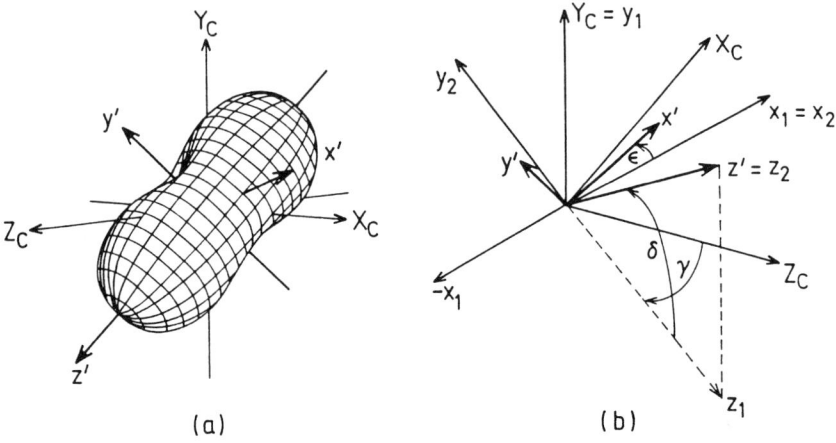

Figure 12(a). Atomic charge cloud after excitation by spin polarised electrons. **(b).** Definition of alignment angles γ, δ, ε (see Raeker et al., 1993).

Figure 13 shows the data of Sohn and Hanne (1992) for the angle δ at an energy of 8 eV. They have demonstrated that the non-zero values observed at all scattering angles is confirmation that the collision system cannot be described in L-S coupling. The behaviour of the data are well reproduced by a Breit-Pauli R-matrix calculation of Raeker et al. (1993).

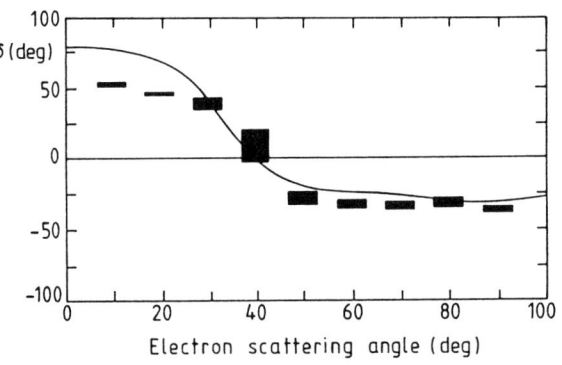

Figure 13. Angle δ for the Hg(6^3P_1) state excited by 8 eV spin polarised electrons. The experimental data of Sohn and Hanne (1992) are compared with the Breit-Pauli R-matrix calculation of Raeker et al. (1993).

FUTURE DIRECTIONS

Despite the fact that only a few groups, both theoretical and experimental, are involved in these studies and that both aspects of the work are time consuming, the large number of recent references cited confirms the continuing vigour of the field. It is clear that a better understanding of excitation processes is emerging for an increasing range of states of different targets. However, many problems remain unsolved and indeed others have been created. Some of these will be best resolved by the merging of spin polarised beam techniques with correlation (Sohn and Hanne, 1992) and superelastic scattering methods as demonstrated by the NIST group. These methods will be particularly important not only in disentangling spin effects in the excitation of heavy atoms but also in providing a complete description of any excitation process which is not fully coherent, including atomic hydrogen.

Considerable work remains to be done with unpolarised beams. Although there are no obvious reasons to doubt the H(2p) data of Williams (1981, 1986), most theoreticians see an independent experimental study as the next logical step in resolving the serious discrepancies between theory and experiment at large scattering angles. In helium, a considerable volume of statistically good experimental data, reliable in the sense that there is substantial agreement between three independent groups, over a range of energies and scattering angles is available for the 3^1D state. Even at small angles no calculation reproduces the experimental data, suggesting little understanding of processes involving transfer of more than a single unit of angular momentum. Discrepancies between theory and experiment for the 3^3P state of helium also suggests a less than complete understanding of exchange processes in simple atoms. Progress in this area may in turn lead to a better description of excitations involving both direct and exchange scattering.

Quite often there is a mismatch between the states and kinematics easily accessible experimentally and using a particular theoretical model. Nevertheless, some areas of the field could benefit from a greater co-ordination of effort. As an example, although interesting in themselves, it is highly unlikely that the detailed resonance structure predicted in the correlation parameters for the near threshold excitation of the 2^1P, 2^3P and 3^3P states of helium (Fon et al., 1993a,b) can be measured in the foreseeable future. Similarly, lack of co-ordination between different experiments has restricted Andersen and Bartschat (1994) in their attempts to obtain a complete description of Na(3^2P) excitation.

Extension of the studies to simultaneous ionisation-excitation could be envisaged. (e, 2e) correlation studies of these processes have already been performed (e.g. Stefani et al., 1990) suggesting that the electron-photon correlation studies should be feasible. Only a single electron-photon correlation experiment has been performed for molecules (McConkey et al., 1986). Despite the considerable difficulties and complexities of both experiments and their analysis arising from the increased degrees of freedom, further applications of the techniques to molecules can be expected.

ACKNOWLEDGMENTS

Correlation work in this laboratory is supported by the UK Engineering and Physical Sciences Research Council (EPSRC).

REFERENCES

Andersen, N., Andersen, T., Dahler, J.S., Nielsen, S.E., Nienhius, G., and Refsgaard, K., 1983, *J. Phys. B.* 16:817.Andersen, N., Gallagher, J.W., and Hertel, I.V., 1988, *Phys. Rep.* 165:1.
Anderson, N., and Bartschat, K., 1994 *Phys. Rev. A.* 49:4232.
Andersen, N., and Hertel, I.V., 1986, *Comments At. Mol. Phys.* 19:1.
Baerveldt, A.W., van Eck, J., and Heidemann, H.G.M., 1994, *J. Phys. B.* 27:1857.
Bartschat, K., and Madison, D.H., 1987, *J. Phys. B.* 20:5839.
Bartschat, K., and Madison, D.H., 1988, *J. Phys. B.* 21:153.
Bartschat, K., and Madison, D.H., 1992, *J. Phys. B.* 25:1361.
Batelaan, H., van Eck, J., and Heideman, H.G.M., 1990, *J. Phys. B.* 23:3993.
Batelaan, H., van Eck, J., and Heideman, H.G.M., 1991a, *J. Phys. B.* 24:L397.
Batelaan, H., van Eck, J., and Heideman, H.G.M., 1991b, *J. Phys. B.* 24:5151.
Becker, K., Crowe, A., and McConkey, J.W., 1992, *J. Phys. B.* 25:3885.
Beijers, J.P.M., Doornenbal, S.J., van Eck, J., and Heideman, H.G.M., 1987a, *J. Phys. B.* 20:5529.
Beijers, J.P.M., Doornenbal, S.J., van Eck, J., and Heideman, H.G.M., 1987b, *J. Phys. B.* 20:6617.
Beyer, H.J., 1988, *in*: Correlation and Polarisation in Electronic and Atomic Collisions, A. Crowe and M.R.H. Rudge, eds., World Scientific: Singapore, p324.
Beyer, H.J., Silim, H.A., and Kleinpoppen, H., 1989, *in*: Proc. 16th Int. Conf. on Physics of Electronic and Atomic Collisions, A. Dalgarno et al. eds., New York, p165.
Blum, K., 1981, "Density Matrix Theory and Applications," Plenum, New York.
Bray, I., and Stelbovics, A.T., 1992, *Phys. Rev. A.* 46:6995.
Bray, I., and McCarthy, I.E., 1993, *Phys. Rev. A.* 47:317.
Cartwright, D.C., and Csanak, G., 1986, *J. Phys. B.* 19:L485.
Cartwright, D.C., and Csanak, G., 1987, *J. Phys. B.* 20:L583.
Chwirot, S., and Slevin, J.A., 1994, *Comments At. Mol. Phys.* 30:15.
Corr, J.J., Wang, S., and McConkey, J.W., 1992, *J. Phys. B.* 25:4929.
Crowe, A., 1990, *in*: Proc. Int. Symp. on Correlation and Polarisation in Electronic and Atomic Collisions (NIST Special Publication 789) P.A. Neill et al., eds., NIST: Washington DC p39.
Crowe, A., 1993, *in*: The Physics of Electronic and Atomic Collisions, T. Andersen et al, eds., AIP Conference Proceedings, 295:286, New York.
Crowe, A., Donnelly, B.P., and McLaughlin, D.T., 1994, *J. Phys. B.* to be published.
Csanak, G., and Cartwright, D.C., 1986, *Phys. Rev. A.* 34:93.
Csanak, G., and Cartwright, D.C., 1987, *J. Phys. B.* 20:L603.
Donnelly, B.P., and Crowe, A., 1989, *Z. Phys. D.* 14:333.
Donnelly, B.P., and Crowe, A., 1988a, *J. Phys. B.* 21:L637.
Donnelly, B.P., and Crowe, A., 1988b, *J. Phys. B.* 21:L697.
Donnelly, B.P., McLaughlin, D.T., and Crowe, A., 1994, *J. Phys. B.* 27:319.
Donnelly, B.P., Neill, P.A., and Crowe, A., 1988, *J. Phys. B.* 21:L321.
Eminyan, M., Mac Adam, K.B., Slevin, J., and Kleinpoppen, H., 1973, *Phys. Rev. Lett.* 31:576.
Eminyan, M., Mac Adam, K.B., Slevin, J., and Kleinpoppen, H., 1974, *J. Phys. B.* 7:1519.
Fano, U., and Macek, J.H., 1973, *Rev. Mod. Phys.* 45:553.
Farrell, P.M., Webb, C.J., Mac Gillivray, W.R., and Standage M.C., 1989, *J. Phys. B.* 22:L527
Fon, W.C., Berrington, K.A., and Kingston, A.E., 1990, *J. Phys. B.* 23:4347.
Fon, W.C., Lim, K.P., and Sawey, P.M.J., 1993a, *J. Phys. B.* 26:4201.
Fon, W.C., Lim, K.P., and Sawey, P.M.J., 1993b, *J. Phys. B.* 26:L747.

Hegemann, T., Oberste-Vorth, M., Vogts, R., and Hanne, G.F., 1991 *Phys. Rev. Lett.* 66:2968.

Herman, H.W., Hertel, I.V., Reiland, W., Stamatovic, A., and Stoll, W., 1977, *J. Phys. B.* 10:251.

Hermann, H.W., Hertel, I.V., and Kelley, M.H., 1980, *J. Phys. B.* 13:3465.

Hermann, H.W., and Hertel, I.V., 1982, *Comments At. Mol. Phys.* 12:61.

Hollywood, M.T., Crowe, A., and Williams, J.F., 1979, *J. Phys. B.* 12:819.

Hood, S.T., Weigold, E., and Dixon, A.J., 1979, *J. Phys. B.* 12:631.

Humphrey, I., Williams, J.F., and Heck, E.L., 1987, *J. Phys. B.* 20:367.

Kelley, M.H., McClelland, J.J., Lorentz, S.R., Scholten, R.E., and Celotta, R.J. *in*: Correlations and Polarisation in Electronic and Atomic Collisions and (e, 2e) Reactions P.J.O. Teubner and E. Weigold eds., IOP Conferences Series 122: 23:Bristol.

Khakoo, M.A., Becker, K., Forand, J.L., and McConkey, J.W., 1986, *J. Phys. B.* 19:L209.

King, S.J., Neill, P.A., and Crowe, A., 1985, *J. Phys. B.* 18:L589.

McConkey, J.W., Trajmar, S., Nickel, J.C., and Csanak, G., 1986, *J. Phys. B.* 19:2377.

McLaughlin, D.T., Donnelly, B.P., and Crowe, A., 1994a, *Phys. Rev. A.* 49:2545.

McLaughlin, D.T., Donnelly, B.P., and Crowe, A., 1994b, *Z. Phys. D.* 29:259.

Mac Gillivray, W.R., and Standage, M.C., 1991, *Comments At. Mol. Phys.* 26:179.

Macek, J., and Jaecks, D.H., 1971, *Phys. Rev. A.* 4:2288.

Madison, D.H., Bray, I., and McCarthy, I.E., 1991, *J. Phys. B.* 24:3861.

Madison, D.H., Bartschat, K., and McEachran, R.P., 1992, *J. Phys. B.* 25:5199.

Mansky, E.J., and Flannery, M.R., 1991, *J. Phys. B.* 24:L551.

Meneses, G.D., da Paixao, F.J., and Padial, N.T., 1985, *Phys. Rev. A.*, 32:156.

Mikoza, A.G., Hippler, R., Wang, J.B., Williams, J.F., and Wedding, A.B., 1994a, *J. Phys. B.* 27:1429.

Mikoza, A.G., Hippler, R., Wang, J.B., and Williams, J.F., 1994b, *Z. Phys. D.* 30:129.

Murray, P.B., Gough, S.F., Neill, P.A., and Crowe, A., 1990, *J. Phys. B.* 23:2137.

Neill, P.A., and Crowe, A., 1984, *J. Phys. B.* 17:L791.

Neill, P.A., and Crowe, A., 1988, *J. Phys. B.* 21:1879.

Neill, P.A., Donnelly, B.P., and Crowe, A., 1989, *J. Phys. B.* 22:1417.

Neitzke, H.P., and Andersen, T., 1984, *J. Phys. B.* 17:1559.

Nic Chormaic, S., Chwirot, S., and Slevin, J., 1993, *J. Phys. B.* 26:139.

da Paixao, F.J., Padial, N.T., and Csanak, G., 1984, *Phys. Rev A*, 30:1697.

Raeker, A., Blum, K., and Bartschat, K., 1993, *J. Phys. B.* 26:1491.

Sang, R.T., Farrell, P.M., Madison, D.H., Mac Gillivray, W.R., and Standage, M.C., 1994, *J. Phys. B.* 27:1187.

Scholten, R.E., Shen, G.F., and Teubner, P.J.O., 1993, *J. Phys. B.* 26:987.

Scholz, T.T., Walters, H.R.J., Burke, P.G., and Scott, M.P., 1991, *J. Phys. B.* 24:2097.

Slevin, J.A., and Chwirot, S., 1990, *J. Phys. B.* 23:165.

Slevin, J., Eminyan, M., Woolsey, J.M., Vassilev, G., Porter, H.Q., Back, C.G., and Watkin, S., 1982, *Phys. Rev. A.* 26:1344.

Sohn, M., and Hanne, G.F., 1992, *J. Phys. B.* 25:4627.

Standage, M.C., and Kleinpoppen, H., 1976, *Phys. Rev. Lett.* 36:577.

Stefani, G., Avaldi, L., and Cannilloni, R., 1990, *J. Phys. B.* 23:L227.

Steph, N.C., and Golden, D.E., 1980, *Phys. Rev. A.* 21:1848.

Teubner, P.J.O., Scholten, R.E., and Shen, G.F., 1989 *in*: Proc. Int. Symp. on Correlation and Polarisation in Electronic and Atomic Collision (NIST Special Publication 789) P.A. Neill, et al., eds., NIST: Washington D.C. p45.

Wang, J.B., Stelbovics, A.T., and Williams, J.F., 1994a, *Z. Phys. D.* 30:119.

Wang, Y.D., Callaway, J., and Unnikrishnan, K., 1994b, *Phys. Rev. A.* 49:1854.

Wang S., van der Burgt, P.J.M., Corr, J.J., McConkey, J.W., and Madison, D.H., 1994c, *J. Phys. B.* 27:329.
Weigold, E., Frost, L., and Nygaard, K.J., 1980, *Phys. Rev. A.* 21:1950.
Williams, J.F., 1981, *J. Phys. B.* 14:1197.
Williams, J.F., 1986, *Aust. J. Phys.* 39:621.
van Wyngaarden, W.L., and Walters, H.R.J., 1986, *J. Phys. B.* 19:929.
Zheng, S.H., and Becker, K., 1993, *J. Phys. B.* 26:517.
Zuo, T., McEachran, R.P., and Stauffer, A.D., 1991, *J. Phys. B.*, 24:2853.

POLARIZATION CORRELATIONS FROM ELECTRON IMPACT EXCITATION OF EARTH ALKALINE ATOMS

Jurgen Beyer

Atomic Physics, University of Stirling
Stirling FK9 4LA, Scotland

INTRODUCTION

With the advance of experimental techniques, the aim of electron scattering experiments has increasingly moved towards so-called complete experiments from which the full information inherent in a given process can be deduced. Such studies are not just curiosity driven, but they lead to a much better understanding of the underlying process and thus a much more meaningful test of the corresponding theories than can be provided by conventional measurements where parameters are averaged for instance over scattering angles, spin polarizations or photon polarizations. The information derived from such complete experiments usually takes the form of a complete set of scattering amplitudes and their relative phases but it may, of course, be expressed by other equivalent parameter sets.

For many electron scattering processes a complete experiment requires some form of spin analysis, i.e. the employment or analysis (or both) of polarized electrons and/or polarized atoms. There are, however, cases such as a fairly substantial group of electron impact excitation processes of atoms where a complete experiment can be achieved solely by the analysis of the polarization of the light emitted by the excited atom provided this is done in coincidence with the scattered electrons which have caused the excitation and which are scattered into a particular direction. The coincidence method thus succeeds by restricting the measurements to a well defined sub-set of all scattering events taking place.

Coincidence studies can take the form of angular or polarization correlation measurements and were first applied to ^{1}P state excitation of He[1,2]. Since then He and the other inert gases have been studied extensively and have stimulated considerable advances in the theoretical treatment of the electron impact excitation of these atoms. (see e.g. the review by Andersen et al.[3]). A limited number of systems other than the inert gases has also been investigated[3], and the current situation for the earth alkali atoms is going to be discussed in this report. In another part of this book A. Crowe is going to look at the past and the future of the coincidence method in a more general way.

There are several reasons why it is of interest to study the electron impact excitation of ^{1}P states of the earth alkaline atoms Be, Mg, Ca, Sr, Ba. In the first place all have a He-like configuration and state system. In fact, they resemble He more closely than the other inert gases do and thus provide a natural extension from He towards more complex and heavier atoms. As in the case of He the excitation from the $^1S_0$ ground state to the n^1P_1 states can be described completely by 2 parameters if - as usual in these particular discussions - the absolute cross section is neglected and if the atoms are not too heavy. The heavier earth alkaline atoms might show spin-orbit coupling in the scattering process itself and the excitation would no longer be completely coherent. In this case the more general 4-parameter description for the excitation of ^1P states would be required necessitating an additional measurement. However, the usual polarization correlation measurements combined with the 2 parameter analysis will reveal the presence of any such effects. The analysis of ^1P state excitation presents no fine structure coupling problems and the prevalence of even-even nuclei in the earth alkaline systems renders the hyperfine coupling effects fairly small too.

The earth alkaline atoms are known to show strong electron correlation effects[4,5] which will influence the scattering process. Compared with He therefore, much more variation is expected of the scattering parameters as a function of the electron scattering angle. From Ca onwards there is an empty (n-1)d shell below the outer ns^2 shell and this should also leave a mark. In this context it would be interesting in some future experiment to compare for instance the Ca data with those of its partner Zn where the 3d shell is filled.

Naturally, experimental coincidence work with metal vapours presents additional problems compared with gas targets. Complications due to the hot beam sources can usually be overcome after a while, but the beam densities remain lower by orders of magnitude compared with gas sources - and have to be low since the experiments require continuous runs of weeks at least in order to provide the minimum of a complete data set for one scattering angle. Resonance trapping is usually not a problem in this case. To obtain a reasonable scattered electron count rate the primary electron beam has to be of the order of $1\mu A$ and since the energy loss of the electrons is only approximately 3eV it is not straight-forward to suppress the strong background of primary electrons. This is aggravated by the condensation of metal vapour on the surfaces near the interaction region which increasingly affects the primary and scattered electrons as well as the resolution of the electron energy analyzer during a run even if the vacuum system is not opened. Regular cleaning of the system is thus mandatory.

Theoretical calculations of the scattering parameters measured in coincidence experiments are now possible in various approximations for such complicated atoms. To be able to judge their predictions it is essential to provide measurements for as wide a range of impact energies and scattering angles as possible.

THE CURRENT STATE

Theoretical results of electron impact coherence parameters for the ^1P states of Mg, Ca, Sr, Ba were reported, notably by groups at Adelaide, Los Alamos and Toronto, often in conjunction with new experimental investigations. Calculations were carried out for the 3^1P state of Mg by Mitroy and McCarthy[6], Meneses et al[7] and Clark et al.[8], for the 4^1P state of Ca by Clark et al.[9,10] and Srivastava et al.[11], for the 5^1P state of Sr by Srivastava et al.[11] and Beyer et al.[12] and for the 6^1P state of Ba by Clark et al.[13] and Srivastava et al.[11]. Most of these calculations are based on fairly "simple" approximations like the first order many body theory[12,14,15] (Los Alamos) or the relativistic distorted wave approximation[11,16] (Toronto) but, nevertheless, represent the experimental results very well in many cases.

Experimental results of the electron impact coherence parameters for earth alkaline atoms have been obtained in two different ways both of which are considered to provide equivalent data: Superelastic electron scattering following laser induced preparation of the excited state and electron-photon polarization correlations using the coincidence technique. Superelastic scattering does not require coincidence measurements, an advantage in signal to noise terms, but the laser excitation and other experimental conditions have to be carefully controlled. It is not subject of this review, but measurements reported so far for the earth alkaline atoms are summarized in Table 1. The coincidence measurements which will be discussed in more detail below are summarized in Table 2. Both tables show the electron impact energies and the range of scattering angles studied.

Table 1. Superelastic scattering experiments on earth alkaline atoms.

Ca 4^1P		^{138}Ba 6^1P	
25.7eV 5°-120°	various	19.48eV 5°-32°	
		34.48eV 4°-32°	
Adelaide[17]	Pasadena/Los Alamos[18,19]	Winnipeg/Pasadena[20]	

Table 2. Coincidence experiments on earth alkaline atoms

Mg 3^1P	Ca 4^1P	Sr 5^{1P}
20eV 5°-25°	no results	30.3eV 20°-130°
40eV 5°-20°	reported	45eV 30°-113°
	so far 25.7eV 15°-40°	58.4eV 20°-100°
	35.7eV 15°-55°	
	45eV 10°-50°	
Adelaide[21,22]	Torun[23] Stirling[24,25]	Stirling[26-28,12]

THE COINCIDENCE APPARATUS

Fig.1 shows a schematic diagram of the experimental arrangement used at Stirling. The electron gun and the 127° energy analyzer for the scattered electrons together determine the scattering plane, here the x-z-plane using the collision coordinate system. The decay photons from the excited ^1P states are detected at right angles to the scattering plane in y-direction through a linear polarizer and an interference filter preceded by a quarter wave plate in the case of circular polarization measurements. A single quartz lens is used to form a magnified image of the interaction region at the photomultiplier cathode. This helps to reduce the divergence of the light beam through the optical components. The Adelaide group[22] placed the lens such that the interaction region was at the focal point. This results in a parallel light beam through the optics and also reduces the area of the interaction region subtended by the detection system thus helping to define the exact scattering plane and scattering angle which is important for the small scattering angles used by this group.

In order to reduce the stray light seen by the photomultiplier from the hot cathode the Stirling gun has an extension tube with another two apertures mounted to its front. Similarly, to reduce the number of stray electrons entering the electron analyzer a narrow cone is attached to the front of the electron lens and steering system.

The atomic beam is directed parallel to the scattering plane along the x-axis. It emanates from a shielded oven, resistively heated to 500-600°C. The electron analyzer,

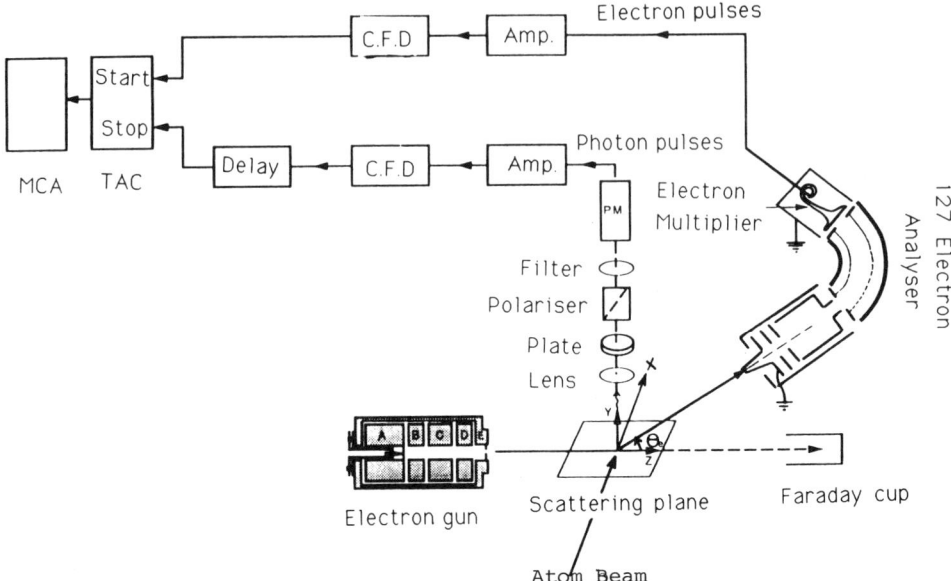

Fig.1 Experimental arrangement.

especially the cone extending close to the interaction region, is exposed to the vapour beam, the more so the closer the electron scattering angle gets to 90° when the analyzer would look straight into the oven. This severely restricted the available range of scattering angles (for the Ca measurements) until (for the Sr measurements) the oven was tilted by about 20° towards the y-axis and the beam forming apertures were reduced further. The Adelaide group[22] directed the atomic beam along the y-axis, i.e. perpendicular to the scattering plane. This does not impose any restriction on the electron scattering angle, but now the beam fires directly onto the optical system and coats the lens. However, due to the low beam density this was found to be less of a problem than would have been expected.

The detectors feed the usual coincidence electronics arrangement and the resulting time correlation spectra were displayed on a multichannel analyzer and further analyzed from there.

MEASUREMENTS AND DATA ANALYSIS

The true coincidence rates, after subtraction of the chance coincidences, were normalized to the electron counts which had started the coincidence circuit. The normalized coincidence rates I_α, where α corresponds to the linear polarizer angle measured from the z-axis towards the x-axis in the same way as θ_e, and I_{RHC}, I_{LHC}, which correspond to right and left hand circularly polarized light in the spectroscopic definition, are combined in the usual way to obtain the Stokes parameters P_1, P_2 for linear polarization and P_3 for circular polarization:

$$P_1 = (I_0 - I_{90})/(I_0 + I_{90})$$

$$P_2 = (I_{45} - I_{135})/(I_{45} + I_{135})$$

$$P_3 = (I_{RHC} - I_{LHC})/(I_{RHC} + I_{LHC})$$

From these the total polarization P can be derived which should be 1 as long as the excitation is fully coherent as in the case of ^{1}P state excitation of He:

$$|P| = \sqrt{P_1^2 + P_2^2 + P_3^2} = 1$$

Most theoretical results are now also presented in terms of the measurable Stokes parameters so that a comparison between theory and experiment is possible at this level. However, the physics of the excitation process becomes more transparent if the results are expressed by other parameters which can be deduced from the 3 Stokes parameters. Two such parameter sets are widely used and discussed in detail in the review by Andersen et al.[3]. The first parameter set λ,χ relates closely to the complex excitation amplitudes of the ^{1}P state. Expressed in the collision coordinate system where the quantization axis z is along the incoming electron beam, λ is the ratio of the cross section for $m_\ell = 0$ to the total cross section and χ the phase angle between the scattering amplitudes for $m_\ell = 0$ and $m_\ell = 1$. In terms of the Stokes parameters, λ and χ are given by:

$$\lambda = \frac{1}{2}(P_1 + 1)$$
$$\chi = \tan^{-1}(P_3/-P_2)$$

The second parameter set $L_\perp$, P_ℓ, γ relates to the charge cloud of the excited ^{1}P state. $L_\perp$ is the angular momentum transfer to the ^{1}P state in the collision, P_ℓ the degree of linear polarization (the two are linked for fully coherent excitation when $|P| = 1$) and γ the alignment angle of the charge cloud with respect to the incoming electron beam. In terms of the Stokes parameters, $L_\perp$, P_ℓ, γ are given by:

$$L_\perp = -P_3$$
$$P_\ell = (P_1^2 + P_2^2)^{1/2}$$
$$\gamma = \tfrac{1}{2}\tan^{-1}(P_2/P_1)$$

RESULTS AND DISCUSSION

Mg. The excitation of the 3^1P state of Mg was studied by the Adelaide group[21,22] and analyzed in terms of the charge cloud parameters as shown in Fig.2. The measurements were carried out for electron impact energies of 20 eV (left) and 40 eV (right) and for scattering angles up to 20° and 25° above which the signal to noise ratio was found to become unsatisfactory. Within this angular range the measurements are very accurate and can be used for comparison with the various theories shown in Fig.2. There is no clear trend as to which theory agrees best with all experimental data. The first order many body theory (FOMBT) seems to differ fairly strongly from the experiment in the Stokes parameter P_1 which works through into the alignment angle γ. Here the close coupling theories perform best while for P_ℓ the first order theories agree marginally better. As expected for a light atom like Mg, the total polarization $|P|$ is found to be one throughout.

Ca. The excitation of the 4^1P state of Ca was studied by the Stirling group[24,25] for electron impact energies of 25.7 eV, 35.7eV and 45eV and scattering angles in the range of 15°-50°. The results for the Stokes parameters P_1 to P_3 are shown in Fig.3 for the excitation energy of 35.7eV. The comparison with the first order many body calculations[9,10,29] and the distorted wave approximation[11,30] shows substantial agreement over this range of scattering angles, possibly with a slight shift of the measured values

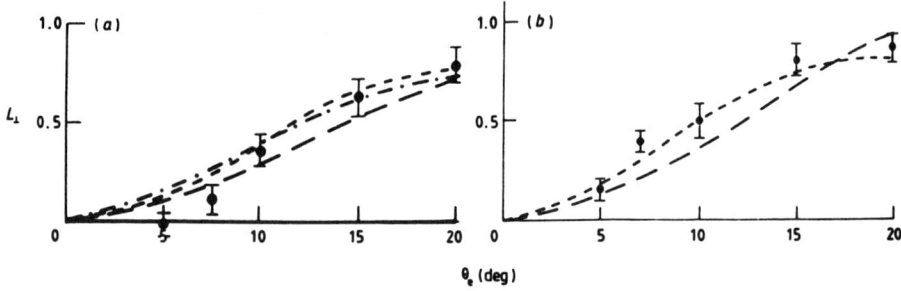

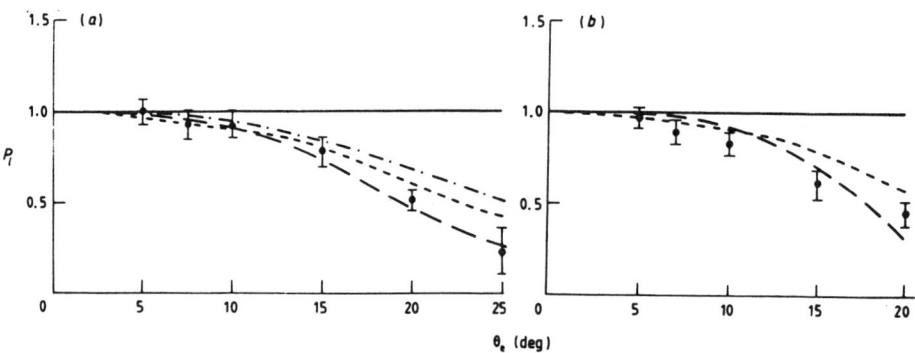

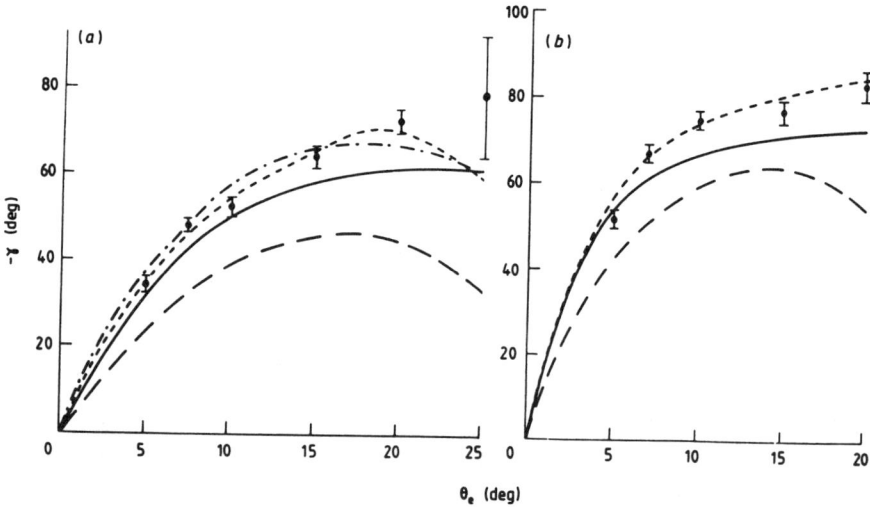

Fig.2. (from Ref.22), Electron impact coherence parameters for the 3^1P state of Mg^{22}.
Top: Angular Momentum transfer, $L\perp$, normal to the scattering plane. Middle: Linear polarization, P_l. Bottom: Alignment angle, γ, of the charge cloud. Electron impact energy 20eV (left) and 40eV (right). Theories: ——— first Born approximation, - - - five state close coupling calculation[6], — — — first order many body theory[7], —·—·— two state close coupling calculation[33].

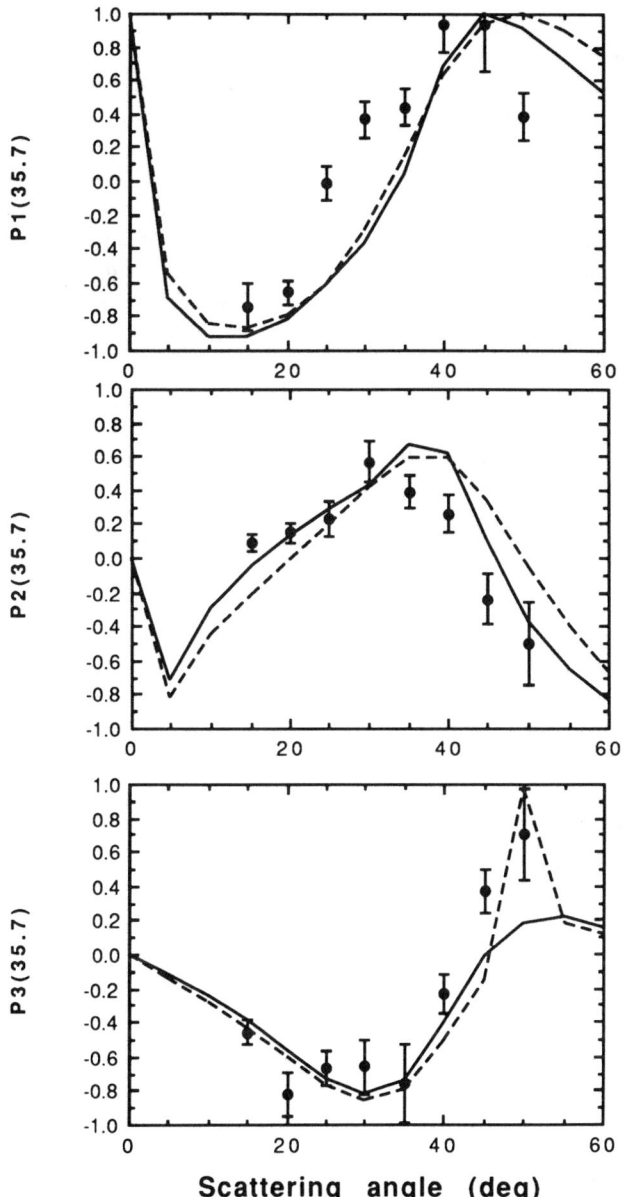

Fig.3. Stokes parameters P_1, P_2, P_3 for the electron impact excitation of the 4^1P state of Ca at 35.7eV. The measured results (Stirling group) are shown in comparison with the first order many body calculations[9,10,29] ——— and the distorted wave approximation[11,30] - - -.

towards smaller scattering angles. For 25.7eV the measured values are limited to the angles $15°-40°$ and P_1 and P_3 show good agreement with the theories while P_2 appears to indicate a different trend even though most values are consistent with the calculations. For 45eV ($10°-50°$) P_1 agrees well, but below $25°$ the measured values are higher than predicted, while P_2 and P_3 show marked differences from theory. In all cases the total polarization $|P|$ is consistent with 1. Another coincidence experiment for Ca is being set up at Torun[23] but no numerical results have been reported so far.

Sr. The excitation of the 5^1P state of Sr was measured by the Stirling group[26-28,12] over a much increased range of scattering angles for the three electron impact energies 30.3eV, 45eV and 58.4eV. Fig.4 shows as an example the experimental and theoretical results for 30.3eV. It is clear that the three Stokes parameters (4a-c) vary greatly as a function of the scattering angle using almost the full range available to them. It is also clear that the high-angle results make an important contribution in spite of their relatively large error bars. As in the case of Ca, the two calculations carried out so far are the first order many body theory[12] and the distorted wave approximation[11,30]. Both theories agree well with each other apart from slight angular shifts, and their agreement with the experimental results is remarkable, certainly for 30.3eV. For 45eV and 58.4eV the general shape of the theories is again in good agreement with the experiment, but some noticeable angular shifts are observed.

The measured values of the total polarization $|P|$ in Fig.4d are on average slightly below 1. However, no corrections have been applied at this stage. Small reductions of the measured polarization are expected as a result of the fairly large acceptance angle of the optical system used and of the hyperfine structure coupling of the odd isotope ^{87}Sr with $I = 9/2$ and an abundance of 7% in natural Sr. The total hyperfine splitting of the 5^1P state of ^{87}Sr is twice the natural width of the state[31] and will, therefore, cause some depolarization. The corresponding corrections would bring the values of $|P|$ fairly close to 1 and thus confirm at least nearly full coherence of the excitation process of Sr in line with both theoretical calculations[11,12,30]. Thus the two-parameter analysis used for the current results represents at least a very good approximation and on this basis the derived parameters $L_\perp$, P_t, and λ, χ are shown in Figs. 4e-i.

Fig.5 shows a comparison of the angular momentum transfer $L_\perp = -P_3$ for the 3 impact energies. The basic shape of the angular dependence is the same for all 3 energies but, in line with the experience from other systems, corresponding structure is shifted towards smaller scattering angles for increasing impact energy. For all 3 energies a clear sign change of the angular momentum transfer is observed at some intermediate scattering angle between $65°$ and $85°$ similar to measurements on the 3^1P state of He (Ibraheim[32] and his results shown in Fig.3.1.13 of the review by Andersen et al.[3]).

The angular dependence of the alignment angle, γ of the charge cloud shows an interesting development over the range of energies studied as can be seen in Fig.6. γ is only determined modulo $180°$ and by convention is usually drawn for the range $-90°$ to $+90°$. Starting from zero scattering angle both theories predict a fast rotation of the charge cloud from $0°$ to just below $-180°$ ($0°$) within the first $30°$ of scattering angle. This applies to all three energies. However, for scattering angles between about $50°$ and $100°$ the theories show a marked difference between 30.3eV and 45eV on the one side and 58.4eV on the other side: For the lower two energies the charge cloud rotates smoothly in a positive sense and then back again to slightly negative values when the scattering angle is increased while for 58.4eV it rotates in a negative sense only, before it reaches a similar value of $\sim -10°$ at a scattering angle of $\sim 80°$ after a rotation by approximately $-180°$. At the scattering angle of $54°$ (distorted wave approximation[30]) γ changes extremely rapidly with a rotation of the charge cloud by $-60°$ over only $2°$ scattering angle. The angular resolution of the experiment is not sufficient to scan such a fast change and all that could be expected would be a step change of γ at this scattering

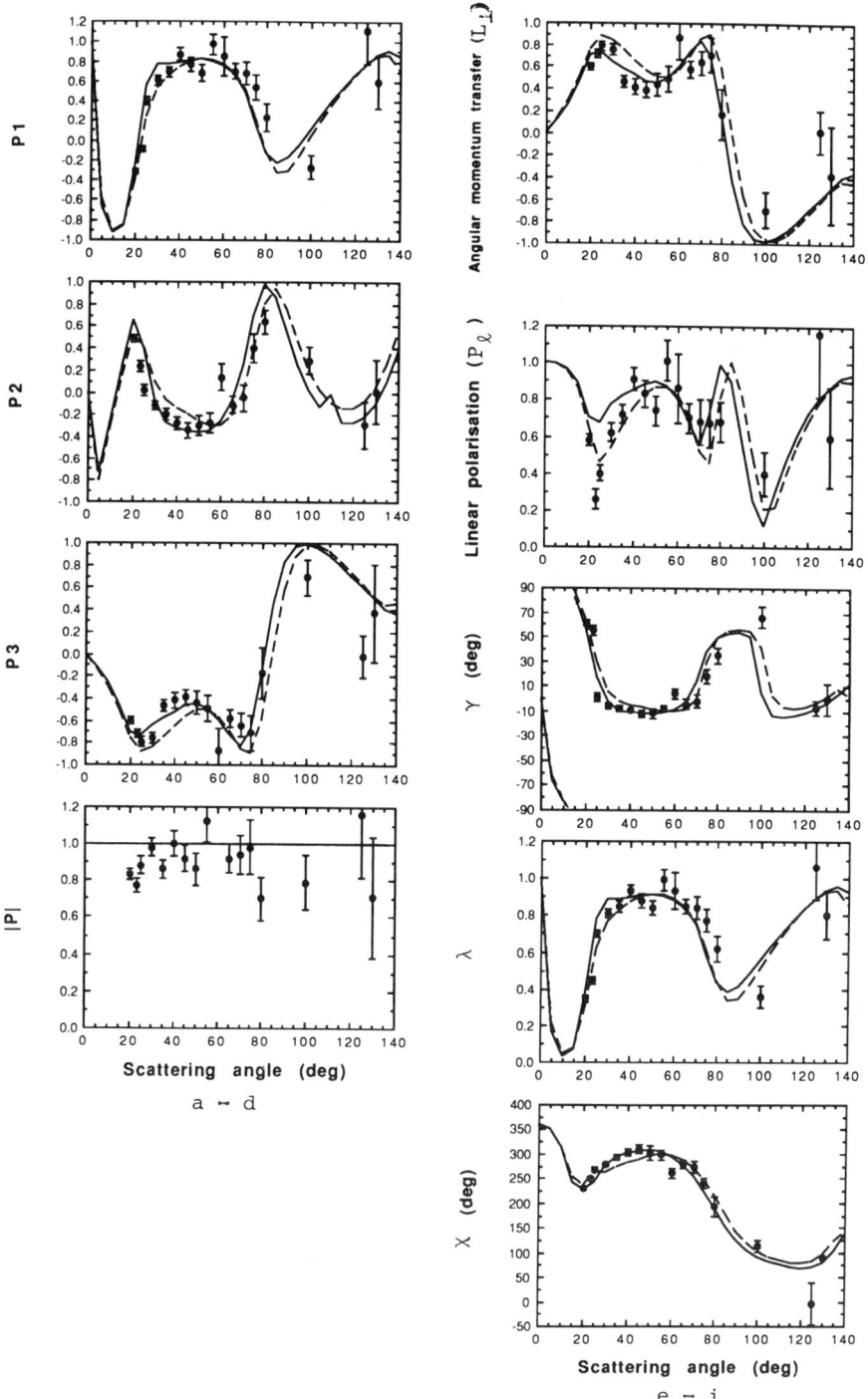

Fig.4. Electron impact excitation of the 5^1P state of Sr at 30.3eV. The measured results[12] are compared with the first order many body theory[12] ——— and the distorted wave approximation[11,30] - - -. a-d: Stokes parameters P_1, P_2, P_3 and total polarisation $|P|$. e-g: Charge cloud parameters $L\perp$, P_ℓ, γ. h-i: scattering parameters λ, χ.

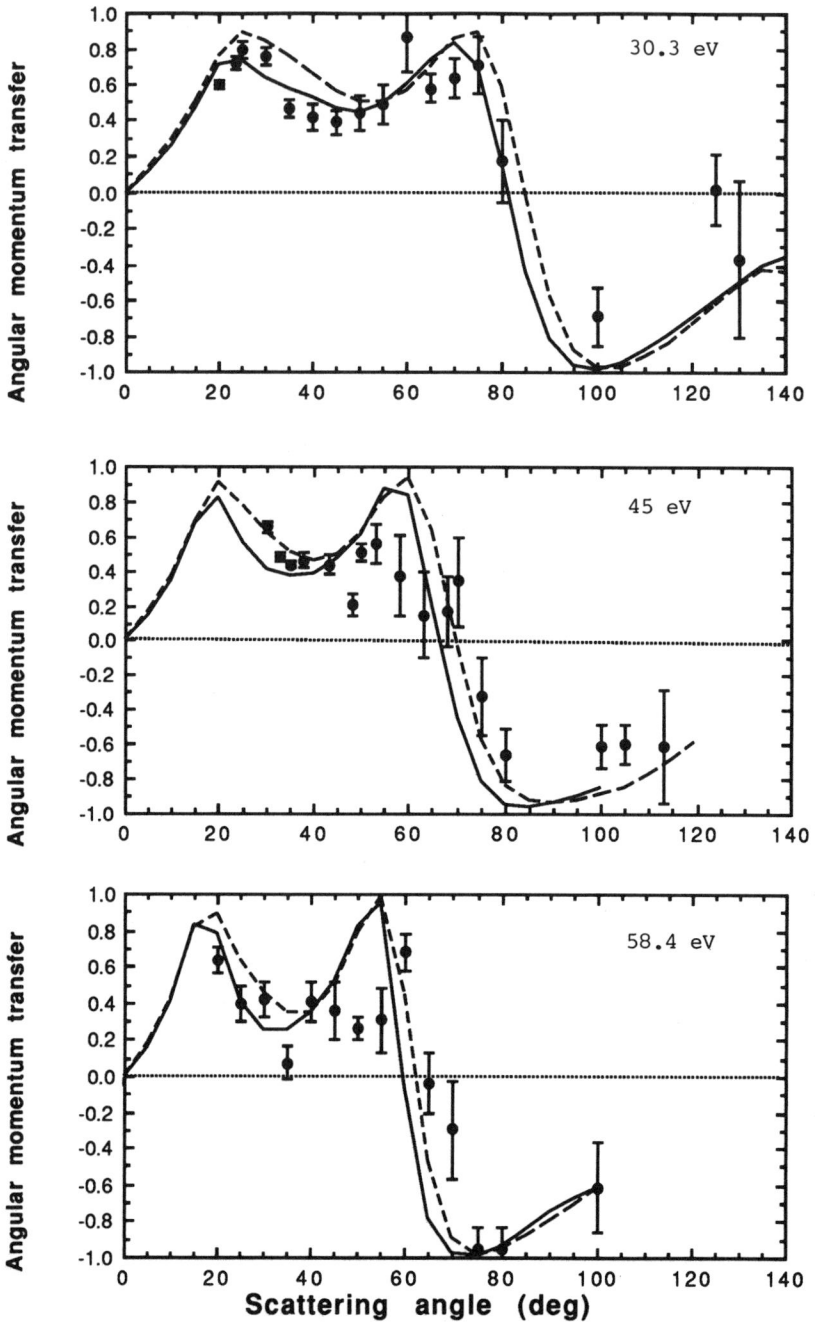

Fig.5. Angular momentum transfer, L⊥, normal to the scattering plane for electron impact excitation of the 5^1P state of Sr at electron impact energies 30.3eV (top), 45eV (centre) and 58.4eV(bottom). The experimental results[28,12] are compared with the first order many body theory[12] ——— and the distorted wave approximation[11,30] - - -. Note the sign change of L⊥ and the shift of corresponding structure towards smaller scattering angles with increasing impact energy.

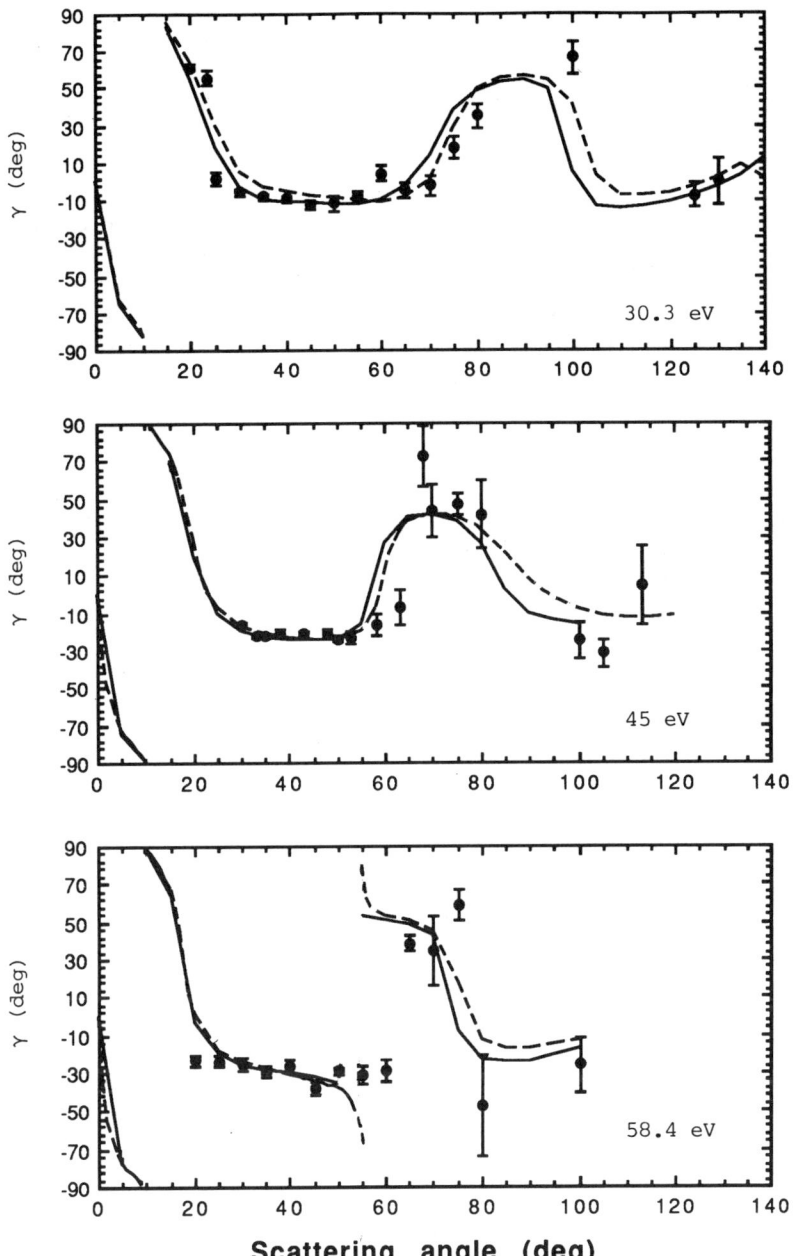

Fig.6. Alignment angle, γ, of the charge cloud with respect to the incoming electron beam for 5^1P state excitation of Sr at electron impact energies 30.3 eV (top), 45 eV (centre) and 58.4 eV (bottom). The experimental results[28,12] are compared with the first order many body theory[12] ———— and the distorted wave approximation[11,30] - - - .

angle. There is indeed a very sudden change of γ between 60° and 65° scattering angle (all measured results for 58.4eV appear to be slightly shifted in angle from the theoretical results) while the results for 30.3eV and 45eV clearly show the onset of the positive rotation of γ. However, corroboration would be required before the rapid negative rotation of the charge cloud for 58.4eV could be accepted as proven.

ACKNOWLEDGEMENTS

For the results of the Stirling group reported here I thank the collaborators H.Hamdy, M.A.K.El-Fayoumi, K.R.Mahmoud, E.I.M.Zohny and H.Kleinpoppen at Stirling and J.Abdallah Jr., R.E.H.Clark and G.Csanak at Los Alamos. I also thank Prof. A.D.Stauffer, Toronto, for making numerical results available. The work carried out at Stirling was financially supported by the Science and Engineering Research Council.

REFERENCES

1. Eminyan, M., MacAdam, K.B., Slevin, J. and Kleinpoppen, H.: *J.Phys.B.* 7:1519 (1974).
2. Standage, M.C. and Kleinpoppen, H.: *Phys.Rev.Lett.* 36:577 (1976).
3. Andersen, N., Gallagher, J.W. and Hertel, I.V.: *Phys.Rep.* 165:1 (1988).
4. Robb,W.D.: *J.Phys.B.* 7:1006 (1974).
5. Sadlej, A.J., Urban, M. and Gropen, 0.: *Phys.Rev.A.* 44:547 (1991).
6. Mitroy, J. and McCarthy, I.E.: *J.Phys.B.* 22:641 (1989).
7. Meneses, G.D., Pagan, C.B. and Machado, L.E.: *Phys.Rev.A.* 41:4610 (1990).
8. Clark, R.E.H., Csanak, G. and Abdallah Jr., J.: *Phys.Rev.A.* 44:2874 (1991).
9. Clark, R.E.H., Csanak, G. and Cartwright, D.C.: *Bull.Am.Phys.Soc.*, 34:1408 (1989).
10. Clark, R.E.H., Cartwright, D.C. and Csanak, G.: *16th Int.Conference on the Physics of Electronic and Atomic Collisions, New York, Book of Abstracts,* p182 (1989).
11. Srivastava, R., Zuo, T., McEachran, R.P. and Stauffer, A.D.: *J.Phys.B.* 25:370 (1992).
12. Beyer H.-J., Hamdy, H., Zohny, E.I.M., Mahmoud, K.R., El-Fayoumi, M.A.K., Kleinpoppen, H., Abdallah Jr., J., Clark, R.E.H., and Csanak, G.: *Z.Phys.D.* 30:91-97 (1994).
13. Clark, R.E.H., Abdallah Jr., J., Csanak, G. and Kramer,S.P.: *Phys.Rev.A.* 40:2935 (1989).
14. Clark, R.E.H., Abdallah Jr. J., Csanak, G. and Kramer, S.P.: *Phys.Rev.A.* 40:2935 (1989).
15. Clark, R.E.H. and Abdallah Jr., J.: *Proceedings of the International Symposium on Correlation and Polarization in Electronic and Atomic Collisions, NIST Special Publication 789, edited by P.A.Neill, K.H.Becker and M.H.Kelley (U.S. Department of Commerce, National Bureau of Standards, Washington, DC)* p22 (1990).
16. Zuo, T.: *PhD thesis,* York University, Toronto, Canada (1991).
17. Law, M.R. and Teubner, P.J.O.: *18th Int. Conference on the Physics of Electronic and Atomic Collisions, Aarhus, Book of Abstracts,* p150 (1993).

18. Register, D.F., Trajmar, S., Csanak, G., Jensen, S.W., Fineman, M.A. and Poe, R.T.: *Phys.Rev.A.* 28:151 (1983).
19. Zetner, P., Trajmar, S., Csanak, G. and Clark, R.E.H.: *Phys.Rev.A.* 40:6022 (1989).
20. Zetner, P., Li, Y. and Trajmar, S.: *J.Phys.B.* 25:3187 (1992).
21. Brunger, M.J., Riley, J.L., Scholten, R.E. and Teubner, P.J.O.: *Proc. 15th Int.Conf. on the Physics of Electronic and Atomic Collisions (Brighton)*, (Amsterdam: North Holland) Abstracts p170 (1987).
22. ----------: *J.Phys.B* 22:1431 (1989).
23. Chwirot, S.: *Private communication.*
24. El-Fayoumi, M.A.K., Beyer, H-J., Shahin, F., Eid, Y. and Kleinpoppen, H.: *Book of Abstracts, 11th Int.Conf. and Atomic Physics, Paris,* XI-2 (1988).
25. Zohny, E.I.M., El-Fayoumi, M.A.K., Hamdy, H., Beyer, H-J., Eid, Y., Shahin, F. and Kleinpoppen, H.: *Proc. XVI Int.Conf. on the Physics of Electronic and Atomic Collisions, New York,* p173 (1989).
26. Hamdy, H., Beyer, H-J., Mahmoud, K.R., Zohny, E.I.M., Hassan, G. and Kleinpoppen, H.: *Book of Abstracts, 17th Int.Conf. on the Physics of Electronic and Atomic Collisions, Brisbane, Australia,* p132, (1991).
27. ----------: *Book of Abstracts, 13th Conference on Atomic and Molecular Physics, Munich,* E23, (1992).
28. Hamdy, H., Beyer, H-J. and Kleinpoppen, H.: *J.Phys.B.* 26:4237-48 (1993).
29. Csanak, G.: *Private communication of numerical results.*
30. Stauffer, A.D.: *Private communication of numerical results.*
31. Kluge, H-J and Sauter, H.: *Z.Physik.* 270:295 (1974).
32. Ibraheim, K.S.: *PhD thesis,* Stirling University, unpublished (1986).
33. Fabrikant, I.I., *Bull.Latvian Acad.Sci.* 5: (1985).

ELECTRON POLARIMETRY ON THE SYNCHROTRON RADIATION SOURCE (SRS)

E.A. Seddon*, I.W. Kirkman and F.M. Quinn

DRAL Daresbury Laboratory
Daresbury
Cheshire
WA4 4AD
UK

INTRODUCTION

Two polarised electron experimental facilities have been developed for use on the SRS. The first is a dedicated beamline and station, Station 1.2, which incorporates a conventional high-energy Mott polarimeter for spin analysis of electrons photoemitted from surfaces under UHV conditions. The second is an electron energy and angle resolving instrument which incorporates a conical retarding potential Mott polarimeter (the "micro-Mott" polarimeter) for use in a variety of configurations on a number of beamlines. These facilities have been the subject of several preliminary reports.[1-3]

STATION 1.2

Station 1.2 is shown schematically in Fig.1. Monochromatic, linearly polarised radiation over the range 5-90eV is focused onto the sample which can be cleaned and characterised by the standard surface science techniques of argon ion sputtering, low energy electron diffraction and Auger electron spectroscopy (Omicron rear view system). The photoemitted electrons are energy analysed in a 100mm radius hemispherical analyser (VSW HA100), which is mounted at 45° to the incident photon beam, and then undergo acceleration (to 3keV) and focusing to a small spot at the entrance to the deflection magnet. After deflection the electrons are electrostatically accelerated to 100keV (National Instruments accelerator tube, type 2FA000530) and focused to a spot approximately 1mm across on a gold scattering target. Although the Mott polarimeter and accelerator vessel are fixed in space, the rest of the station, including the sample, can

---

* Author to whom correspondence should be addressed.

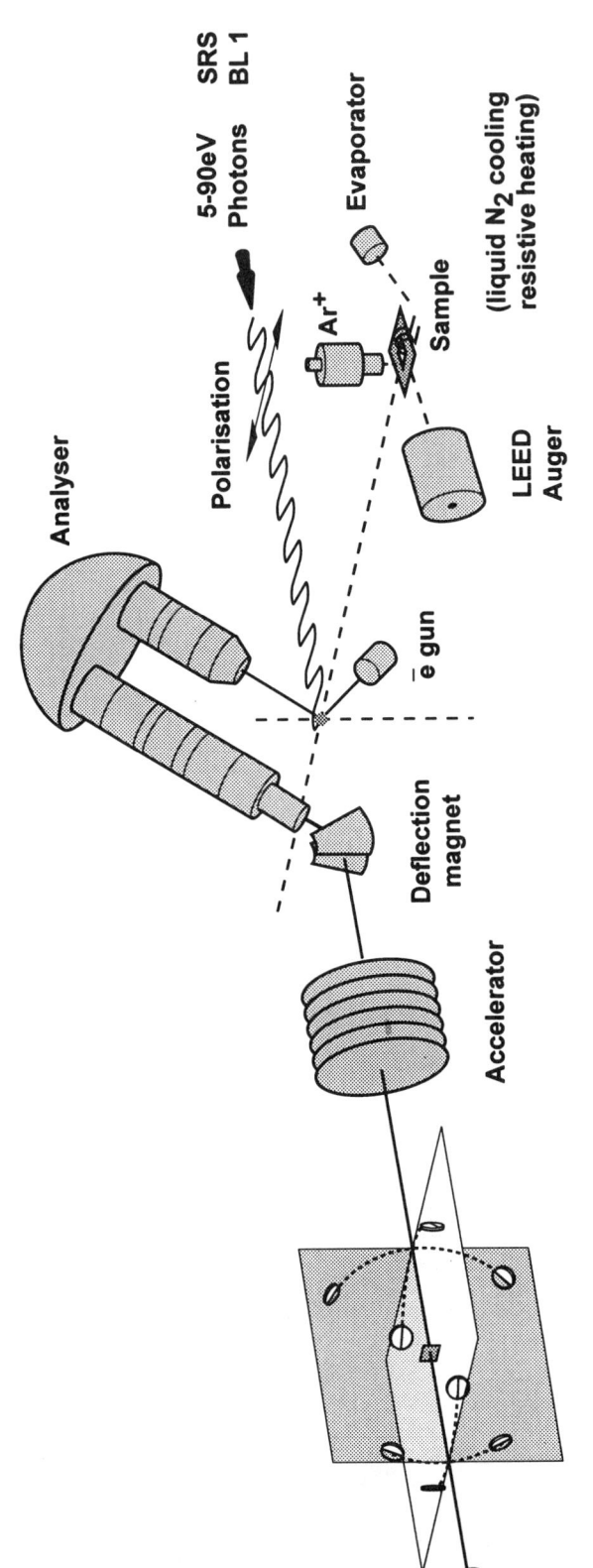

Figure 1 Schematic representation of Station 1.2 on the SRS.

be rotated through 90° allowing alignment of the polarisation vector of the incoming radiation either parallel or perpendicular to the electron emission plane.

The high energy Mott polarimeter shown in Fig. 2 consists of an array of nine solid-state surface barrier detectors (Ortec types H-019-200-100, H-016-100-100 and H-016-050-100). The backward detectors detect electrons scattered over an angular range of 110-130° where the spin dependent scattering asymmetry for 100kV electrons is calculated[4] (assuming single atom scattering) to be at its peak of around 38%. The forward detectors detect electrons over the range 10-14° where the spin dependent scattering asymmetry is calculated to be less than 1%. The electrons are scattered from gold foils held in a motor driven foil holder that currently supports three thicknesses of foil (of 300, 600 and 1000Å), an uncoated Lexan™ backing film and a clear aperture. The whole assembly mounts on a ten inch flange which can be seen to the right of the ion pump port in the centre of the anti-corona housing of the Mott polarimeter depicted in Fig.3. When operating, the chamber housing the polarimeter is at high voltage - routinely 100kV- it is therefore subject to stringent safety precautions.

The 2-3 millivolt signal pulses from the barrier detectors are fed into charge sensitive preamplifier - shaping amplifier (Amptek A225)/discriminator (Amptek A206) units which are mounted directly onto feedthroughs on the end flange of the Mott polarimeter (see Fig. 3) and are therefore also at a potential of 100kV when the polarimeter is in operation. The signals from these are coupled to equipment at ground potential using fibre optic cables. The TTL signal pulses are passed to the counters (Microlink EC8 units) for data logging, processing takes place on an IBM compatible computer (Viglen 386). Full technical details of the data acquisition system are to be reported elsewhere.[5]

The two transverse components of the electron polarisation vector are determined by monitoring the count rate asymmetry in the two pairs of back detectors placed at equal polar but opposite azimuthal angles to the beam direction, as shown in Fig. 1. The spin dependent asymmetries ($A_{x,y}$) arising from the Mott scattering process and the transverse electron beam polarisation components ($P_{x,y}$) are related by the expression:

$$P_{x,y} = A_{x,y} / S_{eff}$$

Where S_{eff} is the effective Sherman function for the scatterer used. The *experimentally* observed asymmetry is however a combination of the true spin dependent asymmetry and any instrumentally derived asymmetries (which may be due, for example, to unequal detector sensitivities and/or misalignment of the apparatus).[6,7]

For ferromagnetic samples the simplest method of eliminating the effects of instrumental asymmetries is to reverse the magnetisation of the sample and hence the polarisation of the electrons photoemitted from it. With the magnetisation in the initial direction the ratio of the counting rates in a pair of detectors is:

$$\frac{I_1}{I_2} = \frac{\eta_1(1 + A)(1 + A_i)}{\eta_2(1 - A)(1 - A_i)}$$

where; I_1 and I_2 represent the detector counts, η_1 and η_2 represent the detector efficiencies, A represents the spin dependent asymmetry and A_i represents the spin independent instrumental asymmetry. Upon reversal of the polarisation direction, and assuming that the beam trajectory at the scattering foil is unchanged, a new ratio of the counts is obtained:

Figure 3 The external view of the high energy Mott polarimeter.

Figure 2 Internal arrangement of the array of nine silicon surface barrier detectors within the high energy Mott polarimeter.

$$\frac{I'_1}{I'_2} = \frac{\eta_1(1-A)(1+A_i)}{\eta_2(1+A)(1-A_i)}$$

where I'_1 and I'_2 are the detector counts after magnetisation reversal. By combining these two equations both the different detector efficiencies and the instrumentally induced asymmetries can be eliminated. The spin dependent asymmetry is then given by:

$$A = \frac{(X-1)}{(X+1)}$$

where $X \equiv \sqrt{I_1 I'_2 / I_2 I'_1}$. The spin resolved spectra are then obtained from the spin integrated data and the polarisation data by applying:

$$I_\uparrow = I_o(1+P) \quad \text{and} \quad I_\downarrow = I_o(1-P)$$

where:

$$I_o = \frac{I_1 + I'_1 + I_2 + I'_2}{4}$$

The expected uncertainty in the calculated asymmetry arising from counting statistics is given by:

$$\Delta A = \frac{2\Delta X}{(X+1)^2}$$

where:

$$\Delta X = \frac{0.5\{I'_2 I_2 I'_1 + I_1 I_2 I'_1 + I_1 I'_2 I'_1 + I_1 I'_2 I_2\}^{1/2}}{I'_1 I_2}$$

Assuming there to be no error in S_{eff}, the uncertainty in the polarisation is given by:

$$\Delta P = \Delta A / S_{eff}$$

More realistically, for an uncertainty in S_{eff} of ΔS_{eff}, the uncertainty in the polarisation is given by:

$$\Delta P = \left[\frac{\Delta A^2}{S_{eff}^2} + \frac{A^2}{S_{eff}^4}\Delta S_{eff}^2\right]^{1/2}$$

In addition to instrumentally derived asymmetries, the measured asymmetries will also be affected by spurious background signals arising principally from higher order radiation in the incident photon beam, electron scattering from the Lexan support and electrons scattered from components within the polarimeter chamber. In the data reported below account has been taken of the first two of these sources of background.

Accurate determination of the Sherman function for this type of polarimeter and an electron beam of unknown polarisation requires measurement of the scattering asymmetries as a function of scattering foil thickness and then extrapolation of the data to zero foil thickness where an experimentally verified theoretical value of S=0.39 may be assumed.[4,7,8] At present the range of foils available in the polarimeter is insufficient to allow an accurate extrapolation to be undertaken. However, comparison of the data described below with equivalent data obtained by Raue *et al.*[9,10] gives a value of S_{eff}=0.26 for the thickest foil.

Some preliminary results on nickel(110) are shown in Figs. 4 and 5. These data were obtained with a nickel picture frame sample mounted so that the electric vector of the incident radiation was parallel to the [110] direction. A photon energy of 16.85eV was used with the beam oriented at 45° to the surface normal. Normal electron emission geometry was used. Typical count rates obtained for the peak of the valence band with a low resolving power were 50,000cps in the straight through detector and 800cps in each of the backward scattering detectors. An accumulation time of approximately two hours was required for moderate counting statistics. Great care was taken with the magnetisation of the nickel as multidomain samples give inconsistent, low, polarisation values. Magnetisation was achieved by passing a current pulse of approximately 120A and 150µs duration through a copper coil wrapped four and a half times around one of the legs of the picture frame. The magnitude of the current required was judged from off-line magneto optical Kerr measurements.

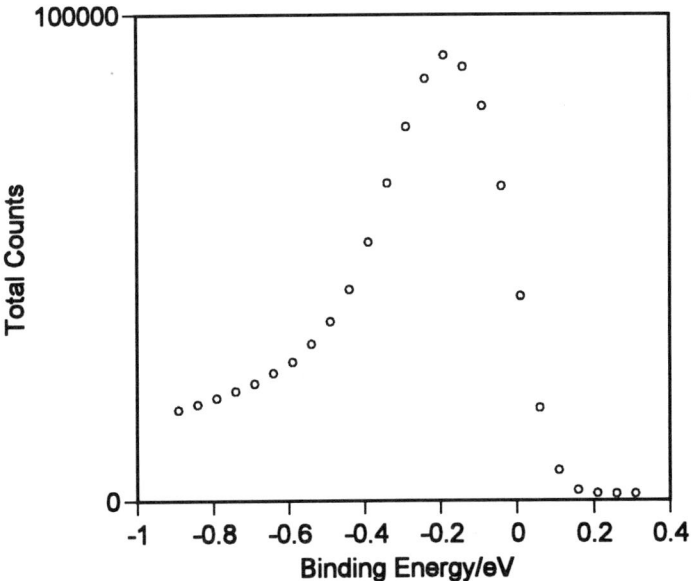

Figure 4 The spin integrated photoemission spectrum for Ni(110).

The spin integrated spectrum (for which a low resolution was used in order to reduce the data accumulation time) is shown in Fig. 4. The spin dependent asymmetries obtained from the two pairs of backward and forward detectors are shown in Fig.5. The backward data clearly show a statistically significant spin dependent scattering asymmetry. The absence of significant asymmetry in the forward direction, in agreement

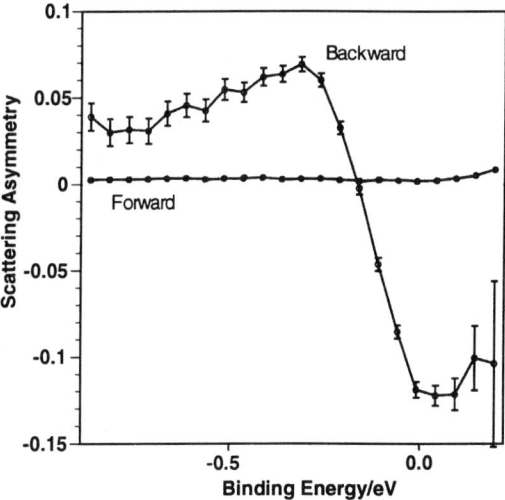

Figure 5 The corrected asymmetries for Ni(110) obtained from the backward and forward detectors.

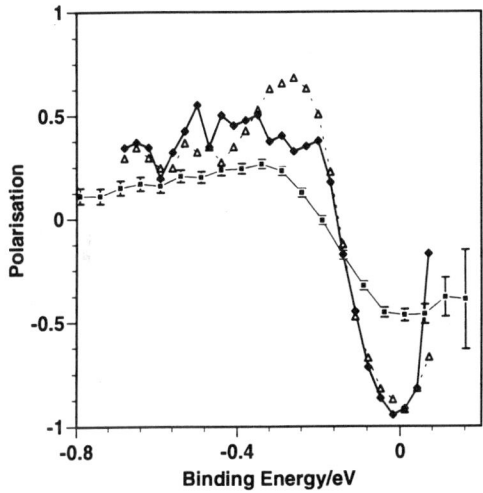

Figure 6 Polarisation data deduced from the asymmetries for Ni(110) obtained on Station 1.2 (–■–) compared with those reported by Raue et al.[9] (–♦–) and Raue[10] (–△–).

with theoretical expectations, leads to a degree of confidence that the procedure adopted for removing the instrumental effects, as outlined above, is valid for the experimental configuration of Station 1.2. The polarisation deduced from the backward asymmetry data displayed in Fig. 5 is compared with that reported by Raue et al.[9,10] in Fig 6; both sets of data were obtained using a photon energy of 16.85eV. The data all show the same trends across the Fermi edge, though the polarisation data obtained from Station 1.2 is slightly lower in overall magnitude which is probably due to a combination of lower resolving power and residual surface contamination. Further work on Ni(110) is planned for the near future.

MICRO-MOTT POLARIMETER

In 1992 it was realised that a more compact polarimeter, which could be used on a number of stations, would be a useful complement to the high energy Mott polarimeter on Station 1.2. Thus funding was obtained to assemble a second polarimetry facility based around a micro-Mott polarimeter. Of the variety of polarimeters currently available,[11] a micro-Mott polarimeter was chosen for use on the SRS for a number of reasons, *viz;*

- compactness
- reliable and reproducible operation
- relative ease of use by a variety of users.

The concept of a retarding potential Mott polarimeter was first suggested by Farago and realised by Dunning and co-workers in 1979.[12] It works on the same principle of spin-orbit induced asymmetric scattering that is the physical basis of high energy Mott polarimeters, but the electron beam is only accelerated to around 20kV before scattering from a heavy element target and the scattered electrons are decelerated to close to ground potential before they are detected. The latest generation of this type of polarimeter have conical geometry and are known as micro-Mott polarimeters owing to their relatively small size.[13,14]

Two priority applications were identified for the facility, those of spin- and angle-resolved photoemission a) from atomic and molecular species under high vacuum conditions and b) from condensed matter under UHV conditions. Both of these applications require energy analysis of the photoelectrons. A four detector polarimeter, depicted in Fig.7, was therefore mounted on a small (50mm radius) hemispherical analyser (shown schematically in Fig. 8) and the whole assembly was mounted either on a rotatable bracket for gas phase work or a two circle goniometer for condensed matter work. This latter option is illustrated in Fig. 9. The polarimeter/analyser assembly and associated wiring *etc.* was constructed from UHV compatible materials so that there would be minimal interconversion downtime between the two applications. A second control and data acquisition system, identical to that used on Station 1.2, was also assembled.

The polarimeter body (Acutech) was modified, as a response to the need for greater compactness, by the incorporation of channel plates (Hamamatsu, non-demountable type F1551-21S) rather than channeltrons as used in the original designs. However, the advantage of reduced size has proved to be offset by the disadvantages of cost and increased sensitivity to arcing which tends to lead to irrevocable damage to the plates.

The data acquisition chain used with the microMott polarimeter is essentially the same as that used on Station 1.2 except that there is no requirement to use optical coupling. This stage was therefore omitted in this instrument. So far no attempt has been made to establish the S_{eff} for the microMott polarimeter, rather the values determined in work reported by Dunning *et al.*[15] have been assumed. These workers found that to the 10% precision level the effective Sherman functions obtained for microMott polarimeters of their standard design were remarkably consistent; at 20kV and zero energy loss window the value of S_{eff} obtained was 0.27.

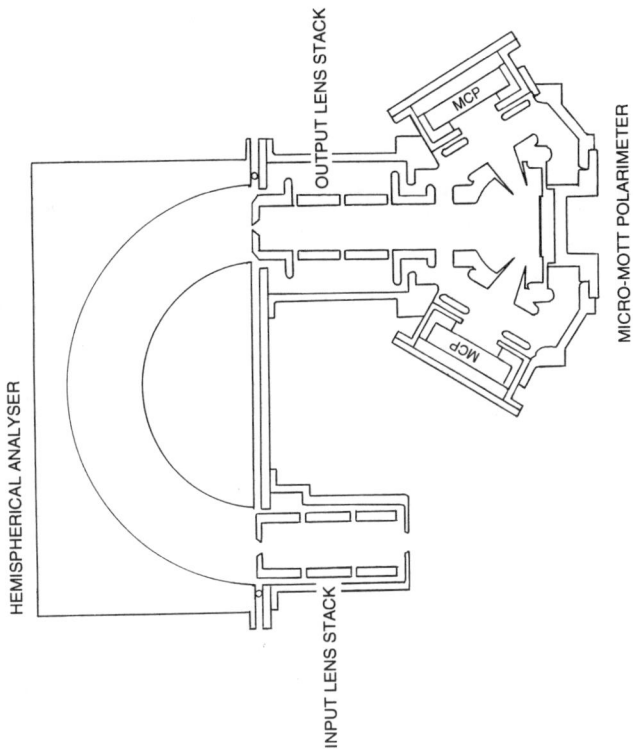

Figure 8 Schematic diagram of the micro-Mott polarimeter/electron energy analyser assembly.

Figure 7 Photograph of the micro-Mott polarimeter.

Figure 9   Photograph of the polarimeter/analyser assembly mounted on a two circle goniometer.

Photoemission of the 4d electrons of xenon was used as a first test of the polarimeter and the data acquisition hardware and software with synchrotron radiation. A gas phase experiment, though more difficult in many respects than a surface experiment, was deliberately chosen for the advantage in turn around time it confers; a choice that proved well founded. Xenon photoemission was achieved using highly plane polarised 105eV radiation from the undulator (Line 5U.1). Although resolution was sacrificed for counts, the two spin orbit components were clearly visible and polarisation measurements were made at angles of 0° and 58° with respect to the E vector of the incoming radiation. Though this initial test yielded a spin dependent asymmetry, more work is needed to obtain precise polarisation data.

Spin polarised photoemission from the valence band of thin films of iron on Cu(110) was used as a first test of the polarimeter in its surface science configuration. Once again spin dependent asymmetries were observed but the quality of the data were limited by poor counting statistics.

Although high quality data have yet to be taken with the micro-Mott polarimeter the route to achieving this is clear - the data acquisition times have to be increased and some instrumentation problems have to be solved. On a more positive note, the work so far has demonstrated that the micro-Mott polarimeter can be configured relatively easily for use in a variety of experimental modes.

FUTURE DEVELOPMENTS

Increasing the magnitude of the Sherman function for the high energy Mott polarimeter on Station 1.2, and its accurate determination, are the highest priorities. To achieve this a range of uranium or thorium foils of thicknesses up to 200Å will be installed, as work by Pappas and Hopster[16] has shown that increases in S_{eff} of some 25% over that for gold can be achieved by the use of passivated uranium. There is also a clear need to be able to determine the longitudinal component of the polarisation - an evaluation exercise is currently underway to establish the best type of spin rotator to use on the station. Over the coming months Kerr effect kit that is currently being tested off-line will be installed on the station and it is anticipated that modifications to the mounting block for the forward barrier detectors will be undertaken. This latter work is in response to detailed studies by Kessler and coworkers[17,18] which clearly showed that the data collected from the forward counters can - if they are appropriately positioned - be used to eliminate *directly* the instrumental, spin-independent, asymmetry in the backward detectors. (This is in contrast to the current situation on Station 1.2 in which the data from the forward counters are independently processed and used *only* as a check of the correction procedure.)

The spin polarised photoemission experimental activities on the SRS are complemented by theoretical work on the electronic structure of solids and photoemission from them. The state-of-the-art codes (for example PHOTON[19]) that have been, and continue to be, developed are however at present rather divorced from the data processing and manipulation activities. Plans are currently well advanced to provide a single platform for the simulation of spectra and manipulation of experimental data. This facility will be based around a UNIX/X-windows environment and is referred to as the Surface Science Shell.

ACKNOWLEDGEMENTS

We would like to thank all those at Daresbury Laboratory who have provided the intensive technical support required for this project. In addition we are happy to acknowledge those who have shared, and continue to share, their expertise with us; in particular discussions with and the help given by Barry Dunning, Mike Hardiman and Murray Campbell have been invaluable. The electron optics program SIMION has been extensively used throughout this project

REFERENCES

1. E.A. Seddon, *Daresbury Laboratory Technical Memorandum*, DL/SCI/TM58E.
2. E.A. Seddon, M.A. Hoyland, H.P. Hughes and R.G. Jordan, *Nucl. Inst. Meth. Phys. Res.,* A319 (1992) 377.
3. F.M. Quinn, E.A. Seddon and I.W. Kirkman, *Rev. Sci. Inst.,* accepted for publication.
4. G. Holzwarth and H.J. Meister, in "Tables of asymmetry, cross-section and related functions for Mott scattering of electrons by screened gold and mercury nuclei", The University of Munich, Munich, 1964.
5. R.A. Hearsey, E.A. Seddon, F.M. Quinn, G. Derbyshire and J. Salvini, *Daresbury Laboratory Technical Memorandum,* to be published.

6. T.J. Gay and F.B. Dunning, *Rev. Sci. Inst.,* 63 (1992) 1635.
7. T.J. Gay, M.A. Khakoo, J.A. Brand, J.E. Furst, W.V. Meyer, W.M.K.P. Wijayaratna and F.B. Dunning, *Rev. Sci. Inst.,* 63 (1992) 114.
8. G. Holzwarth and H.J. Meister, *Nucl. Phys.,* 59 (1964) 56.
9. R. Raue, H. Hopster and E. Kisker, *Rev. Sci. Inst.,* 55 (1984) 383.
10. R. Raue, Ph.D. Thesis, KFA Jülich 1984.
11. See D.M. Campbell and E.A. Seddon, in proceedings of the the SERC workshops on "Polarized Electron/Polarized Photon Physics", Plenum Press, 1994.
12. L.A Hodge, T.J. Moravec, F.B. Dunning and G.K. Walters, *Rev. Sci. Inst.,* 50 (1979) 5.
13. F-C. Tang, X. Zhang, F.B. Dunning and G.K. Walters, *Rev. Sci. Inst.,* 59 (1988) 504.
14. G.C. Burnett, T.J. Munroe and F.B. Dunning, *Rev. Sci. Inst.,* 65 (1994) 1893.
15. F.B. Dunning, F-C. Tang and G.K. Walters, *Rev. Sci. Inst.,* 58 (1987) 2195.
16. D.P. Pappas and H. Hopster, *Rev. Sci. Inst.,* 60 (1989) 3068.
17. A. Gellrich, K. Jost and J. Kessler, *Rev. Sci. Inst.,* 61 (1990) 3399.
18. A. Gellrich and J. Kessler, *Phys. Rev. A,* 43 (1991) 204.
19. For further information on PHOTON or the Surface Science Shell contact B.Searle at DRAL Daresbury Laboratory.

SOURCES AND DETECTORS OF POLARIZED ELECTRONS

D.Murray Campbell

Department of Physics and Astronomy
University of Edinburgh

INTRODUCTION

Twenty five years ago, when the first polarized electron beam source in the United Kingdom was constructed (Campbell et al., 1971), the use of beams of spin-oriented electrons was generally considered to be a highly demanding and expensive technique, of interest only to a few specialist research groups. In the intervening two and a half decades, polarized electron beams have become indispensible tools in many varied fields, including condensed matter and surface studies, atomic and molecular physics, and nuclear and elementary particle collisions.

Numerous technical developments in sources and detectors have contributed to the growth in the use of polarized electron beams; these have been well surveyed and discussed in the textbook by Kessler (1985) and in several more recent reviews (Fletcher et al., 1986; Reichert, 1991; Gay and Dunning, 1992; Pierce, 1995). To attempt a comprehensive survey of the varied and often highly ingenious techniques which have been proposed for the creation and analysis of polarized electron beams would take considerably more space than is available here. Instead, the present article focuses principally on the methods of which the author has personal experience, and which have a well established track record for reliability and relative ease of use: the gallium arsenide polarized electron source, and the Mott polarimeter.

THE POLARIZED ELECTRON SOURCE

Solid versus Gaseous Sources

As early as 1921 Stern and Gerlach created a spin-polarized atomic beam, using an inhomogeneous magnetic field to introduce a deflection whose sense depended on the orientation of the magnetic dipole moment of the atom. This technique cannot be used with an electron beam because of the overwhelming effect of the Lorentz force on the charged electron. However, the use of a polarized atomic beam as an intermediate stage in the polarization of electrons was exploited in several of the early polarized electron beam sources. One obvious possibility was to ionise the spin-oriented atoms in a way which retained the orientation of the electron spin (Baum and Koch, 1969); another was to use spin exchange collisions with spin-oriented atoms to polarize an initially unpolarized electron beam (Farago and Siegmann, 1966). A related technique used an unpolarized atomic beam ionised by circularly polarized radiation; the spin-orbit interaction resulted in a preferential orientation of the electron spin in the continuum state (the Fano effect) (Heinzmann et al., 1970).

All of these approaches suffered from a fundamental limitation on the available beam current, due to the low density of target atoms in the atomic beam (of the order of 10^{18} atoms m^{-3} in optimised sources) and the low collision cross-sections. Although pulses of up to 2×10^9 electrons could be obtained from photoionisation sources, the maximum continuous current output was of the order of 10^{-8} A. The electron polarization was typically in the range 60–85%.

At present, only one type of polarized electron source based on the polarization of atoms in the gaseous state offers levels of performance comparable to that obtained from solid state sources. This is the flowing helium afterglow source, pioneered by the Rice group (Rutherford et al., 1990). In this source, helium atoms are excited in an RF discharge cell and subsequently pumped through a flow tube. Most of the excited atoms in the flow tube are in the relatively long-lived metastable 2^1S_0 and 2^3S_1 states. Circularly polarized laser light is used to optically pump the 2^3S_1 metastables via one of the 2^3P levels. The spin-polarized helium metastables are then chemi-ionised by interaction with carbon dioxide injected into the flow tube; since the ionisation process conserves spin, the resulting free electrons are also polarized. These electrons are extracted from the afterglow and formed into a beam by a system of electrodes.

With a helium pressure of around 0.1 mbar, beam currents of the order of 100 μA can be obtained. However, electron polarizations greater than 80% have been measured only at much lower currents. The major cause of depolarization at high beam current appears to be the reabsorption by helium metastables of the fluorescent radiation emitted during the optical pumping process. The Orsay group have attempted to reduce this radiation trapping effect by using a different pumping scheme, involving the excitation of the 2^3P_0 rather than the 2^3P_1 level, and by modifications in the geometry of the source. By these means they have achieved an electron polarization of 60% with a current of 100 μA (Arianer et al., 1994).

The fact that the polarized electrons emanate from a source region containing gas at a pressure of around 0.1 mbar is a serious drawback for applications in which the beam has to be injected into a target region at ultrahigh vacuum, since it implies that a substantial degree of differential pumping must be used. On the other hand, this drawback can appear as an advantage when the target region is itself only at moderate vacuum, since the source is relatively insensitive to poisoning by contamination backstreaming from the target.

In choosing the optimum source of electrons for a particular experimental arrangement, it is important to consider not only the beam current and polarization, but also the electron optical quality of the beam. This is frequently done by evaluating the product of beam energy (E), cross-sectional area (A) and solid angle (Ω), which is constant for any cross-section of the beam according to the law of Helmholtz and Lagrange. A small value of $EA\Omega$ corresponds to a beam which can be injected into a narrow entrance aperture with a small angle of divergence, even at low energy.

Here again, the gaseous source is at an inherent disadvantage. The density of metastables in the optical pumping region of the flowing helium afterglow source is of the order of 10^{16} atoms m^{-3}; to achieve a large beam current it is necessary to extract the electrons from an ionisation region with a diameter of several centimetres. The corresponding product $EA\Omega$ is of the order of 5×10^{-2}. Although this is significantly better than the corresponding values achieved with earlier types of gaseous source, it is several orders of magnitude worse than the values of $EA\Omega$ offered by sources using solid photoemitters. The crucial factor is the much greater atomic density in the solid.

The Gallium Arsenide Source

Although various types of solid cathode polarized electron sources have been developed, including a successful design using field emission from a tungsten needle coated with ferromagnetic EuS (Kisker et al., 1978), most of the practical polarized electron sources constructed in the last fifteen years have been based on laser-induced photoemission from a planar GaAs cathode. The principle of this source (Garwin et al., 1974, Lampel and Weisbuch, 1975) is that irradiation of the GaAs cathode by circularly polarized photons of energy just greater than the bandgap between valence and conduction bands leads to a preponderance of one spin state in the electrons excited into the conduction band. Referring to Figure 1, it can be seen that, provided the photon energy is between 1.42 eV and 1.76 eV, only electrons from the P$_{3/2}$ state can be excited. The intensity of the transition from the m$_j = -3/2$ sublevel is three times that of the transition from the m$_j = -1/2$ sublevel, so that the theoretical maximum of the conduction band polarization is

$$P_c = (1-3)/(1+3) = -0.5.$$

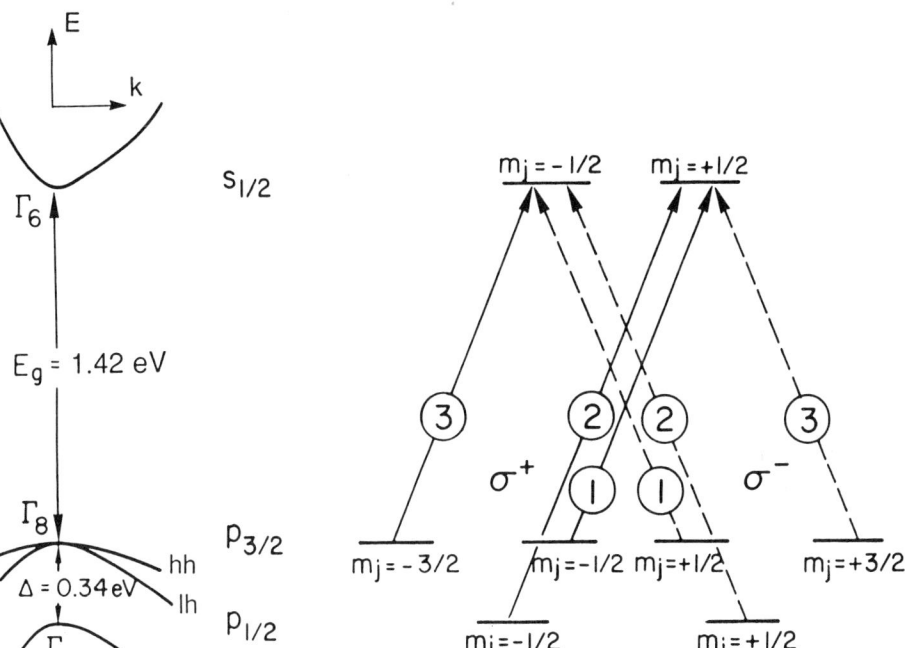

Figure 1. Energy bands and optically induced transitions for gallium arsenide near the centre of the Brillouin zone (from Pierce and Meier, 1976).

The insight which led to the development of the polarized electron source was the recognition that it was possible to extract the conduction band electrons without destroying their spin orientation. The technique of reducing the surface work function of a highly p-doped GaAs crystal by applying minute quantities of Cs and O_2 had already been used to create photocathodes with high quantum yield in the infred; with a Cs coverage of less than one monolayer, the condition of negative electron affinity can be achieved, in which the vacuum level is below the bottom of the bulk conduction band minimum (Bell, 1973). Photoexcited electrons which thermalise to the bottom of the conduction band and diffuse to the surface can then escape into vacuum. The diffusion length for such electrons in bulk GaAs is of the order of $1\mu m$, which is comparable to the absorption depth for the photons; there is thus a fairly high probability that a photoexcited electron will reach the surface. However, during the diffusion process spin relaxation effects can substantially reduce the polarization. For very thin cathodes the time taken to diffuse to the surface is short in comparison with the spin relaxation time, and polarizations approaching the theoretical limit of 0.5 have been achieved. With cathodes thicker than $1\,\mu m$, the maximum polarization is typically around 0.3. Despite this limitation, thick cathodes are frequently used because of their much greater quantum yield.

Almost all GaAs polarized electron sources now in use employ wafers which are cut so as to expose the [100] face of the crystal. A detailed description of such a source, developed at NIST, has been given by Pierce et al. (1980). The achievement of high quantum yield depends on careful cleaning of the crystal surface prior to caesiation. Several procedures for chemical cleaning of the cathode prior to insertion in vacuum have been proposed; most of these involve etching of the crystal surface (see Pierce et al., 1995), although the necessity for this has been questioned (Ciccacci et al., 1995). After the cathode has been installed and the source chamber pumped down, the cleaning process is completed by heating the cathode, typically to around 600°C.

The efficacy of the cleaning process is dependent on many factors, including the previous history of the cathode. In consequence, an air of "cookery" still lingers about this aspect of the GaAs source. Adequate control of the heating cycle is complicated by the difficulty of monitoring the precise temperature of the surface. It is important not to heat the cathode beyond 660°C, at which temperature arsenic evaporates preferentially, leaving a surface with a frosted appearance which can no longer be activated.

One way in which some of these problems can be avoided is by cleaving a crystal in vacuum to give an atomically clean surface. The crystal naturally cleaves to reveal the [110] plane; however, this plane can be activated and used in essentially the same way as the [100] plane (Reihl et al., 1979). A highly compact polarized electron source of this type, which has been in regular use in Edinburgh since 1980, is illustrated Figure 2. The crystal is shown in the cleavage position; after it has been cleaved, it is raised and moved forward to the position at which it is activated and used. The scale of the apparatus can be judged from the 70 mm diameter flanges shown in the photograph.

Once a clean surface has been obtained in vacuum, the activation process is relatively straightforward. Caesium dispensers which are operated by the passage of a current of a few amps are normally used, since the deposition rate can be accurately and quickly adjusted by varying the heating current. In the Edinburgh cleavage source, the dispenser is mounted under the anode of the electron gun, and the Cs atoms reach the cathode through small holes drilled in the anode and deflection electrodes. This arrangement has the advantage that the dispenser can be operated continuously at a very low emission rate during operation of the polarized electron source, replacing Cs atoms which evaporate from the cathode. The stability of the source is greatly improved by this technique, allowing a beam of essentially constant polarization and current to be maintained for several days.

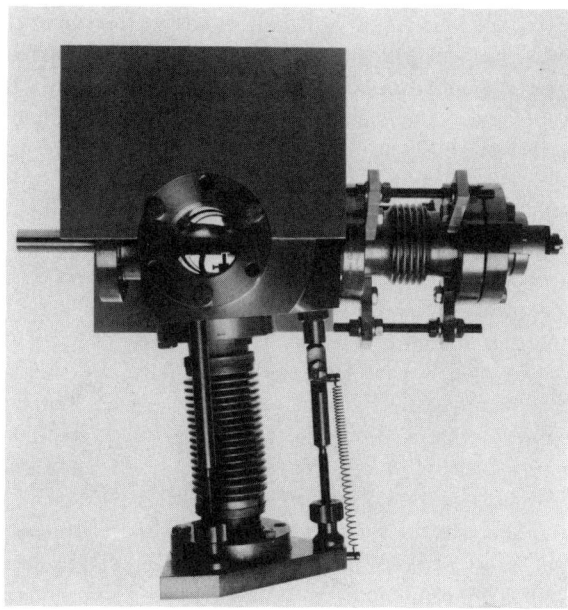

Figure 2. The Edinburgh cleaved crystal polarized electron source.

The admission of oxygen during the activation procedure can be carried out using a UHV leak valve, although in the Edinburgh source a thin-walled silver tube is heated to allow selective permeation of atmospheric oxygen. The creation of a satisfactory cathode requires a base pressure at least in the low 10^{-10} mb region; during activation, the partial pressure of oxygen can be up to two orders of magnitude higher than this. The progress of the activation is monitored by observing the photocurrent generated by either a filtered white light source or a HeNe laser.

One other unusual aspect of the Edinburgh cleavage source is that the HeNe laser is used not only to monitor the activation but also to generate the polarized electron beam. The use of the HeNe laser is very convenient, since the visibility of the beam makes it easy to align and moderately high powers are available cheaply. However, reference to Figure 1 shows that the photon energy of 1.96 eV is high enough to excite electrons from both of the spin-orbit split states at the top of the GaAs valence band. It can therefore be expected that the electrons which thermalise to the bottom of the conduction band have a polarization much less than 0.5. It is indeed true that, when the surface of the crystal is activated to negative electron affinity, the measured polarization of the extracted electron beam is of the order of 0.1. However, if the activation procedure is arrested at a stage at which the electron affinity is still positive, the electrons at the conduction band minimum are unable to escape. Electrons which thermalise into higher minima away from the gamma point can still surmount the potential barrier at the surface, and since these come predominantly from the $P_{3/2}$ state the extracted electron beam polarization is much higher than for the negative electron affinity surface (Reihl et al., 1979). The quantum yield is, of course, correspondingly lower. With a laser power of 1 mW a polarization of 0.3 can be obtained with an emission current of 100 nA, corresponding to a quantum yield of 2×10^{-4}; with a higher work function a polarization of 0.4 can be achieved, but the quantum yield is reduced to 10^{-5}.

Recent Progress in Photoemission Sources

The degeneracy of the heavy and light hole bands at the top of the valence band in GaAs is a consequence of the high degree of symmetry in the crystal. If this symmetry can be reduced, an energy split can be induced between the heavy and light hole bands. At the photoemission threshold, electrons will then be excited only from the upper band, giving a theoretical polarization of either $+1$ or -1 depending on which band is higher.

One way to reduce the symmetry of the crystal is by applying a uniaxial stress. A useful band splitting of around 50 meV requires a stress too great to be achievable by the direct application of external mechanical forces to the crystal (Zorabedian, 1982). The requisite degree of stress can, however, be created by growing a thin layer of GaAs on a substrate which has a slightly smaller lattice constant. The atoms of the GaAs lattice are squeezed together in the plane of the interface, and this lateral compression results in an expansion along the normal to the interface. Tensile strain along this axis raises the heavy hole band above the light hole band, so that photoemission at threshold involves only electrons excited from the $m_j = -3/2$ level in the valence band (see Figure 1). The theoretical polarization limit is then -1.

Successful cathodes have been constructed using this technique, with a thin GaAs layer grown on a substrate of GaAsP (Nakanishi et al., 1991). The degree of strain is dependent on the relative proportions of As and P; a suitable lattice mismatch of just over 1% was obtained from a cathode with a GaAs layer 100 nm thick on a $GaAs_{0.72}P_{0.28}$ substrate (Maruyama et al., 1992). For photon energies very close to the threshold the electron polarization was 0.9, but the quantum yield was of the order of 10^{-4}. Increasing the photon energy gave a rapid increase in the yield, although around 30meV above the threshold electrons from the light hole band started to dilute the polarization. It was possible to attain an electron polarization of 0.8 with a yield of 10^{-3}.

Strained lattice cathodes have also been constructed using GaAs as a substrate and a photoemissive layer of $In_xGa_{1-x}As$ (Maruyama et al., 1991; Meier et al., 1993). An alternative approach has involved the use of superlattices consisting of alternating layers of GaAs and AlGaAs, each layer being a few nanometers thick. Polarizations of around 0.7 have been achieved with cathodes of this type, although the quantum yield is still much lower than that available from bulk GaAs (Omori et al., 1991).

The yield of strained photocathodes is low because the strained layer is too thin to absorb more that a small fraction of the incident light. If the thickness of the layer is increased beyond a critical value, dislocations start to occur; this critical thickness is of the order of 10 nm for the cathodes described above. It has been shown, however, that relaxation of strain only becomes a serious problem for layers about an order of magnitude greater than this critical value (Maruyama et al., 1992). Even so, the limiting thickness of around 200 nm is an order of magnitude less than the photon absorption length, making it impossible to achieve both high polarization and high yield.

Another technique for creating a strained cathode, which does not suffer from the same limitation on layer thickness, involves the deposition of GaAs on a substrate with a different coefficient of thermal expansion. The deposition is carried out at an elevated temperature; strain is induced by the differential contraction of the layer and the substrate on cooling to room temperature. It has been shown that a high degree of strain can be created by this method in a 2 μm thick GaAs layer epitaxially grown on a substrate of CaF_2 (Tessler et al., 1994). A quantum yield of 5×10^{-3} has been obtained with a cathode of this type, although the maximum polarization so far measured is 0.6 (Campbell et al., 1994). An additional advantage of this cathode is the transparency of the substrate, allowing the possibility of a photocathode illuminated from the rear.

MEASUREMENT OF ELECTRON SPIN POLARIZATION

Principles of electron polarimetry

To measure the spin polarization of a particle beam, the particles must clearly be subjected to a physical process whose outcome depends on their spin state. In the case of a beam of photons, the polarimeter may use a dichroic film whose absorption is spin-dependent, a calcite prism whose birefringence leads to spin-dependent refraction, or a pile of plates giving spin-dependent reflection at the Brewster angle.

A major problem in the use of spin-polarized electron beams has been the unavailability of a physical process which can be readily harnessed to separate an electron beam into its component spin states. The equivalent of the optical prism is an inhomogeneous magnetic field, which acts on the magnetic dipole moment of an electron to deflect it in a direction which depends on its spin state. Unfortunately, as was already noted in the discussion of polarized electron sources, it is in principle impossible to design a spin-dependent prism for electrons using a macroscopic magnetic field.

The magnetic field experienced by an electron undergoing a close encounter with an atomic nucleus is, however, capable of generating a substantial spin-dependent scattering asymmetry through the spin-orbit interaction (Kessler, 1985). This phenomenon, known as Mott scattering, has been the basis of most of the polarimeters developed for use with low energy electron beams (Gay and Dunning, 1992). In the simplest case, a beam of electrons collides with a gaseous target whose atomic density is low enough that only single scattering events need be considered. The intensities $I(+\theta)$ and $I(-\theta)$ of electrons scattered through angles $\pm\theta$ in a plane containing the incident beam direction are measured; the scattering asymmetry A is defined by

$$A = \frac{I(+\theta) - I(-\theta)}{I(+\theta) + I(-\theta)}.$$

The polarization P of the electron beam, measured in a direction perpendicular to the scattering plane, is related to the scattering asymmetry by the expression $A = PS$, where S is the Sherman function. S is dependent on the scattering energy and angle, and its detailed form and magnitude are determined by the nature of the scattering potential experienced by the electron. For scattering by a bare nucleus, this is simply the Coulomb potential; in the case of atomic scattering, the screening of the nucleus by the atomic electrons must also be taken into account. Calculated values for S for single scattering by various atoms used in polarimeter targets have been tabulated (Holzwarth and Meister, 1964; Ross and Fink, 1988).

The scattering efficiency μ of a polarimeter may be defined by the expression $\mu = I/I_0$, where I is the total scattered current collected by the detectors and I_0 is the incident beam current. (Some writers use the term sensitivity to describe μ). In practice, the density of target atoms necessary to provide an acceptable scattering efficiency is usually too high to ensure only single scattering. An electron emerging at a large scattering angle from a dense target may have undergone a number of additional small angle collisions (multiple scattering); it is also possible that the measured scattering angle is the result of several large angle collisions within the target (plural scattering). When plural or multiple scattering is significant, the effective value S_{eff} of the Sherman function differs from the calculated single scattering value. Determination of the appropriate value of S_{eff} for a particular polarimeter often involves considerable experimental effort.

Spin-dependent scattering can occur even when the spin-orbit interaction is negligible, through a spin exchange interaction between the incident electron and an electron in the target. This phenomenon has been exploited in polarimeters based on scattering from ferromagnetic crystals and on excitation of inert gas atoms, but neither of these types of polarimeter has achieved widespread use. The use of inert gas excitation in the calibration of polarimeters is described in the final section of this report.

The ideal polarimeter would have the maximum values of μ and S_{eff} (unity in each case). In practice, it is often possible to trade one of these parameters against the other: raising the scattering energy, for example, may increase the Sherman function but lower the scattering efficiency. To discuss the balance of advantage in such a trade off, it is customary to define a figure of merit $M = \mu S_{eff}^2$, which is inversely proportional to the counting time necessary to achieve a given level of statistical uncertainty in a polarization measurement.

Polarimeters in current use

The efficiency of polarimeters based on Mott scattering is limited by the fact that the spin-orbit interaction is normally much smaller than the spin-independent scattering force; it is only in regions of energy and angle for which the overall scattering cross-section is low that substantial scattering asymmetry is found. Since the spin-orbit interaction is proportional to the charge on the nucleus, heavy atoms are usually chosen as targets. A polarimeter based on scattering from a mercury vapour jet has been described by Jost et al. (1981). With a scattering energy of 15 eV the mercury vapour polarimeter was capable of an efficiency $\mu = 2.8 \times 10^{-4}$ with an effective Sherman function $S_{eff} = 0.37$. The corresponding figure of merit was $M = 3.8 \times 10^{-5}$.

The mercury vapour polarimeter can be made very compact, since no high voltages are involved. Its major disadvantage is the vacuum limitation imposed by the nature of the target, which renders it unsuitable for experiments demanding a UHV environment.

The vast majority of Mott scattering polarimeters have used targets consisting of gold films or foils. The atomic density is greatly increased by the use of a solid target, but the effect of multiple scattering on the effective Sherman function becomes a much more serious problem than with a gaseous target. By using a very thin film (of the order of 100 nm) and a very high scattering energy (typically 100 keV), it is possible to approach single scattering conditions, but the scattering efficiency is low. To improve the efficiency it is necessary to use much thicker foils, with a consequent reduction in S_{eff}. Polarimeters in which electrons are scattered at around 100 keV and detected without deceleration are often described as being of the conventional high-energy type. An optimised version of this design, described by Kisker et al. (1982), gave a Sherman function $S_{eff} = 0.26$ with a scattering efficiency $\mu = 1.5 \times 10^{-3}$, resulting in a figure of merit $M = 1 \times 10^{-4}$.

The conventional high energy Mott polarimeter has several practical advantages. The effective Sherman function is insensitive to small changes in scattering energy or contamination of the foil surface. The apparatus does not require ultra high vacuum, but can be made UHV compatible. An important additional advantage is that a self-calibration procedure is available, involving the measurement of asymmetry at several foil thicknesses and an extrapolation to zero foil thickness (Gay et al., 1992). On the other hand, the practical disadvantages of this design are also considerable. The major part of the apparatus, including the scattered electron counters, is at very high voltage, which inevitably results in a somewhat bulky and inflexible measuring instrument. The surface barrier detectors normally employed accept electrons which have lost up to 10 keV, and these inelastically scattered electrons degrade the effective Sherman function.

Many of the disadvantages of the conventional high energy Mott polarimeter are overcome by designs based on the retarding potential concept. The electron beam to be analysed is accelerated to the scattering energy, and the scattered electrons are decelerated before impinging on the counters, which are usually channel electron multipliers. The original design by Farago (1973), employing concentric cylindrical electrodes, was developed by Hodge et al. (1979). Only the inner cylinder, which contains the scattering foil, is at high voltage; it is therefore possible to make this system quite compact, even with a scattering energy of 100 keV. A further advantage is that the cylindrical field helps to focus the incident beam in the scattering plane, while the deceleration of the scattered electrons in the same field provides a high degree of selectivity against inelastically scattered electrons.

Although the cylindrical retarding Mott polarimeter has not been optimised in efficiency, it has found a useful role as an in-line polarization monitor (Campbell and Farago, 1987). A gold coated micromesh target allows most of the incident beam to pass undeviated, while giving adequate scattering for polarization analysis. The transmitted beam is decelerated in the cylindrical field, and extracted through an aperture in the outer cylinder. The effective Sherman function of the mesh can be estimated by interchanging it with a number of standard foils of known thickness. A similar arrangement, with a removable foil, has been used to calibrate another polarimeter (Zhang, 1988).

As the scattering energy of electrons on a gold target is reduced from 100 keV to 10 keV, the cross-section for elastic scattering increases substantially, while the Sherman function decreases. The effect of multiple scattering in further reducing the effective Sherman function is less significant for retarding polarimeters because of their greater energy discrimination, and the optimum figure of merit for this type of polarimeter is found at a scattering energy of between 20 keV and 30 keV (Campbell et al., 1983). A new generation of miniaturised Mott polarimeters has emerged, operating in this energy range. The use of concentric spherical electrodes has the advantage of providing two-dimensional focusing of the incident beam towards the target, while the spherical retarding field provides an efficient transfer of elastically scattered electrons to the counters (Gray et al., 1984). The use of microchannel plates to increase the solid angle of collection has provided a scattering efficiency $\mu = 2.4 \times 10^{-3}$ for a miniature spherical polarimeter operating at 20 keV (Watkins et al., 1988). With an effective Sherman function $S_{eff} = 0.11$, the figure of merit for this polarimeter is $M = 2.9 \times 10^{-5}$. A polarimeter of similar design (Ciccacci et al., 1995), capable of being mounted on a 150 mm internal diameter UHV port, is illustrated in Figure 3.

An alternative design of miniaturised retarding field polarimeter has been developed by the Rice group (Dunning et al., 1987). Although this has been described as a conical polarimeter (Gay and Dunning 1992), the retarding field is not maintained between concentric cones as the description might suggest. A coaxial three element lens is used to accelerate the electron beam to its scattering energy, and the retarding field is created between two apertured planar electrodes. The efficiency, Sherman function and figure of merit for this polarimeter are comparable to those for the spherical polarimeter described above.

Figure 3. The Edinburgh miniaturised spherical Mott polarimeter.

Several other types of electron spin polarimeter have been developed in recent years, which share the practical advantage that they operate with scattering energies of the order of 100 V. The low energy diffuse scattering Mott polarimeter (Scheinfein et al., 1990) uses an evaporated polycrystalline gold target and an incident beam energy of 150 V. Electrons backscattered through a wide range of angles are collected using a carefully shaped electrostatic field and a segmented microchannel plate. Stable operation has been reported with an effective Sherman function $S_{eff} = 0.10$, a scattering efficiency $\mu = 2 \times 10^{-2}$ and a figure of merit $M = 2 \times 10^{-4}$. The principal disadvantages of this very compact polarimeter are its sensitivity to misalignment of the incident beam, and the necessity for regular renewal of the polycrystalline gold surface by in situ evaporation.

A LEED polarimeter making use of scattering from a single crystal of tungsten has been described by Kirschner (1985). In contrast to the continuous angular spread in diffuse scattering from a polycrystalline target, the diffracted beams from a single crystal are concentrated at the angles giving rise to the familiar LEED spot pattern. A highly collimated incident beam is required, and the scattering energy must also be constrained to within a few electron volts to maintain a stable and well resolved diffraction pattern. Under these circumstances, with a scattering energy of 104.5 eV, a Sherman function $S_{eff} = 0.27$, a scattering efficiency $\mu = 2.2 \times 10^{-3}$ and a figure of merit $M = 1.6 \times 10^{-4}$ were reported. The sensitivity to surface contamination is critical; UHV conditions are essential, and the tungsten surface must be cleaned by flashing to a temperature of 2500 K approximately every 20 minutes.

The absorbed current polarimeter operates in a different mode to all of the scattering polarimeters so far discussed, in that the signal which carries the information about the electron polarization is the current absorbed by a polycrystalline gold film (Pierce et al., 1981) or a ferromagnetic glass (Celotta et al., 1981). The electron beam strikes the target at an angle to the normal typically around 30°. At a suitably chosen incident energy, the sum of primary scattering and secondary electron emission is equal to the incident beam current for an unpolarized beam; the absorbed current is then zero. For a polarized beam, the situation is different. The spin-dependence, arising from spin-orbit coupling in the gold and spin exchange in the ferromagnetic glass, introduces an asymmetry in the scattering within the target, while the oblique beam incidence means that not all scattering angles have equal escape probability. For a gold target, the difference in absorbed current between an unpolarized beam and a completely polarized beam can be as much as 1% of the incident beam current (Erbudak and Ravano 1981). Evaluation of the Sherman function and figure of merit is inappropriate here, since the signal is measured by analogue techniques rather than by particle counting; this may be a limitation or an advantage, depending on the experimental circumstances. The quantitative performance of the polarimeter is highly dependent on the state of the target surface and on the angle of incidence.

Calibration and optimisation of polarimeters

The scattering asymmetry A measured using a particular polarimeter is subject to both systematic and random errors. Systematic errors are most frequently associated with instrumental asymmetries arising from beam misalignment; these can be readily eliminated when it is possible to reverse the polarization of the incident beam. It should be borne in mind, however, that some types of systematic error, such as that associated with counter saturation, cannot be cancelled by polarization reversal.

The most obvious source of random error in A is that associated with the counting statistics. Provided that $A \ll 1$, the statistical uncertainty is approximately equal to $N^{-1/2}$, where N is the sum of the counts in the two channels. It is of course possible that there will be additional random errors arising from fluctuations in the properties of the incident beam.

The polarization of the beam is obtained by multiplying the measured asymmetry by the Sherman function S_{eff}. An error in S_{eff} will affect all polarization measurements in the same sense; the overall fractional error in the measured polarization is derived by adding the fractional uncertainties in A and S_{eff}. The uncertainty in S_{eff} is frequently the limiting factor in the precision of a polarization measurement.

The calibration of a polarimeter consists in determining the appropriate value of S_{eff}. For a conventional Mott polarimeter, this can be done with high accuracy using a double scattering experiment. In the most careful study of this type carried out to date, Gellrich and Kessler (1991) were able to establish the value of S_{eff} for a Mott polarimeter operating at 120 kV with a fractional uncertainty of 0.2%. Calibration at this level of accuracy is necessary if the polarimeter is required to measure absolute polarizations with a fractional error of under 1%.

A polarimeter of any type can be calibrated by measurements on an electron beam of known polarization. One method of obtaining such a beam has been described by Uhrig et al. (1989). This is based on an optical technique for electron polarimetry originally suggested by Farago and Wykes (1969), and developed by Eminyan and Lampel (1980). Atoms in a gaseous target are excited by the electron beam, and the measurement of the circular polarization of a suitably chosen line in the radiation spectrum yields the electron polarization. Inert gases have proved most suitable as targets (Gay, 1983). Uhrig et al. made use of the $3^3P \rightarrow 3^3S_1$ transition in helium, and determined the electron polarization with a fractional uncertainty of 1%. Although the helium impact excitation polarimeter is of great value as an in-line calibrator, its efficiency is too low to make it practicable for general use.

The traditional way to calibrate Mott polarimeters has been to measure the scattering asymmetry using several foils of different thickness, and to extrapolate to zero foil thickness. Retarding polarimeters have also been calibrated by a series of asymmetry measurements with different retarding potentials and an extrapolation to zero energy loss (Campbell et al., 1985). Both these methods rely on the assumption that the extrapolations yield asymmetry values corresponding to single scattering, so that the calculated Sherman function can be used. The conditions under which this assumption is valid have been carefully investigated by Gay et al. (1992), who have pointed out that, for thick targets, plural scattering can cause significant depolarization even in the elastic limit. The consequences for extrapolations of both types are illustrated by Figure 4, showing asymmetry measurements for scattering of 20 keV electrons. Curve 1, corresponding to elastic scattering, has a highly non-linear dependence on foil thickness even in the 10 nm range. Curve 3, measured with a 3.4 nm foil, shows a linear dependence on the energy loss ΔE, but the extrapolation to zero ΔE underestimates the true asymmetry. The situation is much more favourable at a scattering energy of 100 keV; in this case, with $\Delta E = 100$ eV, Gay et al. found a linear dependence of asymmetry on thickness with foil thicknesses ranging from 3.4 nm to 68 nm. They could therefore estimate the single scattering asymmetry with a fractional uncertainty of 0.8%. Since the uncertainty in the calculated Sherman function is approximately 2% (Fletcher et al., 1986), the calibration of Gay et al. gave an uncertainty of 3% for absolute polarization measurements.

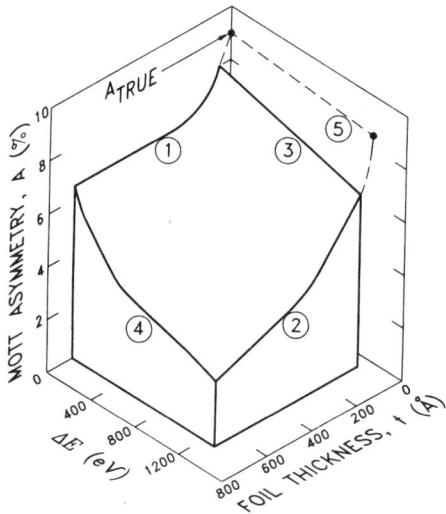

Figure 4. Schematic representation of measured scattering asymmetry A in the $(\Delta E, t)$ plane at 20 keV (from Gay et al., 1992).

Miniaturised retarding polarimeters frequently have a single thick target, and are designed to operate with a maximum accelerating voltage of around 25 kV. A crude self-calibration can be carried out using energy loss extrapolation on such a polarimeter, but it is evident from curve 4 in Figure 4 that the effective Sherman function is likely to be overestimated by around 20%. For precise calibration it is necessary to use a beam whose polarization has been accurately measured using one of the techniques previously described.

The figure of merit of an electron polarimeter can be improved by increasing either the scattering efficiency or the effective Sherman function. One way in which both μ and M can be increased is by replacing the gold target by a material with higher Z. The spin-orbit interaction is stronger, which makes S_{eff} larger; the scattering cross-section also increases, raising the value of μ. The use of a thorium target has increased the value of M by a factor of nearly 2 over that obtained with a gold target (McClelland et al., 1989). However, the reactive nature of thorium makes it more difficult to maintain a stable performance.

The figure of merit of 10^{-4}, characteristic of most of the efficient electron polarimeters in current use, contrasts strongly with the corresponding figure of merit of optical polarimeters, which can be of the order of unity. The relative inefficiency of electron polarimeters remains one of the major obstacles in the investigation of spin-dependent phenomena using polarized electron beams.

ACKNOWLEDGEMENTS

I am grateful to Dr E. Seddon and Dr G. Lampel for helpful discussions during the preparation of this paper. Part of the work was carried out with the support of the C.N.R.S. and the E.P.S.R.C.

REFERENCES

Arianer, J., Cohen, S., Essabaa, S., Frascaria, R. and Zerhouni, O., 1994, Workshop on Polarized Electron Beams and Targets, Les Houches, France, June 7-10 1994, C11.
Baum, G. and Koch, U., 1969, *Nucl.Instrum.Meth.* 71:189.
Bell, R.L., 1973, "Negative Electron Affinity Devices", Clarendon, Oxford.
Campbell, D.M., Brash, H.M. and Farago, P.S., 1971, *Phys.Lett.*36A:449.
Campbell, D.M., Hermann, C., Lampel, G. and Owen, R., 1983, International Symposium on Polarization and Correlation in Electron-Atom Collisions, July 24-26, 1983, University of Munster,22.
Campbell, D.M., Hermann, C., Lampel, G. and Owen, R., 1985, *J.Phys.E* 18:664.
Campbell, D.M. and Farago, P.S., 1987, *J.Phys.B* 20:5133.
Campbell, D.M., Tessler, L.R., Lassailly, Y., De Laharpe, P., Hermann, C., Lampel, G., Puech, P., Pisani, P., Landa, G., Carles, R., Daran, E., Munoz-Yague, A. and Fontaine, C., 1994, Ecole Thematique CNRS-PIRMAT, Marseille, September 12-14 1994.
Celotta, R.J., Pierce, D.T., Siegmann, H.C. and Unguris, J., 1981, *Appl.Phys.Lett.* 38:577.
Ciccacci, F., De Rossi, S. and Campbell, D.M., 1995, submitted to *Rev.Sci.Instrum.*.
Dunning, F.B., Gray, L.H., Ratcliff, J.M., Tang, F.-C., Zhang, X. and Walters, G.K., 1987, *Rev.Sci.Instrum.* 58:1706.
Eminyan, M. and Lampel, G., 1980, *Phys.Rev.Lett.* 45:1171.
Erbudak, M. and Ravano, G., 1981, *J.Appl.Phys.* 52:5032.
Farago, P.S. and Siegmann, H.C., 1966, *Phys.Lett.* 20:279.
Farago, P.S. and Wykes, J.S., 1969, *J.Phys.B* 2:747.
Farago, P.S., 1973, private communication.
Fletcher, G.D., Gay, T.J. and Lubell, M.S., 1986, *Phys.Rev.A* 34:911.
Garwin, E., Pierce, D.T., and Siegmann, H.C., 1974, *Helv.Phys.Acta* 47:393.
Gay, T.J., 1983, *J.Phys.B* 16:L553.
Gay, T.J. and Dunning, F.B., 1992, *Rev.Sci.Instrum.* 63:1635.
Gay, T.J., Khakoo, M.A., Brand, J.A., Furst, J.E., Meyer, W.V., Wijayaratna, W.M.K.P. and Dunning, F.B., 1992, *Rev.Sci.Instrum.* 63:114.
Gellrich, A. and Kessler, J., 1991, *Phys.Rev.A* 43:204.
Gray, L.G., Hart, M.W., Dunning, F.B. and Walters, G.K., 1984, *Rev.Sci.Instrum.* 55:88.

Heinzmann, U., Kessler, J. and Lorenz, J., 1970, *Z.Phys.* 240:42.
Hodge, L.A., Moravec, T.J., Dunning, F.B. and Walters, G.K., 1979, *Rev.Sci.Instrum.* 50:5.
Holzwarth, G. and Meister, H.J., 1964, "Tables of Asymmetry, Cross Sections, and Related Functions for Mott Scattering of Electrons by Screened Au and Hg Nuclei," University of Munich, Munich.
Jost, K., Kaussen, F. and Kessler, J., 1981, *J.Phys.E* 14:735.
Kessler, J., 1985, "Polarized Electrons," 2nd ed., Springer, Berlin.
Kirschner, J., 1985, "Polarized Electrons at Surfaces," Springer, Berlin.
Kisker, E., Baum, G., Mahau, A.H., Raith, W. and Reihl, B., 1978, *Phys.Rev.B* 18:2256.
Kisker, E., Clauberg, R. and Gudat, W., 1982, *Rev.Sci.Instrum.* 53:1137.
Lampel, G. and Weisbuch, C., 1975, *Solid State Comm.* 16:877.
Maruyama, T., Garwin, E.L., Prepost, R., Zapalac, G.H., Smith, J.S. and Walker, J.D., 1991, *Phys.Rev.Lett.* 66:2376.
Maruyama, T., Garwin, E.L., Prepost, R. and Zapalac, G.H., 1992, Phys.Rev.B 46:4261.
McClelland, J.J., Scheinfein, M.R. and Pierce, D.T., 1989, Rev.Sci.Instrum. 60:683.
Meier, F., Grobli, J.C., Guarisco, D., Vaterlaus, A., Yashin, Y., Mamaev, Y., Yavich, B. and Kochnev, I., 1993, *Physica Scripta* T49:574.
Nakanishi, T., Aoyagi, H., Horinaka, H., Kamiya, K., Kato, T., Nakamura, S., Saka, T. and Tsubata, M., 1991, *Phys.Lett.A* 158:345.
Omori, T., Kurihari, Y., Nakanishi, T., Aoyagi, H., Baba, T., Furuya, T., Itoga, K., Mizuta, M., Nakamura, S., Takeuchi, Y., Tsubata, M. and Yoshioka, M., 1991, *Phys.Rev.Lett.* 67:3294.
Pierce, D.T. and Meier, F., 1976, *Phys.Rev.B* 13:5484.
Pierce, D.T., Celotta, R.J., Wang, G.-C., Unertl, W.N., Galejs, A., Kuyatt, C.E. and Mielczarek, S.R., 1980, *Rev.Sci.Instrum.* 51:478.
Pierce, D.T., Garvin, S.M., Unguris, J. and Celotta, R.J., 1981, *Rev.Sci.Instrum.* 52:1437.
Pierce, D.T., 1995, to appear in *Methods of Experimental Physics*.
Reichert, E., 1991, Proc.9th Int. Symp. on High Energy Spin Physics, Springer, Berlin, 1:303.
Reihl, B., Erbudak, M. and Campbell, D.M., 1979, *Phys.Rev.B* 19:6358.
Ross, A.W. and Fink, M., 1988, *Phys.Rev.A* 38:6055.
Rutherford, G.H., Ratliff, J.M., Lynn, J.G., Dunning, F.B. and Walters, G.K., 1990, *Rev.Sci.Instrum.* 61:1460.
Scheinfein, M.R., Unguris, J., Kelley, M.H., Pierce, D.T. and Celotta, R.J., 1990, *Rev.Sci.Instrum.* 61:2501.
Tessler, L.R., Hermann, C., Lampel, G., Lassailly, Y., Fontaine, C., Daran, E. and Munoz-Yague, A., 1994, *Appl.Phys.Lett.* 64:895.
Uhrig, M., Beck, A., Goeke, J., Eschen, F., Sohn, M., Hanne, G.F., Jost, K. and Kessler, J., 1989, *Rev.Sci.Instrum.* 60:872.
Watkins, S., Campbell, D.M. and Farago, P.S., 1988, University of Edinburgh Internal Report.
Zhang, X., 1988, MA Thesis, Rice University.
Zorabedian, P., 1982, SLAC Report 248, Stanford University.

ELECTRON SPIN POLARIMETRY INSTRUMENTATION SURVEY: 1994

E.A. Seddon

DRAL Daresbury Laboratory
Daresbury
Cheshire
WA4 4AD
UK

INTRODUCTION

The following table summarises the world-wide distribution of electron spin polarimeters. It is designed to give those interested in the area; some historical perspective, a 'feel' for the popularity, or otherwise, of the various polarimeters currently used for atomic, molecular and condensed matter physics applications, and finally to highlight which groups are active in roughly which scientific areas. Polarimeters for beta spectroscopy and high energy physics are not (currently) included.

The table has the following organisation. The first two columns give the year of the first substantive literature report of the equipment and its location; clearly, polarimeters used in early reports may have subsequently been moved or dismantled and where the fate is known it is indicated in the Comments column. In a few instances the existence of a polarimeter is known to the author but its specifications have yet to be reported in the literature, where this is the case it is tagged as such in the Comments column. The third and fourth columns indicate the research group leader(s) and the general research area. A list of abbreviations is given at the end of the table.

In the few cases where no reference is given the information is believed to be accurate but unconfirmed. In those cases where the reference is a personal communication the information has come from the group involved or a reliable source close to the group. Every attempt has been made to ensure the accuracy of the entries however it is in the nature of an endeavour such as this that omissions and ambiguities easily occur; any additional information or comments should be sent to the author at the above address (or *via* email to e.seddon@dl.ac.uk).

Year	Location of Equipment	Group	General Area of Interest	Comments	Ref.
CONVENTIONAL HIGH ENERGY MOTT POLARIMETERS					
1959	University of Michigan	Nelson & Pidd	Mott scattering		[1]
1966	University of Groningen	Van Klinken	Mott scattering	No longer in operation.	[2,3]
1966	Karlsruhe University	Jost & Kessler	e/atom and e/molecule scattering	Activities now centred at Münster where four polarimeters of this type are operational.	[4-6]
1968	Rijks University Groningen	Van Duinen & Aalders	triple scattering of electrons	Van Klinken's apparatus with modifications. No longer in operation.	[7]
1969	Yale University	Hughes		No longer in operation.	[8]
1970	ETH, Zurich	Siegmann	SPPE	Dismantled.	[9]
1971	Technical University of Berlin	Boersch, Schliepe & Schriefl	e scattering		[10]
1975	Rice University	Dunning & Walters	SPLEED	No longer in operation.	[11]
1976	FOM Amsterdam	Van der Wiel	gas phase SPUPS	No longer in operation.	[12]
1976	University of München	Chrobok and Hofmann	2^γ e emission		[13]
1976	ETH, Zurich	Meier	e beam characterisation and time resolved SPPE	Still operational.	[14]
1977	Bell Laboratories	Landolt	e source characterisation	Dismantled.	[15]
1977, 1979	SLAC	Garwin	e source characterisation	Originally used for HEP work.	[16,17]
1978	Stirling University	Kleinpoppen	e impact ionisation of atoms	No longer in operation	[18]

Year	Institution	Author	Technique	Notes	Ref
1978	University of Münster	Heinzmann	gas phase SPUPS	Activities now centred at Bielefeld.	[19]
1978	University of Bielefeld	Raith and Baum	e source characterisation		[20]
1979	KFA Jülich	Kisker	SPARUPS, 2^y e spectroscopy	Activities now centred at Dusseldorf where two polarimeters of this type are operational. Polarimeter dismantled.	[21]
1979	ETH, Zurich	Erbudak	e source characterisation		[22]
1981	KFA Jülich GmbH	Campagna & Alvarado	e source characterisation		[23]
1983	Central Research Laboratories, Hitachi Ltd, Tokyo	Koike	SPLEED		[24]
1984	KFA Jülich GmbH	Kisker	SPARUPS		[25]
1984	BESSY	Heinzmann	surface SPUPS		[26]
1985	BESSY	Kisker	SPARUPS		[27]
1985	ETH, Zurich	Landolt	SPAES, SPSES	Spherical focusing field. Another under construction for eventual use at Trieste.	[28]
1986	BESSY	Heinzmann	gas phase SPUPS		[29]
1986	Johannes Gutenberg University, Mainz	Reichert	e source characterisation		[30]
1986	Nagoya University	Nakanishi	e source characterisation		[31]
1987	KEK, Photon Factory	Ishii	SPARUPS		[32]
1990	IBM, San Jose	Poppa & Landolt	SEMPA	Polarimeter moved to Stanford and then to LURE (currently used by Landolt & Rossi, see below).	[33]
1991	University of Münster	Merz	e source characterisation		[34]
1992	SRS	Seddon	SPUPS		[35]
1994	University of Mainz	Schönhense	SPARUPS	Spherical focusing field.	[36]

	Institution	Person	Technique	Comments	Ref.
-	University of Tübingen	Nakel	e source characterisation		[37]
-	University of California	Hopster	SPEELS	Design based on that reported by Raue et al. [25].	[38]
-	R-WT Hochschule Aachen	Guntherodt	SPUPS,STM		[37]
-	Hokkaido University	Hayakawa	SPEM		[37]
-	IBM, Zurich	Allenspach	SEMPA	Spherical focusing field.	[37]
-	LURE	Landolt & Rossi	time-resolved photoemission	Spherical focusing field.	[37]

SPHERICAL RETARDING POTENTIAL MOTT POLARIMETERS

Year	Institution	Person	Technique	Comments	Ref.
1984	Rice University	Dunning and Walters		No longer in operation.	[39]
1989	University of Münster	Kessler and Hanne	e/atom and e/molecule scattering	Two operational.	[40]
1990	University of Western Australia	Williams	e/atom scattering		[41]
-	NIST	Pierce and Celotta		Copy of the Dunning and co-workers design [39].	[37]
-	Edinburgh University	Campbell	e/atom scattering		[37]
-	Naval Research Laboratory, Washington	Prinz	SPSES	Design based on [39].	[37]
-	LURE	Hricovini		Under construction.	[37]
-	Fritz-Haber Institute, Berlin	Becker	gas phase SPUPS		[37]
-	KFA Jülich GmbH	Eberhardt	SPARUPS	Under construction	[37]

CYLINDRICAL RETARDING POTENTIAL MOTT POLARIMETERS

Year	Institution	Designer	Application	Status	Ref
1979	Rice University	Dunning and Walters		No longer in operation.	[42]
1985	Ecole Polytechnique	Lampel	e source characterisation		[43]
1985	Edinburgh University	Campbell	e scattering		[44]
1989	NIST	Pierce & Celotta	e scattering	Currently used for calibration purposes.	[45]
1994	Inst. for Nuclear Physics, Orsay	Brissau	e source characterisation		[46]
-	University of Bielefeld	Raith & Baum	e source characterisation		[37]
-	University of Missouri-Rolla	Gay	e/atom collisions		[37]

CONICAL RETARDING POTENTIAL MOTT POLARIMETERS

Year	Institution	Designer	Application	Status	Ref
1987, 1988	Rice University	Dunning and Walters	SPEELS, atom de-excitation spectroscopy		[47,48]
1989	University of California	Hopster	SPEELS,SPUPS, SPSES	Design based on that reported in [48] but with U scattering foil. Equipment currently undergoing modification for SEMPA work.	[49]
1993	State Technical University, St. Petersburg	Mamaev	gas phase and surface SPSES	Currently under construction.	[50]
1993	University of Texas at Austin	Erskine	SPXPS & SPAES		[51]
1994	SRS	Seddon	gas phase and surface SPUPS & SPXPS	Undergoing commissioning tests.	[52]
-	University of Nebraska	Gay		At the planning and design stage.	
-	Institute for Nuclear Physics, Orsay	Brissau	e source characterisation	Expected to be operational by 1995.	[37]

125

		CAMD	Erskine	SPXPS,SPAES		[37]
		Sussex University	Hardiman	SEMPA,SPAES		[53]
		Rice University	Rau			

MERCURY VAPOUR MOTT POLARIMETERS

1961	University of Mainz	Deichsel				[54]
1965, 1981	University of Münster	Kessler		No longer operational.		[55]
1967	University of Mainz	Eckstein	e scattering			[56]
1976	University of Mainz	Reichert	e scattering			[57]

ABSORBED CURRENT POLARIMETERS

1981	ETH Zürich	Erbudak and Müller				[58]
1981	NIST	Siegmann, Pierce and Celotta		No longer operational.		[59-61]
1986	State Technical University of St. Petersburg	Mamaev				[62]
1990	IBM, Zurich Research Lab	Grentz	SPARIPES			[63]
1992	Politecnico di Milano	Ciccacci	e source characterisation			[64]
1993	Advanced Research Lab. Hitachi Ltd.	Koike	SEMPA			[65]

LOW ENERGY DIFFUSE SCATTERING POLARIMETERS

1986, 1989	NIST	Pierce & Celotta	SEMPA and electron tunnelling	Two incorporated into a SEMPA apparatus, one in a scanning electron microscope and one for field emission work.	[66,67]
1989	MIT	O'Handley	SPSES		[68]
1992	NSLS	Johnson	SPUPS, SPXPS		[69]
1993	Lehigh University, Pennsylvania	Klebanoff	SPXPS	Based on the design reported by Unguris et al. [66]. Two operational.	[70]
-	New York University	Sinkovic		Based on the design reported by Unguris et al. [66]	[37]
-	Rice University	Rau			
-	Gakushuin University	Mizoguchi		Under construction.	[37]
-	Institute for Molecular Sciences, Okazaki	Kamada	SPUPS	Under construction.	[37]

LEED POLARIMETERS

1979	KFA Jülich GmbH	Kirschner & Feder	SPLEED	Currently used by Oepen for SEMPA at Jülich.	[71]
1986	State Technical University of St. Petersburg	Mamaev	e source characterisation		[62]
1991	McMaster University, Ontario	Venus	SPES		[72]
-	BESSY	Kirschner	SPPE		[73]
-	Halle	Kirschner	e scattering from surfaces		[37]
-	Tokyo University	Kakizaki	SPPE		[37]
-	Osaka University	Suga	SPIPES		[37]

MAGNETIC SCATTERING POLARIMETERS

1989, 1992	University of Düsseldorf	Kisker	SPSXPS, SPES	One currently in use, two more under construction.	[74,75]

OPTICAL POLARIMETERS

1980	Ecole Polytechnique	Lampel		e source characterisation	[76]
1983, 1987	University of Münster	Kessler and Hanne		e source characterisation	[77,78]
1983, 1993	University of Missouri-Rolla	Gay		First polarimeter constructed in 1988. Activities now centred at the University of Nebraska.	[79,80]
-	Sophia, Japan	Suzuki			

Abbreviations

LEED	low energy electron diffraction
SEMPA	scanning electron microscopy with polarisation analysis
SPAES	spin polarised Auger electron spectroscopy
SPARIPES	spin polarised angle resolved inverse photoemission spectroscopy
SPARUPS	spin polarised angle resolved ultraviolet photoemission spectroscopy
SPEELS	spin polarised electron energy loss spectroscopy
SPES	spin polarised electron spectroscopy
SPIPES	spin polarised inverse photoemission spectroscopy
SPLEED	spin polarised low energy electron diffraction.
SPPE	spin polarised photoemission
SPSES	spin polarised secondary electron spectroscopy
SPUPS	spin polarised ultraviolet photoemission spectroscopy
SPXPS	spin polarised X-ray photoemission spectroscopy
STM	scanning tunnelling microscopy

Acknowledgements

I would like to express my gratitude to all those who have given up their valuable time and freely contributed to the information contained in this survey.

References

1. D.F. Nelson and R.W. Pidd, *Phys. Rev.*, 114 (1959) 728-735.
2. J. van Klinken, *Thesis,* The University of Groningen, 1965.
3. J. van Klinken, *Nucl.Phys.*, 75 (1966) 161-188.
4. K. Jost and J. Kessler, *Z. Phys.*, 195 (1966) 1-12.
5. G.F. Hanne and J. Kessler, *J. Phys. B: At. Mol. Phys.*, 9 (1976) 791-804.
6. R. Möllenkamp, W. Wübker, O. Berger, K. Jost and J. Kessler, *J. Phys. B: At. Mol. Phys.*, 17 (1984) 1107-1121.
7. R.J. van Duinen and J.W.G. Aalders, *Nucl. Phys.*, A115 (1968) 353-363.
8. W. Raith in "Atomic Physics", ed. B. Bederson, V. Cohen and F.M.J. Pichanick, Plenum Press, New York, 1969.
9. G. Busch, M. Campagna and H.C. Siegmann, *J. Appl. Phys.*, 41 (1970) 1044-1051.
10. H. Boersch, R. Schliepe and K.E. Schriefl, *Nucl. Phys.*, A163 (1971) 625-636.
11. M. Kalvisvaart, M.R. O'Neill, T.W. Riddle, F.B. Dunning and G.K. Walters, *Phys. Rev. B*, 17 (1978) 1570-1578.
12. E.H.A. Granneman, *Thesis*, 1976, Amsterdam.
13. G. Chrobok and M. Hofmann, *Phys. Lett.*, 57A (1976) 257-258.
14. D.T. Pierce and F. Meier, *Phys. Rev. B*, 13 (1976) 5484-5500.
15. M. Landolt and M. Campagna, *Phys. Rev. Lett.*, 38 (1977) 663-666.
16. C.K. Sinclair, E.L. Garwin, R.H. Miller and C.Y. Prescott, *Proceedings of the Argonne Symposium on High Energy Physics with Polarized Beams and Targets*, ed. M.L. Marshak, AIP Conf. Proc. No. 35 (AIP, New York, 1977), p. 424.
17. M.J. Alguard, J.E. Clendenin, R.D. Ehrlich, V.W. Hughes, J.S. Ladish, M.S. Lubell, K.P. Schüler, G. Baum, W. Raith, R.H. Miller and W. Lysenko, *Nucl. Inst. Meth.*, 163 (1979) 29-59.
18. D. Hils and H. Kleinpoppen, *J. Phys. B: At. Mol. Phys.*, 11 (1978) L283-L287.
19. U. Heinzmann, *J. Phys. B: At. Mol. Phys.*, 11 (1978) 399-412.
20. E. Kisker, G. Baum, A.H. Mahan, W. Raith and B. Reihl, *Phys Rev. B*, 18 (1978) 2256-2275.
21. E. Kisker, M. Campagna, W. Gudat and E. Kuhlmann, *Rev. Sci. Inst.*, 50 (1979) 1598-1601. E. Kisker, R. Clauberg and W. Gudat, *Rev. Sci. Inst.*, 53 (1982) 1137-1144.
22. B. Reihl, M. Erbudak and D.M. Campbell, *Phys. Rev. B*, 19 (1979) 6358-6366.
23. S.F. Alvarado, F. Ciccacci and M. Campagna, *Appl. Phys. Lett.*, 39 (1981) 615-617.
24. K. Koike and K. Hayakawa, *Jpn J. Appl. Phys.*, 22 (1983) 1332.
25. R. Raue, H. Hopster and E. Kisker, *Rev. Sci. Inst.*, 55 (1984) 383-388.
26. A. Eyers, F. Schäfers, G. Schönhense, U. Heinzmann, H.P. Oepen, K. Hünlich, J. Kirschner and G. Borstel, *Phys. Rev. Lett.*, 52 (1984) 1559-1562.
27. E. Kisker, K. Schröder, W. Gudat and M. Campagna, *Phys. Rev. B*, 31 (1985) 329-339.
28. M. Landolt, R. Allenspach and D. Mauri, *J. Appl. Phys.*, 57 (1985) 3626-3631.

29 Ch. Heckenkamp, A. Eyers, F. Schäfers, G. Schönhense, U. Heinzmann, *Nucl. Inst. Methods Phys. Res.*, A246 (1986) 500-503.
30 N. Ludwig, A. Bauch, P. Naß, E. Reichert and W. Welker, *Z. Phys. D - At. Mol. Clust.*, 4 (1986) 177-183.
31 T. Nakanishi, K. Dohmae, S. Fukui, Y. Hayashi, I. Hirose, N. Horikawa, T. Ikoma, Y. Kamiya, M. Kurashina and S. Okumi, *Jpn J. Appl. Phys.*, 25 (1986) 766-767.
32 K. Soda, H. Sugawara, S. Suga, A. Kakizaki, M. Taniguch, T. Mori, M. Fujisawa, S. Suzuki, Y. Kamiya, T. Miyahara and T. Ishii, *Activity Report of the Synchrotron Radiation Laboratory,* (1987) 70-71.
33 R. Browning, T. VanZandt, C.R. Helms, H. Poppa and M. Landolt, *J. Electron. Spectrosc. Relat. Phenom.*, 51 (1990) 315-320.
34 Ch. Bathe, M. Rissmann and H. Merz, *Surf. Sci.*, 251/252 (1991) 276-280.
35 E.A. Seddon, M.A. Hoyland, H.P. Hughes and R. Jordan, *Nucl. Inst. Meth. Phys. Res.*, A319 (1992) 377-382.
36 M. Getzlaff, J. Bansmann and G. Schönhense, *J. Mag. Mag. Mat.*, 131 (1994) 304-310.
37 personal communication.
38 D.P. Pappas, *Ph.D. Thesis*, The University of California, 1990.
39 L.G. Gray, M.W. Hart, F.B. Dunning and G.K. Walters, *Rev. Sci. Inst.*, 55 (1984) 88-91.
40 M. Uhrig, A. Beck, J. Goeke, F. Eschen, M. Sohn, G.F. Hanne, K. Jost and J. Kessler, *Rev. Sci. Inst.*, 60 (1989) 872-878.
41 C. Ranganathaiah, J.L. Robins, A.L. Yates, W.C. Macklin, R.A. Anderson and J.F. Williams, *J. Electron Spect. Relat. Phenom.*, 51 (1990) 331-338.
42 L.A. Hodge, T.J. Moravec, F.B. Dunning and G.K. Walters, *Rev. Sci. Inst.*, 50 (1979) 5-8.
43 H-J. Drouhin, C. Hermann and G. Lampel, *Phys. Rev. B*, 31 (1985) 3872-3886.
44 D.M. Campbell, C. Hermann, G. Lampel and R. Owen, *J. Phys. E: Instrum.*, 18 (1985) 664-672.
45 J.J. McClelland, M.R. Scheinfein and D.T. Pierce, *Rev. Sci. Inst.*, 60 (1989) 683-687.
46 Anon., *CERN Courier*, 34 (1994) 12-13.
47 F.B. Dunning, L.G. Gray, J.M. Ratliff, F.-C. Tang, X. Zhang and G.K. Walters, *Rev. Sci. Inst.*, 58 (1987) 1706-1708.
48 F.-C. Tang, X. Zhang, F.B. Dunning and G.K. Walters, *Rev. Sci. Inst.*, 59 (1988) 504-505.
49 D.P. Pappas, *Ph.D. Thesis,* 1990, The University of California.
50 Yu.A. Mamaev, L.P. Ovsyannikova, V.N. Petrov, T.Ya. Fishkova and E.V. Shpak, *Tech.Phys.*, 38 (1993) 827-829.
51 D-J. Huang, J-Y. Lee, J-S. Suen, G.A. Mulhollan, A.B. Andrews and J.L. Erskine, *Rev. Sci Inst.*, 64 (1993) 3474-3479.
52 E.A.Seddon, D. Eastham, M. Siggel, F.M. Quinn, M.C. Grossel, I.K. Lewis, W. Schwarzacher, M.C. Dowling, Appendix to the *Daresbury Laboratory Annual Report,* (1993/94) 238.
53 M.S. Bhella, *D.Phil. Thesis,* The University of Sussex, 1991.
54 H. Deichsel, *Z. Phys.*, 164 (1961) 156-165.
55 K. Jost and J. Kessler, *Phys. Rev. Lett.*, 15 (1965) 575-577. K. Jost, F. Kaussen and J. Kessler, *J. Phys. E: Sci. Instrum.*, 14 (1981) 735-741.
56 W. Eckstein, *Z. Phys.*, 203 (1967) 59-65.
57 M. Düweke, N. Kirschner, E. Reichert and S.Schön, *J. Phys. B, Atom Molec. Phys.*, 9 (1976) 1915-1921.

58 M. Erbudak and N. Müller, *Appl. Phys. Lett.,* 38 (1981) 575-577.
59 D.T. Pierce, S.M. Girvin, J. Unguris and R.J. Celotta, *Rev. Sci. Inst.,* 52 (1981) 1437-1444.
60 H.C. Siegmann, D.T. Pierce, R.J. Celotta, *Phys. Rev. Lett.,* 46 (1981) 452-455.
61 R.J. Celotta, D.T. Pierce, H.C. Siegmann and J. Unguris, *Appl. Phys. Lett.,* 38 (1981) 577-579.
62 Yu.A. Mamaev, B.S. Makarov, A.N. Mishin, V.N. Petrov, V.N. Yakovlev and Yu.P. Yashin, *Bull. Acad. Sci. USSR, Phys. Ser.,* 50 (1986) 89-91.
63 W. Grentz, M. Tschudy, B. Reihl and G. Kaindl, *Rev. Sci. Inst.,* 61 (1990) 2528-2533.
64 F. Ciccacci, E. Vescovo, G. Chiaia, S. de Rossi and M. Tosca, *Rev. Sci. Inst.,* 63 (1992) 3333-3338.
65 T. Furukawa and K. Koike, *Jpn J. Appl. Phys. Lett.,* 32 (1993) 1851-1854.
66 J. Unguris, D.T. Pierce and R.J. Celotta, *Rev. Sci. Inst.,* 57 (1986) 1314-1323.
67 M.R. Scheinfein, D.T. Pierce, J. Unguris, J.J. McClelland, R.J. Celotta and M.H. Kelley, *Rev. Sci. Inst.,* 60 (1989) 1-11.
68 J. Woods, M. Tobise and R.C. O'Handley, *Rev. Sci. Inst.,* 60 (1989) 688-692.
69 P.D. Johnson, N.B. Brookes, S.L. Hulbert, R. Klaffky, A. Clarke, B. Sinkovic, N.V. Smith, R. Celotta, M.H. Kelly, D.T. Pierce, M.R. Scheinfein, B.J. Waclawski and M.R. Howells, *Rev. Sci. Inst.,* 63 (1992) 1902-1908.
70 L.E. Klebanoff, D.G. Van Campen and R.J. Pouliot, *Rev. Sci. Inst.,* 64 (1993) 2863-2871.
71 J. Kirschner and R. Feder, *Phys. Rev. Lett.,* 42 (1979) 1008-1011.
72 J. Sawler and D. Venus, *Rev. Sci. Inst.,* 62 (1991) 2409-2418.
73 C.M. Sneider, J.J. de Miguel, P. Bressler, P. Schuster, R. Miranda and J. Kirschner, *J. Electron Spectrosc. Relat. Phenom.,* 51 (1990) 263-274.
74 D.Tillmann, R. Thiel and E. Kisker, *Z. Phys. B - Cond. Matt.,* 77 (1989) 1-2.
75 R. Jungblut, Ch. Roth, F.U. Hillebrecht and E. Kisker, *Surf. Sci.,* 269/270 (1992) 615-621.
76 M. Eminyan and G. Lampel, *Phys. Rev. Lett.,* 45 (1980) 1171-1174.
77 J. Goeke, G.F. Hanne, J. Kessler and A. Wolcke, *Phys. Rev. Lett.,* 51 (1983) 2273-2275.
78 J. Goeke, J. Kessler and G.F. Hanne, *Phys. Rev. Lett.,* 59 (1987) 1413-1415.
79 T.J. Gay, *J. Phys. B: At. Mol. Phys.,* 16 (1983) L553-L556.
80 J.E. Furst, W.M.K.P. Wijayaratna, D.H. Madison and T.J. Gay, *Phys. Rev. A,* 47 (1993) 3775-3787.

A BOLT-ON SOURCE OF SPIN POLARISED ELECTRONS FOR STUDIES OF SURFACE MAGNETISM

Fredrik Schedin, Ranald Warburton, and Geoff Thornton

Interdisciplinary Research Centre in Surface Science and
Chemistry Department
University of Manchester
Manchester M13 9PL
United Kingdom

INTRODUCTION

Experiments with spin-polarised electrons are rapidly gaining importance due to their ability to selectively measure the exchange-split majority- and minority-spin electronic states. By employing the inverse photoemission process with its small probing depth, information about surface magnetism can additionally be obtained. In this chapter a description will be given of a versatile and transportable spin-polarised electron gun for angle-resolved spin-polarised inverse photoemission spectroscopy (ARSPIPES) experiments. A GaAs(001) single crystal is employed to obtain spin-polarised electrons by photoemission with circularly polarised light. To obtain a transversely polarised beam, which has the objective of optimising the spin-dependent interaction with a sample magnetised in the surface plane, a 90° electrostatic spherical sector is used to deviate the direction but not the spin of the electrons. A set of electrostatic lenses are employed in order to transport and focus the electrons onto a target of investigation. A brief description of results obtained with this apparatus will be given.

ANGLE RESOLVED SPIN POLARISED INVERSE PHOTOEMISSION

The phenomenon of emission of radiation caused by electrons impinging on a solid surface has been known since the last century (e.g. Röntgen, 1898). With the development of UHV systems and energy selective photon detectors in recent years it has become possible to measure the emission of UV radiation. The method of measuring the emission of a particular energy of UV photons as a function of impinging electron energy is known as inverse photoemisson spectroscopy (IPES) (e.g. Dose, 1985). The incident electron enters an unoccupied electronic band above the vacuum level and decays from there via a radiative transition into another, lower lying unoccupied band (see Figure 1). If the impinging electrons have a well defined energy, E, and momentum, k, information is also gained about the dependence of the parallel component of the electron wave vector, $k_\parallel$, on the emitted radiation intensity, and angle resolved IPES can be performed. With this technique the energetic position of the unoccupied electronic bands as a function of $k_\parallel$ is obtained.

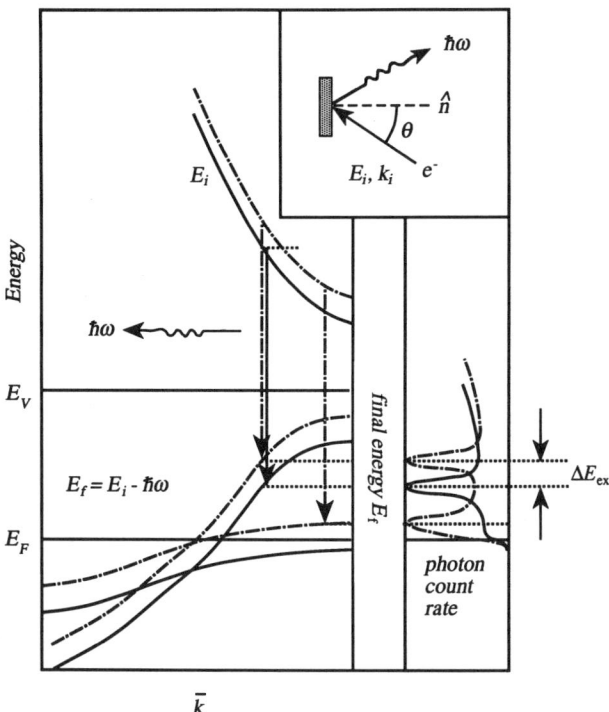

Figure 1. Illustration of the inverse photoemission process, indicating direct radiative transitions. The dot-dashed lines represent minority-spin bands and the full lines represent majority-spin bands. Two minority-spin transitions and one majority-spin transition for a certain photon energy are shown. These lead to maxima in the measured spectra (e.g. Dose, 1985; Donath, 1994).

In ferromagnetic materials the electronic bands for electrons with spin-up and spin-down are shifted in energy with respect to each other by the exchange interaction. This results in a higher density of occupied electronic states in that set of bands which is lower in energy and thereby has more available states below the Fermi level. A preferential spin orientation of the electrons in the ferromagnetic material is obtained and is known as the majority-spin direction. With the use of a spin polarised electron gun in the IPES experiment, one can selectively access the two sets of spin-split bands and the spin-polarised unoccupied electronic band structure is obtained.

ELECTRON GUN DESIGN

A number of constraints have affected the design of the electron gun. Due to the sensitivity to contamination of the photocathode, it is desirable to keep it under vacuum continuously, especially when the substantial problems of cleaning it are considered. It has also been necessary to optimise the appropriate distance between the end of the gun and the target in order to accommodate other experiments in an arbitrary vacuum chamber. Moreover, it is necessary to provide a means of illuminating the photocathode with circularly polarised light during operation of the gun. It is also necessary to enable illumination of the photocathode during the activation procedure. The connection to the measurement chamber is through a 114 mm O.D. conflat type flange.

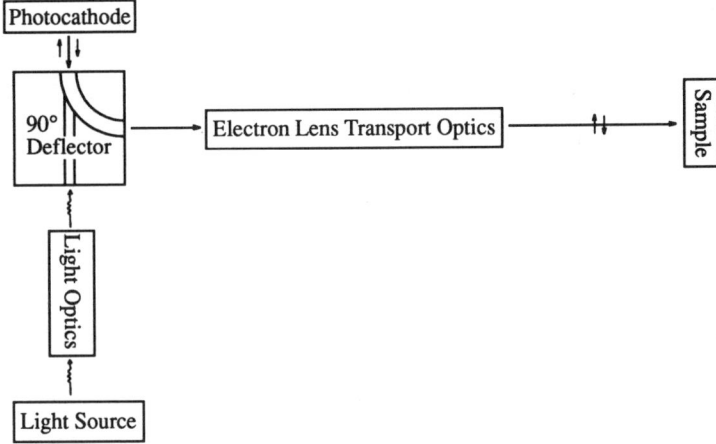

Figure 2. Block diagram of the spin-polarised electron gun.

The Photocathode

The photocathode serves as a source of spin-polarised electrons. A GaAs(100) photocathode (Pierce et al., 1980) has been employed. Polarised electrons are obtained by means of photoemission with the exciting light being circularly polarised and of energy slightly higher energy than the band gap of GaAs (Pierce and Meier, 1976). The photoemission process can be described using the three step model (Spicer, 1958; 1977), which involves *photo-excitation, transport to the surface* and *emission into vacuum*.

Due to the direct band gap and the spin-orbit splitting of the 3p valence band in GaAs, and the selection rules for transitions excited with circularly polarised light, a differential population in the $S_{1/2}$ conduction band level of electrons with spin-up and spin-down can be obtained as shown in Figure 3. If σ^+ circularly polarised light is used the selection rules are $\Delta l = \pm 1$, and $\Delta m_j = +1$. The relative transition probabilities can be calculated (Pierce and Meier, 1976) and yield a three times higher result for the $m_j = -3/2$ to $m_j = -1/2$ transition compared to the $m_j = -1/2$ to $m_j = +1/2$ transition. Thus the upper $S_{1/2}$ level becomes populated with three times as many electrons in the state with $m_j = -1/2$ as in the state with $m_j = +1/2$.

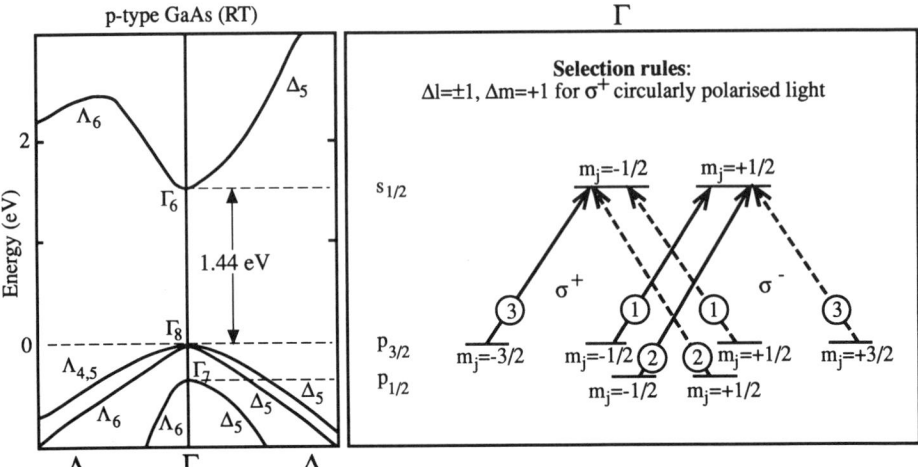

Figure 3. Section of the band structure of GaAs around the Γ point (after Chelikowski and Cohen, 1974) and selection rules with transition probabilities for optical excitation with circularly polarised light from the valence band maximum (Γ_8) to the conduction band minimum (Γ_6).

The spin polarisation, P is defined as:

$$P = (n_\uparrow - n_\downarrow)/(n_\uparrow + n_\downarrow), \tag{1}$$

where $n_\uparrow$ and $n_\downarrow$ are the numbers of electrons with spin-up and spin-down respectively. According to Figure 3, the maximum obtainable polarisation for σ^+ light is $(1-3)/(1+3)$ = -50%.

In order to enable escape of the excited electrons into vacuum, the ~4 eV work function has to be lowered. This is achieved with the formation of a negative electron affinity (NEA) surface accomplished by dosing the clean GaAs(100) surface with cesium and oxygen (Turnbull and Evans, 1968). In this manner the vacuum level is lowered to a position below that of the bulk conduction band minimum and the electrons photoexcited to the conduction band have no potential barrier to overcome to escape into vacuum. The emission is therefore limited mainly by the electron diffusion length, which measures about 0.5 µm for bulk material (James and Moll, 1969).

Light Optics

A GaAlAs laser diode has been used as the infrared light source. The band gap of GaAlAs is slightly larger than that for GaAs, the exact value depending on the concentration of Al. The energy of the emitted laser light is hence sufficient to photoexcite electrons to just above the photothreshold for GaAs. Here, a diode of energy 1.56 eV and a maximum power of 20 mW (Sony SLD201V) has been used. The laser diode is mounted on a water-cooled reservoir to ensure a constant operational temperature. Two lenses are used to obtain a parallel beam and subsequently focus it onto the cathode. Circularly polarised light is generated by a combination of a linear polariser and a Pockels cell. The helicity of the circularly polarised light is defined by the particle angular momentum with respect to the particle momentum and can easily be switched by reversing the polarity of the voltage applied to the Pockels cell.

The 90° deflector

The use of a 90° deflector serves to produce a transversely polarised electron beam at the target, by arranging that the photon momentum is at a 90° angle to the final beam direction. An electrostatic spherical sector with a pass energy of 250 eV was chosen to deflect the beam. If suitable voltages are applied to the 90° sector it forms a real image at a point on the exit beam where a line drawn from an object on the optical axis at the entrance, through the centre of curvature of the sector, crosses the optical axis at the exit (Purcell, 1938). In the GaAs cathode system the object is a virtual object formed by the optical system comprising of the photocathode and the anode. It is located at ~3 mm behind the photocathode surface (Pierce et al., 1980). The image is formed at ~50 mm beyond the exit of the 90° deflector at which point the electrons have an energy of 250 eV relative to the photocathode. This point also functions as the object for the transport lens system.

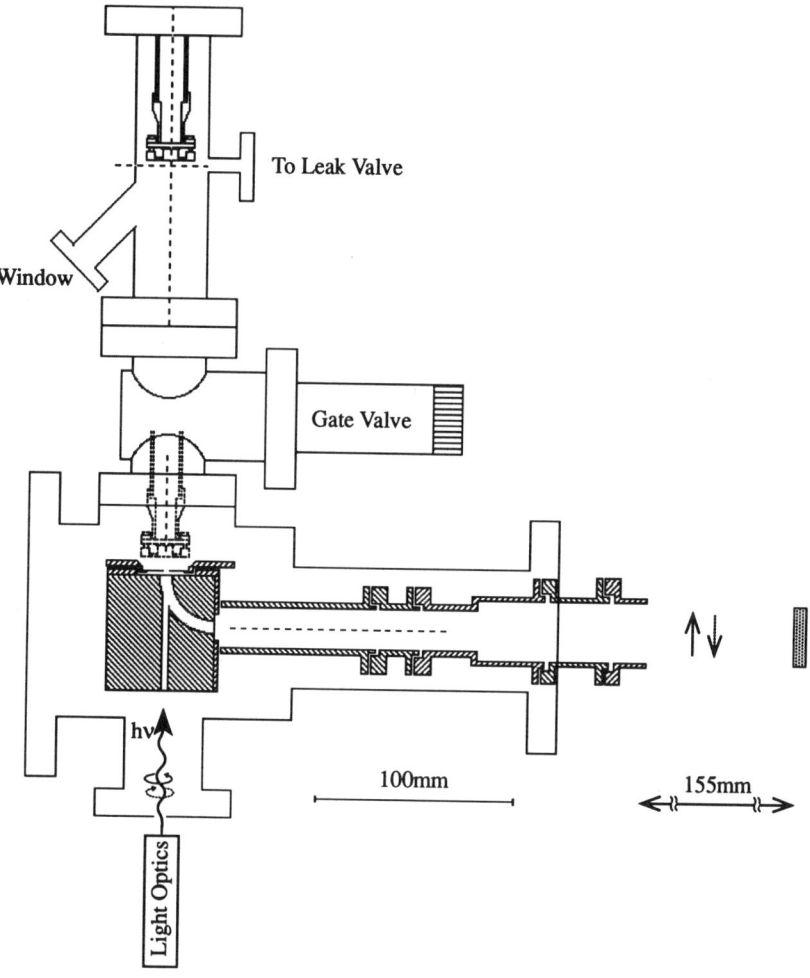

Figure 4. Spin-polarised electron source chamber with GaAs cathode in operation position. The aperture, the 90° deflector and the electron transport optics are also shown.

The Electron Lens Transport Optics

In order to transport the beam to the target and provide a maximum retard ratio of ~50, a twin three-element lens system was used, where the length of the middle electrode is equal to the diameter and the gaps between the lens elements are a tenth of the diameter. The point where the electrons are focussed by the spherical sector functions as the first object plane. The first lens is designed for an image plane at an equal distance away from the reference plane on the other side of the lens. A set of four steering plates are placed at the object and image planes on either side of this first lens. The second lens serves to generate an as near to parallel and as narrow beam as possible at the target.

The voltages on the lens elements were initially calculated from lens curves (Harting and Read, 1976). In reality, the focussing voltages in particular those on the second three element lens need to be adjusted to account for the uncertainty in the position of the electrons at which they are focussed by the 90° sector, and to account for the distance between the end of the gun and the sample.

Construction Details

The lenses and the 90° spherical deflector are made of aluminium alloy and the internal surfaces are covered with a coating of colloidal graphite in order to reduce the effect of patch fields and to achieve a homogeneous conducting surface with a constant work function. The apertures are made of molybdenum for similar reasons. All the lens elements and the 90° sector are electrically isolated from ground and each other.

The electron optics are contained in a nonmagnetic stainless steel tube, as shown in Figure 4. The circularly polarised light is admitted into the source chamber vertically upwards through a vacuum window. Thereafter, the light beam travels through a hole in the spherical sector to hit the photocathode which is mounted on a liquid nitrogen reservoir. The photocathode may be retracted upwards by a linear drive from the operating position to a position above the gate valve where the surface activation takes place. While the cathode is in this higher position the gate valve may be shut to isolate this section from the main vacuum system.

The GaAs Electron Source

All the procedures of cleaning, activating, and maintaining the GaAs photocathode require ultra high vacuum. The vacuum in the cathode region is maintained by an ion pump. During cleaning and activation the cathode is positioned in the higher position above the gate valve. A vacuum window allows illumination of the cathode during activation. A stainless steel pipe has been added for admission of oxygen to the GaAs surface, the pressure being controlled by a variable leak valve.

The GaAs(100) photocathode is clamped against a heater block to enable cleaning of it by heating in vacuum. The temperature is monitored by a thermocouple spot-welded to the heater block. The heater block is pressed against a stainless steel disk which is welded onto the bottom of a stainless steel tube. A sapphire disc is placed between the heater and the stainless steel disk to ensure electrical isolation while enabling a high thermal conductivity

(Holland, 1962) between the heater block and the stainless steel disc at low temperatures. The stainless tube may be filled with liquid nitrogen to cool the GaAs cathode. Before insertion into the vacuum chamber the cathode is cleaned by a chemical etching procedure based on a 4 : 1 : 1 mixture of concentrated H_2SO_4, 30% H_2O_2, and H_2O by volume as described by Pierce et al (1980).

Lens Voltages

The voltages for the five lens elements, the cathode, the aperture, and the two spherical sectors are generated by a computer controlled power supply. The target is held at ground potential and the accelerating voltage controlling the beam energy is set by biasing the cathode with a negative potential. The optimal operating conditions were investigated experimentally by adjusting the lens element voltages manually and monitoring the beam with a custom made fluorescent imaging screen. A spot size of ~7 mm diameter at the screen was achieved in this fashion. These functions were then incorporated into the computer program so that at the target a focussed beam could be obtained for each beam energy in the range 5-20 eV.

OPERATION AND PERFORMANCE OF THE ELECTRON GUN

When UHV conditions ($\sim 1 \times 10^{-10}$ mbar) are achieved after bakeout for ~48 hours at 450 K, the GaAs surface can be further cleaned by heat treatment in vacuum. The cathode is heated to temperatures just below the congruent evaporation temperature (Foxon et al., 1973; Goldstein et al., 1976).

Activation of the Photocathode

After heat treatment the GaAs is left to cool down. If liquid nitrogen is injected into the stainless tube the cooling time can be decreased. Once the crystal is cooled down to ≤ 320 K the activation procedure can take place. The activation is performed in the position, ca. 20 cm above the operating position and is started by passing a current through a Cs dispenser positioned ca. 50 mm from the photocathode. The GaAs cathode is illuminated during the activation by a white projector lamp . The photocurrent is monitored by means of measuring the drain current between the crystal and ground with the crystal held at a potential of -10 V. This is used as an aid to maximise the photocurrent.

After typically 5 min of Cs deposition the emission current starts rising, reaches a peak and starts dropping again. While maintaining Cs deposition, oxygen is then let in through a leak valve and the O_2 leak rate is adjusted so as to obtain a maximum emission current. The system is left at this stage for 10-15 min, conditioning being completed by first increasing the oxygen pressure until the photocurrent drops to half of the maximum value and then closing the oxygen valve to allow an excess of Cs onto the photocathode surface. The current increases back up to the maximum value and starts decreasing again. When the photocurrent has decreased back to half of the maximum value the current through the Cs getter is switched off. The photocurrent will then slowly increase back up to near the maximum value as Cs diffuses away from the surface.

Some of the activations have been completed by treatment with an excess of O_2 rather than Cs. The most effective way to produce a good photocathode has varied between different activations. The method used in the present work has produced photocathodes with a long lifetime (~24 h). Pierce et al. (1980) and Drouhin et al. (1985) have measured the quantum yield for a GaAs photocathode with similar treatment to be a few percent at threshold.

Transmission of the Electron Optics and Energy Resolution of the Beam

The maximum emission current obtained from the GaAs with the laser diode operated at approximately 10 mW was 60 µA. This was achieved for an optimised NEA surface, shortly after an activation of the cathode. This optimised state of the surface, however, was found to be rather unstable due to the vacuum level increasing as, e.g. cesium diffuses away from the surface or residual gases are adsorbed. To ensure stable operating conditions in the very time-consuming ARSPIPES experiments, the cathode was operated in less optimised NEA conditions. This resulted in constant emission currents for up to 24 hours. The spin polarised electron gun was normally operated at an emission current of 2 µA.

The transmission through the electron optics and the beam energy resolution was measured with a custom made four parallel grid electron beam analyser. The transmission was measured to be larger than 70% for the energy range of 7 - 20 eV used in the experiment. The energy resolution of the spin polarised electron beam was measured using the custom made beam analyser in retarding field mode (RFA). A small oscillation (10 mV p-p) was superimposed on a retarding ramp voltage applied to the middle two grids in the same fashion as in an RFA Auger experiment. Since the retarding field analyser integrates the beam current intensity over all energies above that of the retarding voltage, the first derivative of the collected current at the screen with respect to retarding voltage is proportional to the beam intensity as a function of beam energy. The measured upper limit of 0.27 eV (FWHM) as the electron beam energy resolution at operating conditions with $I_E = 2$ µA, suggests that the broadening of the total resolution the inverse photoemission experiment due to electron energy spread is insignificant in comparison to the optical resolution of the photon detector (0.73 eV FWHM for the combination of CaF_2 and KCl).

Polarisation of the Electron Beam

The degree of polarisation of the emitted electrons from the spin polarised electron source is one of the main criteria in the performance of the electron gun described in this paper. In addition, to represent the true shape and position of the spin resolved features in ARSPIPES experiments one needs to extrapolate the measured data for a hypothetical 100% polarised beam to obtain any physical meaning by reference to the minority- and majority spin states in the crystal. It is therefore necessary to know the polarisation of the electron beam, as an incorrect estimate affects both the energy position and peak heights in the re-calculated spectra.

Unfortunately, in the experimental chamber there was no conventional means of detecting the degree of spin polarisation, such as a Mott detector or an absorbed current detector. A variant of the method described by Donath (1989); Donath et al. (1990); and

Grentz et al. (1990) was used in this work to determine the electron beam spin polarisation. In this method, the IPES spectra themselves serve as a spin detector. Here we use this method, employing a Ni(110) sample. Ni is a strong ferromagnet, meaning that the majority spin d-bands are completely filled at $T = 0$ K. The uppermost minority d-band, however, extends above the Fermi level. As a consequence, optical IPE transitions involving the d-band as a final state can occur at T = 0 K only in the minority channel. The method of re-calculating the data, which largely follows Donath's description, is performed as follows:

Let $N_{\uparrow,\downarrow}$ be the number of counts from the photon detector per unit incident charge in either spin direction for a hypothetical 100% polarised beam, with $\uparrow$ ($\downarrow$) being defined as parallel (antiparallel) to the majority spin direction, i.e., antiparallel (parallel) to the magnetisation direction, M. The observed number of counts $n_{\uparrow,\downarrow}$ produced by an incompletely polarised electron beam is then related to $N_{\uparrow,\downarrow}$ by:

$$N_{\uparrow,\downarrow} = \frac{N}{2}(1 \pm A), \qquad (2)$$

where N is the total number of counts $N = N_\uparrow + N_\downarrow = n_\uparrow + n_\downarrow$ and A is the asymmetry in the number of photon counts, defined as:

$$A = \frac{N_\uparrow - N_\downarrow}{N_\uparrow + N_\downarrow} = \frac{n_\uparrow - n_\downarrow}{n_\uparrow + n_\downarrow} \cdot \frac{1}{P \cdot \cos 35.3°}. \qquad (3)$$

The factor cos 35.3° arises from the angle between the spin vector direction, $[1\bar{1}0]$, and the magnetisation direction, $[1\bar{1}1]$ of the Ni(110) sample, and P is the spin polarisation of the electron beam, defined in eq.1. P can be varied as a parameter until a sensible physical representation of the majority-spin spectrum is obtained. Thereby a polarisation of 25±5% has been concluded (Schedin, 1992).

PHOTON DETECTION

Photons are counted using a solid state band pass photon detector with a KCl photocathode high-pass filter and a CaF_2 window low-pass filter, giving an optical energy resolution of approximately 0.73 eV (FWHM). Although the energy position of the peak maximum in the asymmetric distribution of the detection efficiency of this combination was measured to be at a photon energy of 9.91 eV, the mean energy value of the response curve is 9.62 eV. The effective energy of the detected photons has therefore been approximated to ~9.8 eV. The photon detector was mounted at an angle of approximately 71° off the electron beam direction.

EXPERIMENTAL ARSPIPES RESULTS

Experimental details

The spin polarised electron gun is employed to supply a beam of spin-polarised electrons to the Ni(110) sample. The sample is cut to the shape of a picture-frame with legs along the $\langle 111 \rangle$ directions of easy magnetisation. The spin vectors are aligned parallel or antiparallel to the $[1\bar{1}0]$ direction of the crystal lattice, implying that the angle between the spin vector and the magnetisation direction $[1\bar{1}\bar{1}]$ is constant at 35.3° as the polar angle of incidence with respect to the surface normal, θ is varied with the $k_\parallel$ vector lying along the $\overline{\Gamma Y}$ line of the surface Brillouin zone. A coil was wound around one of the legs of the sample to enable magnetisation. The sample was magnetised employing a number of short pulses (~1 ms) of high current (~100 A) through the coil.

The two spin channels were measured quasi-simultaneously, the spin direction being flipped by changing the helicity of the circularly polarised laser light, accomplished by changing the sign of the voltage applied to the Pockels cell, while keeping all the other parameters (lens voltages, accelerating potential and angle of incidence, etc.) constant for each data point. The photons were counted during a time interval corresponding to a constant accumulated charge of incident electrons at the sample (integrated drain current) for each data point. Both the cathode and the Ni(110) sample were held at room temperature (~300 K) during measurements. The base pressure was $< 2 \times 10^{-10}$ mbar.

The series of spin-resolved spectra in this report have been normalised using the value $P = 25\%$ for the polarisation. The measurements were carried out with a remanently magnetised sample. All of the four possible permutations for spin- and magnetisation-directions were investigated. No experimental asymmetry was detected on flipping both electron beam spin- and sample-magnetisation directions.

Sample Preparation

The sample was mechanically polished to a surface roughness of 1 μm. Laue diffraction revealed the surface normal to be within 0.5° of [110]. As sulphur is a common contaminant in Ni, repeated cycles of Ar$^+$ sputtering (10 min, 500 V, ~3 μA) and annealing (~870 K, 20 min) were employed to obtain a depletion of sulphur in the surface layer. Carbon contamination was removed towards the end of the annealing period by exposing the crystal to 1×10^{-7} mbar of O_2 for 5-60 s, depending on the degree of contamination. Subsequent Auger spectra corresponded to a clean surface with small traces of S, Ar, and C (≤ 0.01 ML). A sharp 1x1 LEED pattern characteristic of the (110) face of Ni was observed.

RESULTS AND DISCUSSION

Exchange splitting of bulk and surface states

Figure 5 shows ARSPIPES spectra of Ni(110) in the $\overline{\Gamma Y}$ azimuth. The left hand panel represents the normalised data for a hypothetical 100% polarised beam. The right hand panel represents the sum of the two spin channels. Four features are clearly visible in the spectra and are labelled B_1, B_2, S_1 and S_2. B_1 is a result of transitions into the minority-spin 3d bands just above the Fermi level. Since the corresponding majority-spin part of these bands lies almost completely below the Fermi level it is not available for IPES transitions. A spin splitting of the features B_2 and S_1 is clearly resolved for all angles where the intensity of these features is non-vanishing. B_2 represents a bulk sp-like band and S_1 is a Schockley surface state occurring in a bulk band gap (Altmann et al., 1985).

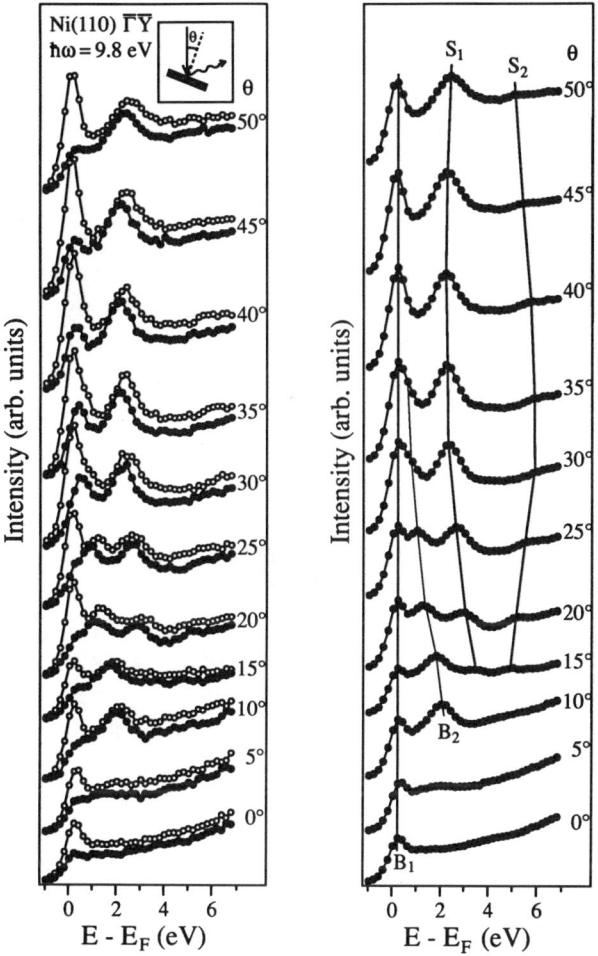

Figure 5. ARSPIPES measurements of Ni(110) in the $\overline{\Gamma Y}$ azimuth, recorded at 300 K. The empty circles represent minority-spin states and the filled circles represent majority-spin states. The left-hand panel shows spin-resolved data and the right-hand panel shows spin-averaged data. The lines connecting the data points are included as a guide to the eye.

The peaks in the minority-spin spectrum consistently occur at a higher energy than the corresponding majority-spin peaks, as would be expected from band structure calculations

(Eckardt and Fritsche, 1987; Wang and Callaway, 1977; Weling and Callaway, 1982). This is also in agreement with the results of Donath (1989) and Donath et al. (1990) for the $\overline{\Gamma X}$ azimuth of Ni(110).

In order to estimate the magnitude of the spin-splittings, a function was fitted to each spectrum. The peaks were represented by Gaussians and the background by a linear function multiplied by an error function (integrated Gaussian) to account for the step representing the onset of available unoccupied states at the Fermi level. This procedure yielded spin splittings of the surface state of about 0.2 eV for the Schockley surface state and about 0.3 eV for the sp-like bulk band.

The peaks in the two spin channels normally occur not only at a different energy position but as a consequence also at a different momentum. If the dispersion is steep the spin splitting in the recorded spectra does not accurately represent the exchange splitting. It should be kept in mind that the experimental splitting will also be affected by any, although generally much smaller, exchange splitting of the initial bands (Donath et al., 1990). At 40° electron incidence angle, the Schockley surface state S_1 appears close to the $\overline{Y}$ point in the surface Brillouin zone At this point the dispersion is flat and the measured splitting of (0.21 ± 0.05) eV can be interpreted as corresponding to the exchange splitting.

The peak positions for the feature B_2, are readily determined only for the lower angles of incidence. Due to the finite experimental resolution, for higher angles in the minority spin channel it is masked by the $3d$ peak (B_1) and is not resolvable. Although no definite conclusions can be drawn about the exchange splitting of this feature for angles > 30°, the general trend for both B_2 and S_1 appears to be that the spin-splitting increases with decreasing energy towards the Fermi level. In agreement with Donath et al (1990) the linewidth of the spin-down (minority-spin) states measured in this work is slightly larger than that for spin-up states. This is expected from the high density of spin-down states at the Fermi energy, E_F (Donath et al., 1990). The detailed nature of the exchange splitting is still to be determined.

The image potential induced state S_1 is detectable as a weak step-like structure in the spin-averaged data in Figure 5. A spin-splitting of this feature may be suggested from the spin-resolved spectra at, e.g. 15° and 35°. The statistics and the resolution in the spectra discussed here are, however, not good enough to draw any conclusions about the splitting. No effort has been made in this work to improve the statistics around this feature. Future studies with better resolution and statistics are planned to investigate any spin-splitting of the image potential feature.

REFERENCES

Altmann, W., Donath, M., Dose, V., and Goldmann, A., 1985, Dispersion of empty surface states on Ni(110), *Sol. State Commun.* 53:209.

Chelikowsky, J.R., Cohen, M.L., 1974, Electronic structure of GaAs, *Phys. Rev. Lett.* 32:674.

Donath, M., 1989, Spin-resolved inverse photoemission of ferromagnetic solids, *Appl. Phys. A* 49:351.

Donath, M., Dose, V., Ertl, K. and Kolac, U., 1990, Polarisation effects in inverse-photoemission spectra from Ni(110), *Phys Rev B* 41:5509.

Donath, M., 1994, Spin-dependent electronic structure at magnetic surfaces: the low-Miller-index surfaces of nickel, *Surf. Sci. Reports* 20:251.

Dose, V., 1985, Momentum-resolved inverse photoemission, *Surf. Sci. Reports* 5:337-378.

Drouhin, H.-J., Hermann, C., and Lampel, G., 1985, Photoemission from activated gallium arsenide. II. Spin polarization versus kinetic energy analysis, *Phys. Rev. B* 31:3872.

Eckardt, H., and Fritsche, L., 1987, The effect of electron correlations and finite temperatures on the electronic structure of ferromagnetic nickel, *J. Phys. F* 17:925.

Foxon, C.T., Harvey, J.A., and Joyce, B.A., 1973, The evaporation of GaAs under Equilibrium and Non-equilibrium conditions using a modulated beam technique, *J. Phys. Chem. Solids* 34:1693.

Goldstein, B., Szostak, D.J., Ban, V.S., Langmuir evaporation from the (100), (111A), and (111B) faces of GaAs, 1976, *Surf. Sci.* 57:733.

Grentz, W., Tschudy, M., Reihl, B., and Kaindl, G., 1990, Angle-resolved inverse photoemission spectroscopy with longitudinally spin-polarized electrons, *Rev. Sci. Instrum.* 61:2528.

Harting, E. and Read, F.H., 1976, "Electrostatic Lenses," Elsevier, Amsterdam.

Holland, M.G., 1962, Thermal conductivity of several optical Maser Materials, *J. Appl. Phys.* 33:2910.

James, L.W. and Moll, J.L., 1969, Transport properties of GaAs obtained from photoemission measurements, *Phys. Rev.* 183:740.

Pierce, D.T. and Meier, F., 1976, Photoemission of spin-polarized electrons from GaAs, *Phys. Rev. B* 13:5484.

Pierce, D.T., Celotta, R. J., Wang, G.C., Unertl, W.N., Galejs, A., Kuyatt, C.E., and Mielczarek, S.R., 1980, GaAs spin polarized electron source, *Rev. Sci. Instrum.* 51:478.

Purcell, E.M., 1938, The focussing of charged particles by a spherical condenser, *Phys. Rev. B* 54:818.

Röntgen, W.C., 1898, Röntgen rays, *Annal. Phys. Chem.* 64:18.

Schedin, B.F., 1992, "Angle-resolved Spin-polarised Inverse Photoemission Spectroscopy Studies of Ni(110)," PhD Thesis, Liverpool University.

Spicer, W.E., 1958, Photoemissive, Photoconductive, and optical adsorption studies of alkali-antimony compounds, *Phys. Rev.* 112:114.

Spicer, W.E., 1977, Negative affinity 3-5 photocathodes: Their physics and technology, *Appl. Phys.* 12:115.

Turbull, A.A., and Evans, G.B., 1968, Photoemission from GaAs-Cs-O, *Brit. J. Appl. Phys. (J. Phys. D) Ser. 2* 1:155.

Wang, C.S., and Callaway, J., 1977, Energy bands in ferromagnetic nickel, *Phys. Rev. B* 15:298.

Weling, F., and Callaway, J., 1982, Semiempirical description of energy bands in nickel, *Phys Rev. B* 26:710.

SPIN POLARIZED ELECTRON DETECTORS FOR SURFACE MAGNETISM

M. Hardiman, I. R. M. Wardell, M. S. Bhella, M. Whitehouse-Yeo, P. Gendrier,
C. J. Harland, G. Roussel, C.-K. Lo, S. Lis, D. König and J. Agernon

Physics and Astronomy Division, University of Sussex, Brighton BN1 9QH, U.K.

I INTRODUCTION

The study of magnetic processes at surfaces and in ultra thin layers has seen significant advances during the past ten years as a result of a wide variety of experiments involving spin polarized electrons. The majority of these experiments have measured the polarization of electrons emitted from a magnetic sample, and a variety of polarimeter types have been used. The use of polarized electron sources, however, is less widespread. A comprehensive review was given in reference 1.

The phenomenon of Spin Polarized Secondary Electron Emission (SPSEE) [2] occurs when an unpolarized primary electron beam is incident on a magnetized region of a (clean) ferromagnetic surface. This effect has been used to develop Scanning Electron Microscopy with Polarization Analysis (SEMPA) [3,4,5] which can image surface domain structure independently of surface topography. These techniques are sensitive at the monolayer level and, with appropriate polarimeters, can be used to infer the vector magnetization.

Spin and energy resolving detectors for Electron Spectroscopy with Polarization Analysis (ESPA) have been used for spin resolved UV photoemission [6], Spin Polarized Auger Electron Spectroscopy (SPAES) [7] and most recently Spin Polarized X-ray Photoemission Spectroscopy (SPXPS) [8,9]. These techniques, particularly the latter two, can probe element specific magnetization, at least indirectly, and are hence of interest for systems containing several magnetic species, whether at an interface or in an alloy.

In this paper we briefly discuss two areas within surface magnetism for which spin polarized techniques can yield unique information. We then describe two polarimeters that we have developed and give some examples of their current capabilities.

II SURFACE MAGNETISM

Much of the current interest in magnetism and novel magnetic materials is directly related to effects associated with surfaces or interfaces [10,11]. Atoms at a surface are in sites of reduced coordination and different symmetry compared to the bulk. The surface layer of a bulk magnetic material can show differences in moment, surface transition temperature and magnetic anisotropy from the bulk values. The variations in these parameters seen in an Ultra Thin Film (UTF) on a non-magnetic substrate can be even more striking, particularly the switching of the anisotropy axis from in-plane to perpendicular and the accompanying changes in domain structure [12,13]. In such cases the roles of the structures of the interface and of the film, on both the atomic and the microscopic scales, have yet to be clearly determined.

Interfaces between two magnetic materials offer more challenges. Again, it is expected that the particular structural arrangements will determine both the exchange coupling and the surface anisotropy term [14]. In addition, there are now several reports of "induced ferromagnetism" in a UTF of a normally non-magnetic material deposited on a ferromagnetic substrate [6].

Perpendicular Anisotropy and SEMPA

A number of UTF systems have a magnetic phase diagram in which there is a ferromagnetic region, Ferro$_\perp$, where the easy axis is perpendicular to the film plane, as shown in Fig.1a The occurrence of this phase is due to a favourable surface anisotropy contribution [15] which, with increasing thickness, is eventually overcome by the usual shape anisotropy of a thin film. Fig.1b shows typical hysteresis loops at various points in the phase diagram [16] for fields oriented both parallel and perpendicular to the film plane. The existence of the transition from the perpendicular to the parallel state, Ferro$_{//}$, at $T_{SW}(t)$ immediately poses questions as to the manner in which the magnetization changes and whether any domain structure develops.

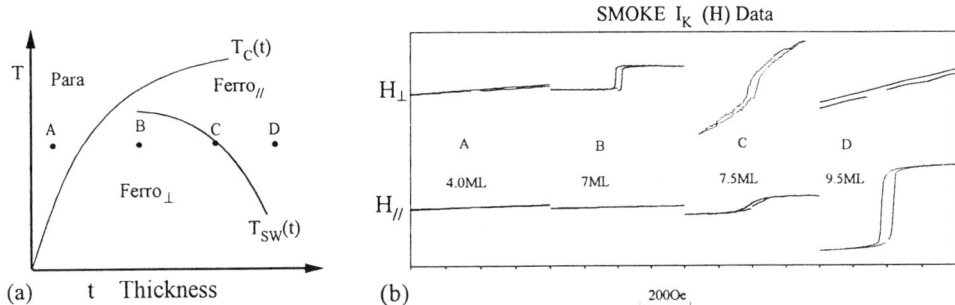

Fig.1 *(a) Characteristic magnetic phases shown by a number of 3-d UTF systems. (b) Representative hysteresis loops of the behaviour in the different phases. These were obtained from in-situ SMOKE on Fe:Si(111) grown and maintained at 100K. Loop (A): Paramagnetic state. Loop (B): 7.0ML deposit, field perpendicular - hysteresis only observed with field perpendicular to surface. Loop (C): 7.5ML deposit - tilted hysteresis loop for field parallel and perpendicular. Loop (D): field parallel- hysteresis only for field parallel.*

The easy axis might rotate smoothly over some temperature (or thickness) interval, leading to "canted" magnetization, or it could switch abruptly. Fritzsche et al [14] have recently concluded that in principle either case is possible, depending on the first and second order surface anisotropy constants. The 7.5 monolayer (ML) data shown in Fig.1, obtained using Surface Magneto Optical Effect (SMOKE) from a highly disordered Fe film grown at 100K on Si(111), appear to show a canted easy axis. However, the SMOKE experiment probes a relatively large area and it is not yet clear if the observed effect is due to a single canted axis or to sample inhomogeneity resulting in a distribution of perpendicular and in-plane easy axes from different regions.

Of the systems exhibiting the above phase diagram, Fe/Cu(100) has been studied the most. In their elegant work on this system using SPSEE, Pappas et al [12] appeared to find a complete loss of remanent magnetization around T_{SW}. However, by using SEMPA, Allenspach and Bischof [13] were able to show conclusively that this was due to the development of an up-down strip pattern in the Ferro$_\perp$ state, as had been predicted by Yafet and Gyorgy [17].

The ability of the SEMPA technique to probe the domain structure and the magnetic inhomogeneities present in monolayer deposits is expected to be crucial in understanding their behaviour.

Induced Ferromagnetism, Exchange Coupling at Interfaces and ESPA

The interface between a bulk 3-d ferromagnet and an otherwise non-magnetic d-series overlayer is currently generating much interest [6]. Given that all d-series ions carry full spin moments in insulators, the effects of d-band hybridization and, in the case of monolayer deposits, of reduced coordination, might be expected to result in ferromagnetism of the overlayer in at least some cases. This has indeed been observed in Pd [18] and Rh [6] using spin resolved UPS and where the hybridization is the dominant effect. The sole application to date of SPAES to study induced magnetism of an overlayer has been for the Ru/Fe(100) system by Landolt and co-workers [19]. This group made an earlier pioneering experiment using the element specificity of SPAES to measure the sign of the coupling across the Gd/Fe(100) interface [20]. The interface between a non-magnetic d-series element and a ferromagnetic Rare Earth (RE) is inherently more complicated but, intriguingly, an enhanced moment at low temperature has been reported recently in sputtered Gd:W multilayers [21].

III   POLARIMETERS FOR SURFACE MAGNETISM

We present here details of two polarimeters that we have developed for surface magnetism studies. The same design of medium voltage (25kV) Mott detector is used in each to measure the spin polarization. The energy and spin resolving "ESPA detector" is optimized for SPAES and incorporates a large commercial Concentric Hemispherical Analyser (CHA). Although the ESPA detector can produce energy selected SEMPA images, we have also developed an energy integrating "dedicated SEMPA detector" which maximizes the SPSEE signal for faster imaging. Extensive use was made of the SIMION [22] simulation package when designing both sets of electron optics.

The Retarding Field Mott Detector

Figure 2 shows the mechanical arrangement of the Retarding Field Mott Detector (RFMD), the optics of which are based on a design by Dunning and co-workers [23]. In our case the incident electron beam has kinetic energy 3keV at MT1 and between 18 and 28keV (depending on operating conditions) at the thick gold scattering target in MT3. Those electrons which are scattered in the x-z plane by ±120° pass through exit holes in MT3 and can be detected by the channeltrons.

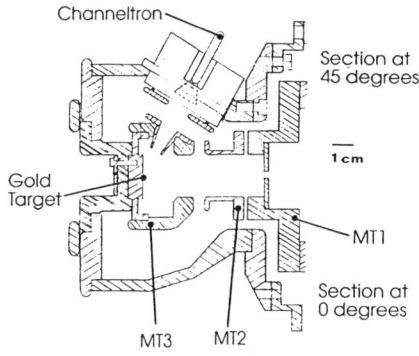

Fig.2.   *Cross section of the Retarding Field Mott Detector*

The y-component, P_y, of the incident beam polarization is related to the asymmetry, A_y, measured in such a Mott polarimeter by:

$$P_y \cdot S_{eff} = A_y = (N_u - N_d)/(N_u + N_d)$$

where $N_{u,d}$ are the number of electrons counted by up and down detectors in the x-z plane. Hence, the effective Sherman function, S_{eff}, [24] gives the asymmetry which would be measured for a fully polarized incident beam.

If the up and down detectors are at ground potential, only elastically scattered electrons are counted. In practice, a variable energy window, ΔE_w, between the channeltrons and ground is used to trade off between scattering asymmetry and count rate. As a progressively greater number of inelastically (and hence multiply) scattered electrons are let through, S_{eff} (and hence A_y) decreases. The Figure of Merit (FOM) for a Mott detector is defined as $S_{eff}^2 \times I_d/I_o$, where I_d is the detected current (in the channeltrons) and I_o is that incident on the scattering target. The characteristics of our detector are similar to those reported by Tang et al [23]. We deduce the S_{eff} to be ≈ 0.12 at 25keV scattering energy and with ΔE_w equal to 500eV. This value for the energy window corresponds to the maximum FOM we can presently achieve.

Although we have operated this design of RFMD up to 28keV scattering energy, all the data reported in Section IV were taken at 19keV and with $\Delta E_w = 500eV$. On the basis of previous measurements [25], this configuration corresponds to $S_{eff} \approx 0.10$.

The ESPA Detector

In the ESPA detector the RFMD is attached to a three-element transfer lens of 60mm bore which contains an x and y double deflection steering assembly. This transfer lens is then mounted directly onto the backplate of the large CHA (VSW model HA 150).

The analyser is operated at constant pass energy (typically 90eV) and the transfer lens accelerates the electrons from the selected pass energy to the required 3keV at element MT1. The potentials of all elements in the transfer and RFMD optics are referenced to the analyser local earth potential, V_{le}, where $-eV_{le}$ is the initial kinetic energy of the electrons to be detected. In this arrangement, apart from the first two elements of the input lens to the analyser, the relative potentials of all other elements in the electron optics remain fixed.

The transfer optics were optimized for a pass energy of 90eV which, with a 5mm analyser aperture, gives a nominal 1.5eV energy resolution appropriate to SPAES measurements. However, since the simulations indicate that the transmission of the transfer optics will not be degraded at lower pass energies, further optimization of the analyser pass energy and aperture size should permit the ESPA detector to be operated with an improved resolution.

The SEMPA Detector

The SEMPA detector consists of a 60mm bore four-element collection optics onto which the RFMD is mounted. The entire assembly connects to the sample chamber by a 150mm od flange. Figure 3 shows a cross section of the optics and sample and the electron trajectories obtained using SIMION.

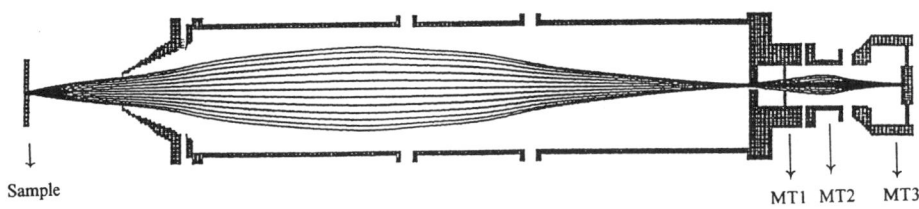

Fig.3. *Cross-section of the SPSEE/SEMPA detector optics and SIMION simulation. Emitted electron energy 4eV; emitted electron cone half angle 60°. The trajectories are calculated at 10° intervals. Mott target voltage 25kV.*

Since SPSEE usually shows an enhanced polarization below 10eV, the transmission of the collection optics was optimized for secondary electron energies in the range 0 to 20eV and designed to fall to zero at about 100eV. A moderately large extraction field between the sample and the first element leads to a very large collected solid angle for the lowest energy electrons.

In SEMPA the raw measured asymmetry, $A_y(x,y)$, is used as the contrast signal. Therefore, any position dependent instrumental term, $A_{yinst}(x,y)$, will be superimposed on the image. Such a term is inherent in the technique since, as the primary beam is scanned across the sample, a corresponding movement of the incident beam on the scattering target in general must occur. This leads to a difference in the scattering angles subtended by the detectors and hence to a significant $A_{yinst}(x,y)$. Considerable problems due to this have been encountered in some SEMPA instruments [3,26], particularly for large scanned areas. In our optics we have attempted to minimize this effect by arranging that lateral source movement produces both a shift and a tilt in the beam at the scattering target, such that the tilt induced change in asymmetry compensates that caused by the shift.

Experimental Configurations and Operating Modes

The two detectors are currently mounted on separate UHV chambers, each of which is equipped with a low resolution ($\approx$0.2mm) 5keV electron gun and sample cleaning facilities. Experimental control and all data acquisition, including imaging, is carried out by 386 PC computers running common C-language software.

In both cases the measured parameter is the total asymmetry, A_y, including any instrumental terms. For the ESPA detector

$$A_y(E,x,y,\underline{H}) = S_{eff} \cdot P_y(E,x,y,\underline{H}) + A_{yinst}(E,x,y)$$

where P_y is the polarization of electrons with kinetic energy E emitted from position (x,y) on a sample which has field $\underline{H}$ applied to it.

In the "Polarized Spectroscopy" mode, the primary beam coordinates (x,y) are constant and any energy dependent instrumental asymmetry, $A_{yinst}(E)$, is eliminated by taking

$$P_y(E) = \{A_y^+(E) - A_y^-(E)\}/2.S_{eff}$$

where the $A_y^{\pm}(E)$ are the asymmetries measured with the sample equivalently magnetized in the positive and negative y-directions.

By scanning the primary beam position, SEM images or linescans can be produced from either detector. In a "Polarization Linescan", A_{yinst} is eliminated and

$$P_y(x) = \{A_y^+(x) - A_y^-(x)\}/2.S_{eff}$$

whereas in an "Asymmetry Linescan" and in a SEMPA image the raw asymmetry is displayed. Hysteresis loops can be obtained in either apparatus by measuring $A_y(\underline{H})$ at fixed E and (x,y). Since A_{yinst} is independent of $\underline{H}$, these loops are correctly termed "P-H" or "P-I" loops.

IV RESULTS

ESPA Detector

We have characterized the imaging, linescan and spectral capabilities of our ESPA system using a $Fe_{81}B_{13.5}Si_{3.5}C_2$ metglass ribbon sample [27]. This is arranged as a loop, the ends of the ribbon being clamped together. The sample can be magnetized parallel to the detection axis of the RFMD by passing a current through a small coil wound on the ribbon, 0.1A being sufficent to obtain saturation. All data reported here were taken at a pressure of 10^{-9}mbar after cleaning the sample by Ar bombardment.

Figure 4 shows various energy selected SEMPA images, asymmetry and deduced polarization linescans obtained with 45eV electrons from the sample in different states of remanent and saturated (±0.2A) magnetization. Also shown is a series of P-I loops taken at various positions on the linescan. This series demonstrates that whilst the sample is always magnetically saturated by less than 0.05A current, the polarization varies from point to point.

Asymmetry scans along the same line for the positive and negative remanent magnetic states are shown in the lower left panel of Fig.4. The mean of these two signals is the position sensitive instrumental asymmetry, $A_{yinst}(x,y)$, which varies between 38.2% and 40.5% across the 5.4mm scan. The large offset from zero is attributed to a small mechanical misalignment which will be compensated for by re-commissioning the beam steering plates incorporated into the transfer optics. However, the relatively small variation in $A_{yinst}(x.y)$ across such a large scan is particularly pleasing.

The polarization linescan, deduced from the difference between the two remanent asymmetry scans, is predominantly independent of instrumental effects. The sample polarization varies between 0% and 22% along the scan, confirming the magnetic inhomogeneity suggested by the P-I loops. As expected, there is a high degree of correspondence between the measured remanent asymmetry-differences and the P-I loop amplitudes measured at the same points, as shown in Table 1.

If we assume, as is usual, that the local polarization, $P_y(x,y)$, is proportional to the local magnetization, then an uncorrected SEMPA image corresponds to $M_y(x,y)$ superimposed on any prevailing $A_{yinst}(x,y)$. The 64x64pixel SEMPA images presented here are of the raw asymmetry and hence still contain any such instrumental contribution. They have been processed such that black (M_y^-) to white (M_y^+) corresponds to a change of 4% in A_y or (40% in P_y). The scanned area is 5.4x6.1mm^2 and the positions of the linescan and the selected P-I loop points are indicated on the remanent images.

The SEMPA images in Fig.4 show the expected change in general brightness level corresponding to the magnetization being reversed from negative (Figs.4a and 4b) to positive (Figs.4c and 4d). As is expected from the square shape of the P-I loops the remanent negative (Fig.4a) and the saturated negative (Fig.4b) images are very similar to each other as are the remanent positive (Fig.4c) and saturated positive (Fig.4d) images. It is worthy of note that the scanned area here is large by usual SEMPA standards and that the overall instrumental asymmetry variations are relatively weak compared to the variations due to the sample. These SEMPA images illustrate clearly that the magnetization is not uniform across the sample surface, even in the saturated state, as might be deduced from a single point P-I loop.

The spatial resolution of a polarization imaging system can defined as the distance separating 20% and 80% of the polarization change between a peak and a trough in a polarization linescan. In our case the value of 0.15mm we obtain is limited by the (large) spot size of the primary beam.

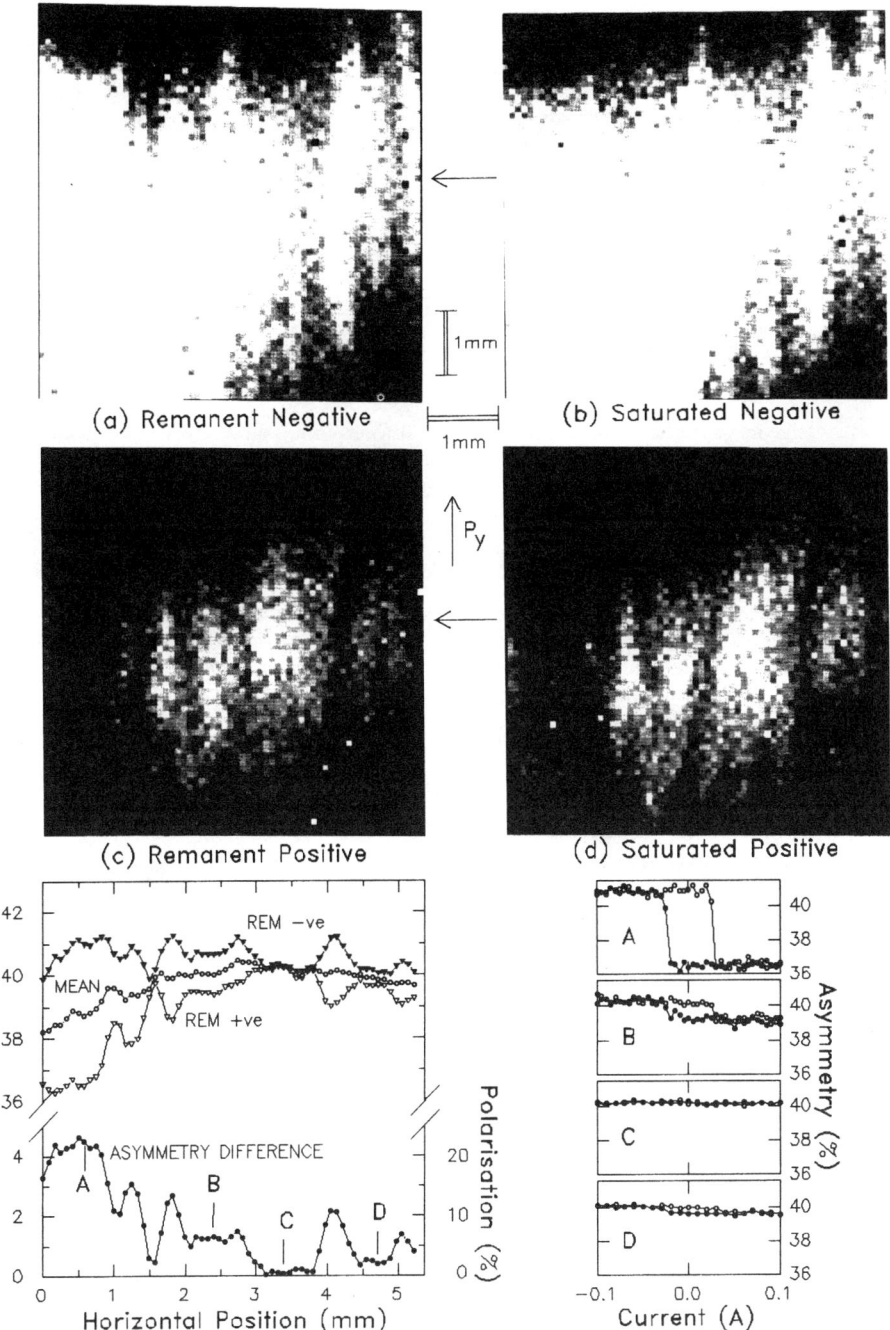

Fig. 4. *Energy selected SEMPA images obtained at 45eV from a region of $Fe_{81}B_{13.5}Si_{3.5}C_2$ metglass in the remanent and saturated states showing complimentary contrast of negative and positive magnetization. Asymmetry linescans, deduced polarization linescans and P–I loops measured along the linescan show sample inhomogeneity.*

The effect of stray magnetic fields due to the sample on the emitted low energy electrons is an important consideration in these experiments. However, to within the resolution quoted, we have yet to find any discernable image shift on changing the sample magnetization. We also note that the saturated regions of the P-I loops are horizontal, implying that the polarization measurements are not influenced by stray fields. This is the case even in a region such as point A in Fig.4 which is close to the edge of the sample and where the measured stray fields are particularly high.

The reason for the large magnetic inhomogeneity of this sample is not clearly understood; it may be inherent in the sample or possibly related to the Ar bombardment cleaning technique to which the sample has been subjected. This may have caused segregation or selective sputtering. To check this and also to demonstrate the potential applicability of the ESPA detector to correlate magnetic and chemical information, we measured the Auger spectrum at the points A to D shortly after the corresponding P-I loops were obtained. The signal-to-background ratios of Auger peaks measured at the various points are collated in Table 1, together with the corresponding asymmetry differences and P-I loop amplitudes.

Unfortunately no specific conclusions can be drawn from these particular measurements, except to note that the carbon intensity seems anomalously large at the point yielding the largest changes in magnetization. Otherwise there are only small Auger intensity differences between regions exhibiting large and small magnetization changes.

Table 1. *Asymmetry and Auger measurements at the points A, B, C and D indicated in Fig.4. The remanent asymmetry difference from the linescan data of Fig.4 is compared with the saturation asymmetry difference from the P-I loop data. Also compared are the signal-to-background ratios of the Fe $M_{23}M_{45}M_{45}$ (45eV), Fe LMM (645eV), B (172eV) and C (263eV) Auger peaks.*

Position	A	B	C	D
Asymmetry Difference %				
Remanent (Linescan)	4.5	1.3	0.1	0.4
Saturated (P-I Loop)	4.1	1.2	0.0	0.4
Auger Signal/Background %				
Iron (45eV)	20	21	20	21
Iron (645eV)	70	58	57	53
Boron (172eV)	6	7	9	10
Carbon (263eV)	13	9	7	6

The ESPA detector has been developed principally for SPAES measurements and it should be noted that such experiments are significantly more demanding in terms of stability and background count rate than those already discussed. Figure 5 shows the $P_y(E)$ polarization spectrum associated with the Fe $M_{23}M_{45}M_{45}$ 45eV Auger peak, as well as the usual N(E) spectrum. The data are from a point which is 0.7mm below the point A shown in Fig.4, and which exhibits a P-I loop very similar to that for point A. For reference the data of Landolt [28] on a similar metglass sample ($Fe_{83}B_{17}$) are shown. Our data show a 3% SPAES structure superimposed on the expected 15% secondary electron polarization. It is this relatively large background signal which is used to produce the linescan and image data discussed above.

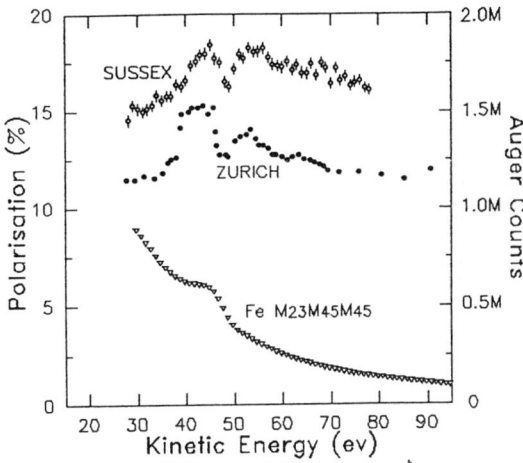

Fig.5. *Spin Polarized Auger Electron Spectrum from the spectral range containing an Iron $M_{23}M_{45}M_{45}$ Auger peak. The spectrum was recorded from a point on the metglass sample shown in Fig.4 0.7mm below point A which exhibits a similar amplitude P-I loop characteristic. The Auger spectrum is also shown. Error bars correspond to the expected statistical noise based on the square root of the counts in each of the Mott channels. For reference purposes the data of Landolt, ETH Zurich, are included.*

The SPAES data was accumulated over a 5.5hour period at a base pressure of 10^{-9}mbar. The sample was cleaned by Ar bombardment at one hour intervals since contamination build-up during this period attenuated the P-I loop amplitude by approximately 10%. The sharply decreasing secondary electron background in the usual N(E) spectrum leads to a decreasing signal-to-noise ratio (SNR) with increasing energy, if a constant dwell time is used at each energy. To obtain a uniform SNR across the polarization spectrum we vary the dwell time at each energy by always acquiring the same (preset) number of counts in one channel of the RFMD. The effect of any temporal variations in $A_{yinst}(E)$ are minimized by reversing the magnetization state of the sample at 75 second intervals.

The SPAES data of Fig.5 represent 1.3×10^7 total counts per energy interval and the error bars shown of $\delta P = \pm 0.3\%$ correspond to a $\pm \sqrt{N}$ variation in each acquisition channel. Generally the resultant SNR in the polarization spectrum is consistent with this expected statistical limit although there is some evidence for excess noise, particularly at higher energies. In order to obtain good SPAES from this magnetically inhomogeneous sample it has been found important to use the various imaging, linescan, Auger and P-I loop capabilities of the ESPA detector.

SEMPA Detector

We have characterized the imaging and linescan capabilities of the dedicated SEMPA detector using a $Fe_{97}Si_3$ sample known to have very large domains [29]. All data presented here were obtained at 10^{-9}mbar using a 5keV electron beam at 3nA sample current. With these primary beam conditions and the RFMD set up as described in Section III, count rates of $10^5 s^{-1}$ were obtained in each channel. The long axis of the domains was aligned with the detection axis of the RFMD. This corresponds to the vertical axis of the images in Fig.6 which shows a selection of the results that have been obtained.

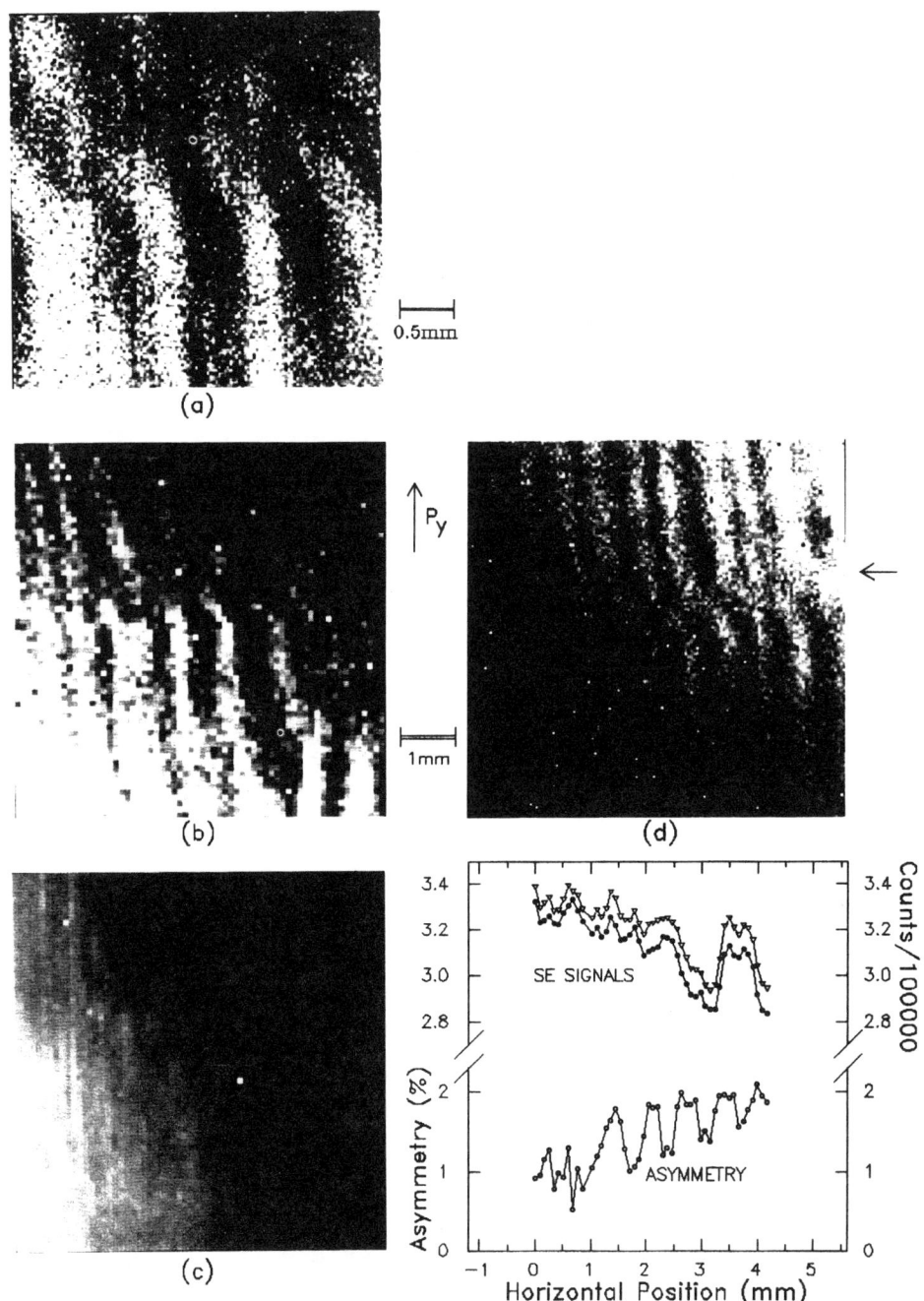

Fig.6. *SEMPA asymmetry and SEM images and linescans from an Iron–Silicon ($Fe_{97}Si_3$) domain sample. The data were recorded with a 3nA, 5kV electron beam and show strong magnetic contrast in the asymmetry signal and no significant magnetic contrast in the SEM signal.*

Figure 6a is a raw asymmetry SEMPA image from a 3.4x3.25mm^2 area of the sample. This 128x128pixel image was generated with a dwell time of 0.5s/pixel. Figure 6b is a 64x64pixel image taken at 1.8s/pixel centered at the same point on the sample but at half the magnification. Both images indicate that the SEMPA detector has adequate sensitivity to produce clear domain structure with currents of the order 1nA.

Figure 6c is an SEM image using one channel of the RFMD and taken over the the same area as in Fig.6b. This image has been considerably contrast enhanced in order to show up detail which is mainly topographical. There is very little contrast attributable to the magnetic structure. A 128x128pixel SEMPA image from an adjacent 6.8x6.5mm^2 area is shown in Fig.6d. The final panel of Fig.6 shows the two secondary electron linescans obtained via the different RFMD channels taken along the line shown in Fig. 6d. The resultant asymmetry linescan also is shown in the final panel of Fig.6. Again the secondary electron signal shows little correlation with the clear domain contrast shown by the asymmetry signal. The asymmetry variation between domains is about 0.8% which corresponds to relatively small polarization changes of ±4%. It is not known at present if this unexpectedly low measured polarization is sample or instrument related.

These SEMPA images and linescans demonstrate that $A_{yinst.}(x,y)$ is minimal for this detector in relation to relatively small values of sample polarization, even over large scanned areas. This result is notable in the light of the problems with position dependent asymmetry encountered by the NIST group [3]. Since the effects of instrumental asymmetry will decrease with increasing magnification, we are confident in the ability of this detector to resolve fine scale domain structure. At the moment, the spatial resolution of the system is limited by the large spot size of the electron gun used. The asymmetry linescan of Fig.6 indicates the present system resolution to be ≈ 0.2mm.

IV CONCLUSIONS

We have demonstrated the performance of an energy and spin resolving ESPA detector by observing the SPAES structure associated with the 45eV Fe $M_{23}M_{45}M_{45}$ transition. The ability of this instrument to provide energy selected polarization linescans, images and hysteresis loops has also been shown. In addition to the polarized spectroscopy measurements of magnetic interfaces for which it was developed, the ESPA instrument may find future application to inhomogeneous magnetic materials in which it is desired to correlate the spatial variations of the magnetic and compositional information.

Large area magnetic domain images have been obtained with the dedicated SEMPA detector which shows minimal instrumental background and in which the spatial resolution is limited by the primary beam spot size. This detector will soon be combined with a high resolution and high brightness electron gun. The resulting instrument will be used to study domain structure and anisotropy in ultra-thin magnetic layers, particularly those growing in the intermediate layer plus island mode.

Acknowledgements

The Mott detectors and transfer optics described above were constructed with great skill by Keith Nie. Melvyn Moore and Dick White designed and built the associated electronics. Invaluable support to start this project was provided by Don Whitehead of VSW. We should like to thank Elaine Seddon and Murray Campbell for many useful conversations. The constant encouragement and assistance of John Venables is gratefully acknowledged. This work has been supported by SERC Grants GR/H71475 and GR/K32197.

REFERENCES

1. "Polarized Electrons in Surface Physics", Ed. R Feder, World Scientific (Singapore) 1985.
2. M Donath, Surf. Sci. 287/288, 722 (1993).
3. M R Scheinfein, J Unguris, M H Kelley, D T Pierce and R J Celotta, Rev. Sci. Instrum. 61, 2501 (1990).
4. H Matsuijama and K Koike, Rev. Sci. Instrum. 62, 970 (1991).
5. R Allenspach, J. Magn. Magn. Mat. 129, 160 (1994).
6. T Kachel, W Gudat, C Carbone, E Vescovo, S Blügel, U Alkemper and W Eberhardt, Phys. Rev. B46, 12888 (1992).
7. O Paul, S Toscano, W Hürsch and M Landolt, J. Magn. Magn. Mat. 84, L7 (1990).
8. R Jungblut, Ch Roth, F U Hillebrecht and E Kisker, Surf. Sci. 269/270, 615 (1992)
9. L E Klebanoff, D G van Campen and R J Pouliot, Phys. Rev. B49, 2047 (1994).
10. For a recent review, see: U Gradmann in Handbook of Magnetic Materials 7 pp 1-96, Ed. K H J Buschow, Elsevier, (1993).
11. L M Falicov et al, J. Mater. Res. 5, 1299 (1990).
12. D P Pappas, K-P Kamper and H Hopster, Phys. Rev. Lett. 64, 3179 (1990).
13. R Allenspach and A Bischof, Phys. Rev. Lett. 69, 3385 (1992).
14. H Fritzsche, J Kohlhepp, H J Elmers and U Gradmann, Phys. Rev. B49, 15665 (1994).
15. S Pick, J Dorantesdavila, G M Pastor and H Dreysse, Phys. Rev. B50, 993 (1994).
16. C K Lo, D.Phil. Thesis, University of Sussex (1994).
17. Y Yafet and E M Gyorgy, Phys. Rev. B38, 9145 (1988).
18. W. Weber, D A Wesner, D Hartmann and G Güntherodt, Phys. Rev. B46, 6199 (1992).
19. K Totland, P Fuchs, J C Grobli and M Landolt, Phys. Rev. Lett. 70, 2487 (1993).
20. M Taborelli, R Allenspach, G Boffa and M Landolt, Phys. Rev. Lett. 56, 2869 (1986).
21. A Heys, P E Donovan, A K Petford-Long and R Cywinski, J. Magn. Magn. Mat. 131, 265 (1994).
22. D A Dahl, J E Delmore and A D Appelhans, Rev. Sci. Instrum. 61, 607 (1990).
23. F-C Tang, X Zhang, F B Dunning and G K Walters, Rev. Sci. Instrum. 60, 3068 (1989).
24. D M Campbell, C Hermann, G Lampe and R Owens, J. Phys. E. 18, 664 (1985).
25. M S Bhella, D.Phil. Thesis, University of Sussex (1991).
26. R Allenspach, A Bischof, M Stampanoni, D Kerhman and D Pescia, Appl. Phys. Lett. 60, 1908 (1992).
27. Goodfellow Cambridge Ltd., Cambridge, UK.
28. M Landolt and D Mauri, Phys. Rev. Lett. 49, 1783 (1982).
29. Orb Electrical Steel Ltd., Newport, UK.

RESONANCE LINE RADIATION FROM SPIN POLARIZED SODIUM AND POTASSIUM ATOMS

Muhammad Afzal Chaudhry[1-3] and Hans Kleinpoppen[1]

[1]Atomic Physics Unit
University of Stirling
Stirling FK9 4LA, Scotland

[2]Department of Physics
Edinburgh University
James Clerk Maxwell Building
Mayfield Road, Edinburgh EH9 3J2

[3]*Permanent Address:* Centre for the Advanced Study of Physics
Government College
Lahore, Pakistan

INTRODUCTION

There is a growing interest in the study of spin effects in atomic collision processes as it can help to analyze the different interactions involved and also provide a stringent test for theoretical models. However, due to the rather complicated experimental techniques employed for measurements of the spin states of collision systems before and after a collision, only relatively simple cases have so far been investigated. For a more complete description of the research work devoted to such problems reviews of Moisewitch and Smith (1968), Bederson (1969), Blum and Kleinpoppen (1975, 1976, 1983), Chen and Gallagher (1978), Heddle and Gallagher (1989), Kessler (1991), Teubner and Weigold (1992) and research papers by Kleinpoppen (1971), Moores and Norcross (1972), Tripathi and Mathur (1973), Jitschin et al. (1984) and Anderson and Bartschat (1993) may be consulted.

This report briefly describes experimental investigations of resonance lines excited in collisions between unpolarized electrons and spin polarized sodium and potassium atoms. The results discussed include:
i) The Stokes polarization parameters of the first resonance lines of sodium and potassium atoms which can be related to Coulomb direct and exchange interactions during the impact process.
ii) The electron-photon coincidence technique which can be used to measure more accurately the Stokes polarization parameters for the resonance radiation, and
iii) to extend, in future, these measurements to the use of spin polarized electrons as

projectiles and thus be able to make the so called 'complete experiment' for the study of such collision processes.

Experimental techniques

Figure 1 shows schematically the apparatus used to measure the Stokes polarization parameters. Briefly, alkali metal is heated in an oven to produce an atomic beam which is collimated with the help of slits and is allowed to pass through a hexapole magnet which acts as a spin polarizer for the atoms. The hexapole magnet distinguishes between the electron spins of the atom which are radially oriented either towards the centre line, i.e. the axis of the hexapole magnet or in the opposite direction. The result is that atomic states with $m_s = +\frac{1}{2}$ are focussed by the hexapole magnet while atomic states with $m_s = -\frac{1}{2}$ are defocussed and lost from the beam. However, due to the hyperfine structure coupling of the ground state the spin polarization of the atomic beam diminishes from 100% at the exit of the hexapole magnet to about 25% when the beam enters into a weak magnetic field (Hils et al. 1981). Practically all polarized Na and K atoms have polarizations of approximateluy 21% after passing adiabatically through guiding magnetic fields of which the final one provides the quantization axis for the atoms at the interaction region. The atomic beam passes a Rabi magnet for the measurement of the atomic polarization at the variable positions of the Langmuir-Taylor detector.

The electron source consists of an electron gun which uses a directly heated tungsten wire as cathode. It is a multistage gun (Simpson and Kuyatt, 1963) by which electrons are extracted at higher energy to overcome the space charge effect and then decelerated to the required energy. Fig. 1 shows the various parts of the electron gun.

For the measurement of Stokes polarization parameters the resonance line radiation of the excited alkali atoms is collected from the interaction region and allowed to pass through a suitable interference filter and a polarization detection optics consisting of a linear polarizer and a quarter wave plate. The radiation is detected by cooled photomultipliers EMI D624 and EMI 9883QA for sodium and potassium radiation, respectively. If $I(\alpha°)$ is the intensity of the radiation after passing through the linear polarizer set at an angle α (measured from the z-axis towards the x-axis as in Fig.1), and I(RHS), I(LHC) are the intensities of he right and left-handed radiation then the Stokes polarization parameters P_1, P_2 and P_3 can be measured using the formulae given by Born and Wolf (1970):

$$P_1 = \frac{I(0°)-I(90°)}{I(0°)+I(90°)}$$
$$P_2 = \frac{I(45°)-I(135°)}{I(45°)+I(135°)}$$
$$P_3 = \frac{I(RHC)-I(LHC)}{I(RHC)+I(LHC)}$$

the polarization data (Bukhari et al. 1995) is corrected for the depolarization effects due to the finite solid angle of the detected light and the divergence of the electron eam.

Fig.2 shows, schematically, the apparatus used to find coincidences between the inelastically scattered electrons detected in a particular direction, having lost the threshold energy for the excitation of the D-lines and the resonance radiation passing through a polarization measurement optics as in Fig.1. The atomic beam source, the electron gun and the polarization optics are the same as in Fig.1. A 30° parallel plate electrostatic analyzer

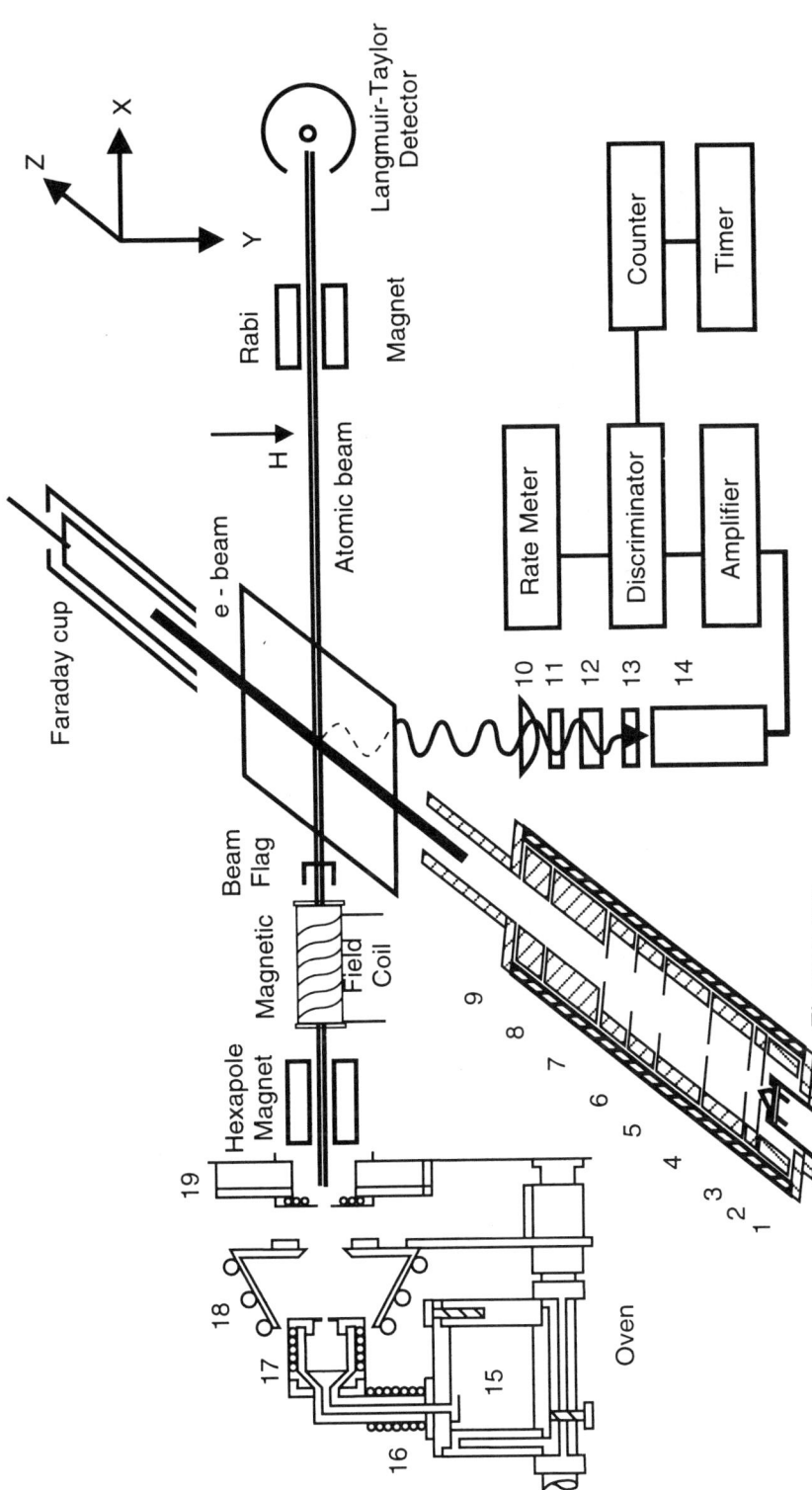

Figure 1. The experimental arrangement. Numbers 1 to 9 are the various parts of the electron gun. (10) plano-convex lens; (11) quarter-wave plate; (12) light filter; (13) linear polarizer; (14) photomultiplier tube. Numbers 15 to 19 are the parts of the atomic beam source (Bukhari *et al.*, 1994).

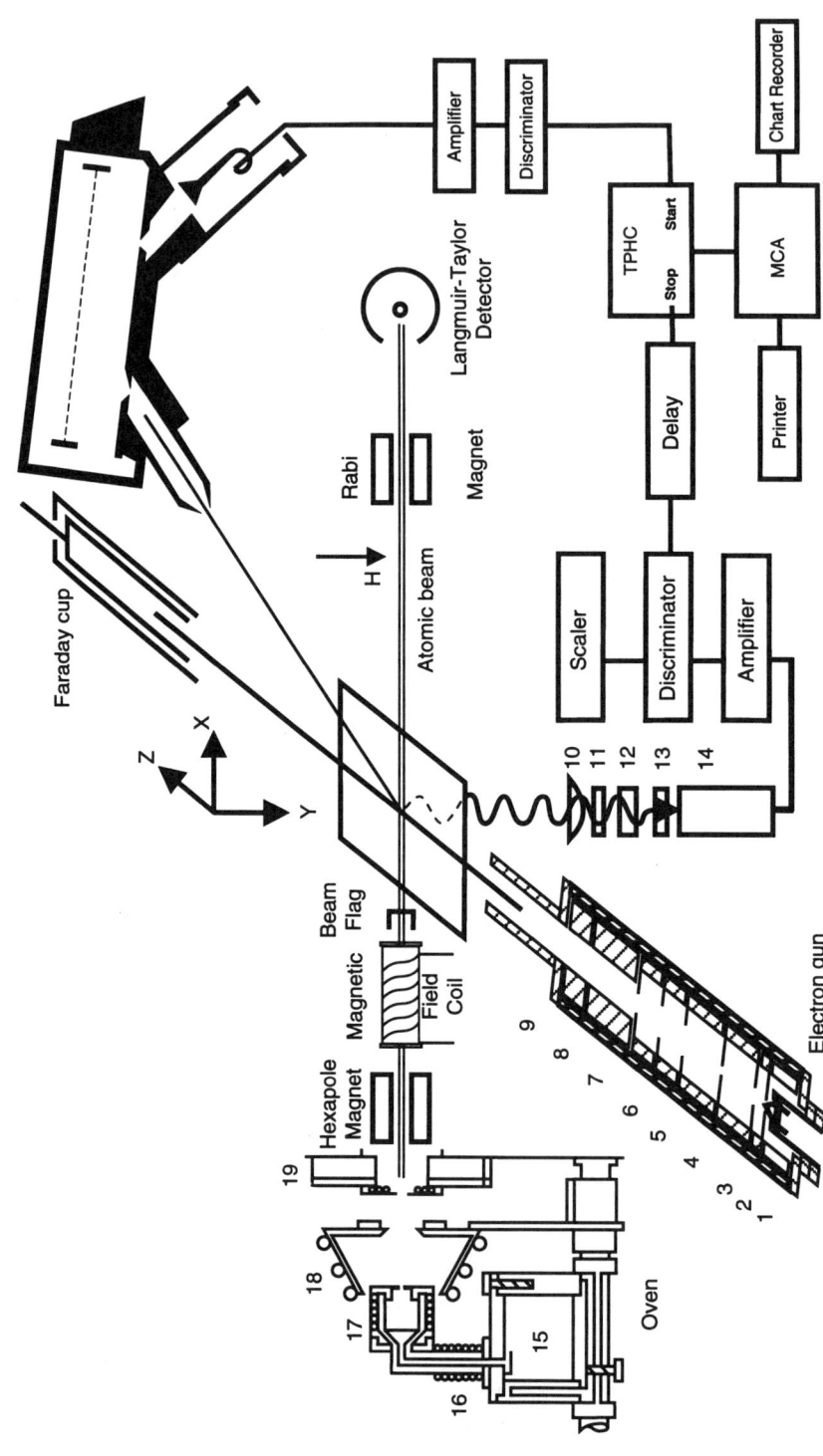

Figure 2. The experimental set up. Numbers 1 to 9 are the various parts of the electron gun, 10 to 14 are the components of polarization detection optics, and 15 to 19 are the parts of the atomic beam source. A 30° parallel plate electrostatic analyzer is applied. A hexapole magnet polarizes, a Rabi magnet analyzes, and a Langmuir–Taylor detector detects the atomic beam.

is used to detect, by a channeltron, the scattered electrons with an energy loss equal to the excitation potential of the D-lines of the alkali atoms. The photons of the resonance lines which have passed through the polarization optics and a light filter are detected by a photomultiplier. Coincidences are then found between the detected electrons and the photons using standard electronic equipment.

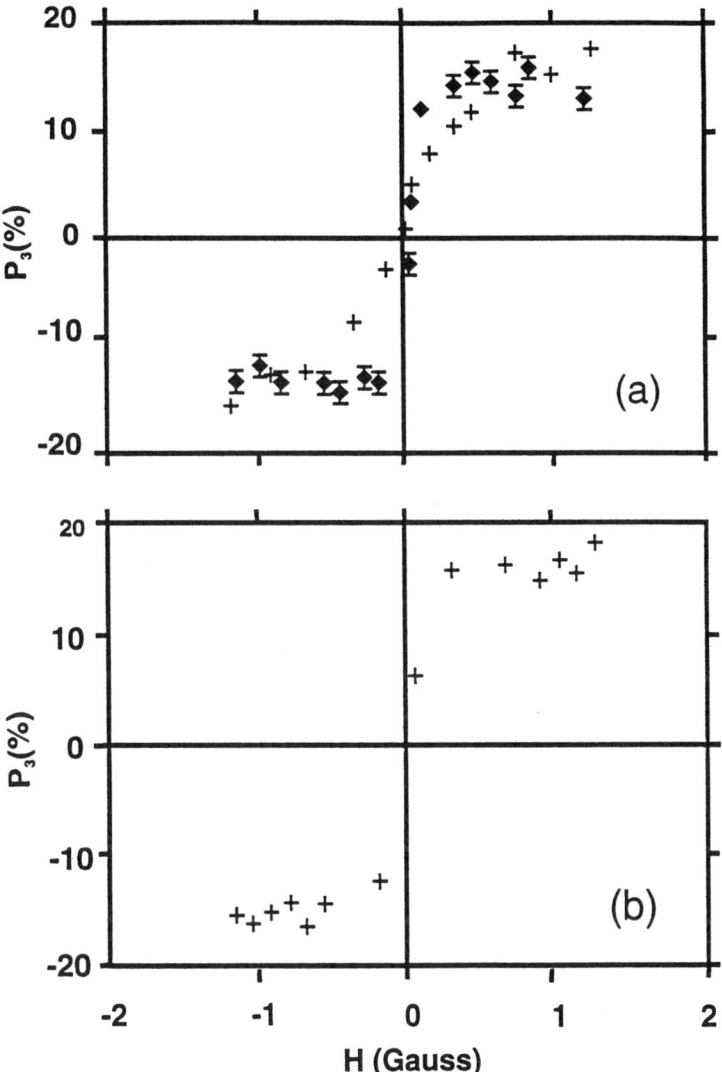

Figure 3. Graphs (a) and (b) are, respectively, the measurements (Bukhari et al.1994) of the polarization Stokes parameter P_3 for sodium resonance radiation ($3^2P_{3/2,1/2} - 3^2S_{1/2}$) and the potassium resonance radiation ($5^2P_{3/2,1/2} - 4^2S_{1/2}$) as a function of the magnetic field H in the interaction region. The symbol (+) and (♦) show, respectively, the measurements of Bukhari et al. (1994) and Osimitsch (1983).

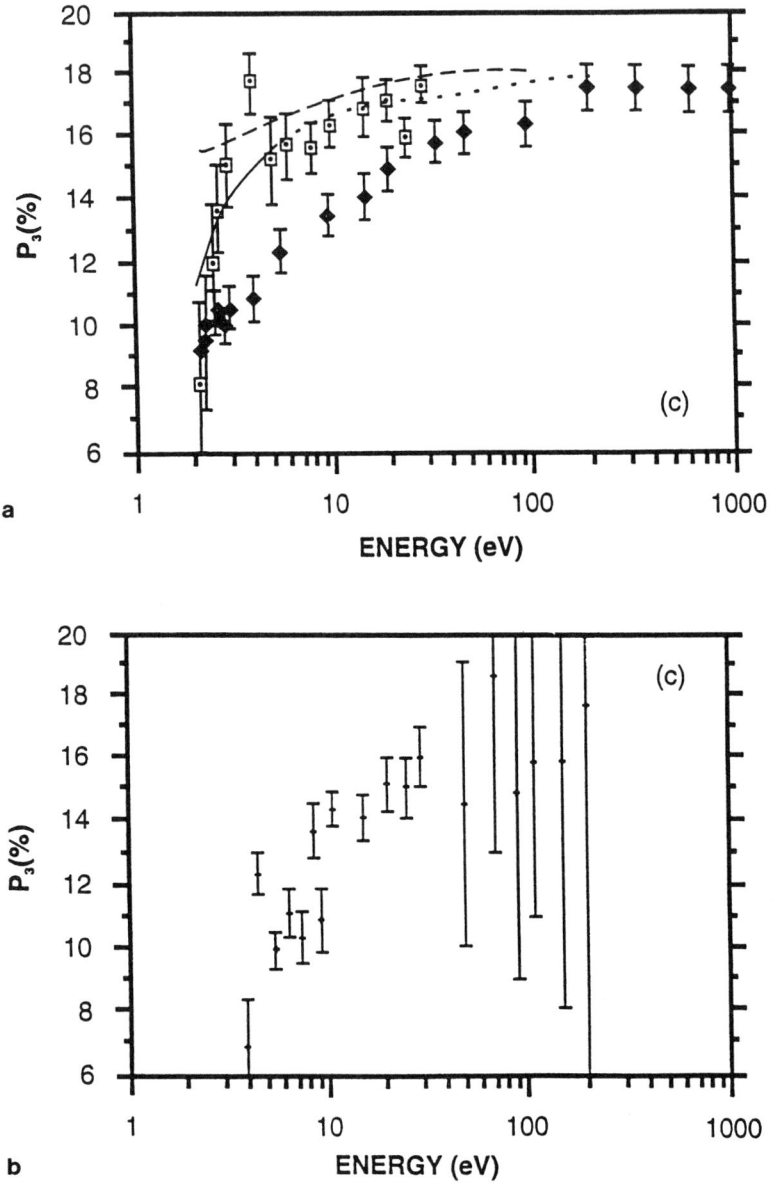

Figure 4. Graphs (a) and (b) are, respectively, the polarization Stokes parameter P_3 for the sodium resonance radiation ($3^2P_{3/2,1/2} - 3^2S_{1/2}$) and for potassium lines ($S^2P_{3/2,1/2} - 4^2S_{1/2}$) as a function of the incident electron energy. Experimental data (□) Bukhari et al. (1994) and (♦) Jitschin et al. (1984) while the theoretical calculations are (—) Moores and Norcross (1972), (- - -) Tripathi (1973) and (---) Kennedy et al. (1977).

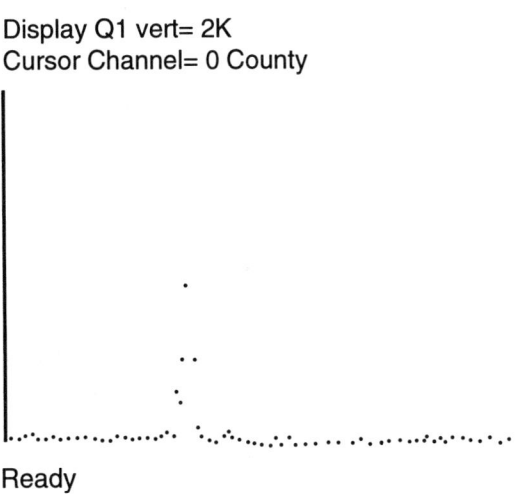

Figure 5. This graph shows electron-photon coincidence counts versus delay time from the excitation of the sodium resonance radiation. (See text).

Results and discussion

Figures 3(a) and 3(b), respectively, show measurements (Bukhari et al. 1995) of the polarization Stokes parameters P_3 for sodium resonance lines ($3^2P_{3/2,1/2}$ - $3^2S_{1/2}$) and potassium resonance lines ($5^2P_{3/2,1/2}$ - $4^2S_{1/2}$) as a function of the magnetic field (H) in the interaction region. These data show positive and negative values of P_3 which seem fairly symmetric and furthermore P_3 reaches saturation values at magnetic fields of approximately 0.5G and 0.2G for sodium and potassium, respectively. Accordingly the measurements (Bukhari et al. 1994) of the Stokes polarization parameters P_1, P_2 and P_3 were carried out in the presence of magnetic fields for 0.5G for sodium and 0.2G for potassium in a direction parallel to the photon emission.

The Stokes polarization parameters P_1, P_2 and P_3 for the sodium D-lines ($3^2P_{3/2,1/2}$-$3^2S_{1/2}$) and potassium resonance radiation ($5^2P_{3/2,1/2}$ - $4^2S_{1/2}$) have been measured (Bukhari et al. 1995) as a function of the incident electron energy.

Figures 4(a) and 4(b), respectively, show the measurements of Stokes polarization parameter P_3 for sodium and potassium. Fig.4(a) includes theoretical calculations of Moores and Norcross (1972), Tripathi (1973) and Kennedy et al. (1977) and experimental results of Osimitsch (1983) and Jitschin et al.(1984). The measurements of Bukhari et al. (1995) for sodium show better agreement with the theoretical data than the previous experimental results of Osimitsch (1983) and Jitschin et al. (1984). For potassium there are no other experimental or theoretical data available for comparison. The P_3 values for potassium, however, show a trend similar to sodium.

Figure 5 shows a coincidence signal between the inelastically scattered electrons which have lost energy equal to the excitation potential of sodium D-lines ($3^2P_{3/2,1/2}$ - $3^2S_{1/2}$) and photons of the resonance radiation which have been collected from the interaction region and have been allowed to pass through a light filter and a polarization optics and are finally detected by a photomultiplier. The number of coincidence counts is reasonably large and encouraging for making such measurements even with spin-polarized electrons as projectiles.

A suitable source of spin-polarized electrons for such measurements have been described by D M Campbell and E A Seddon in another section of this book. The spin-polarized electron source is ready to be used for such measurements with both spin-polarized electrons as projectiles and spin-polarized alkali and hydrogen atoms as targets. An appropriate geometry for the observation of the coincident electrons and photons with Stokes parameter analysis will result in extracting the excitation amplitudes f_m ($m_t = 0, \pm 1$) for Coulomb direct interactions and (g_m ($m = 0, \pm 1$) for Coulomb exchange interactions. Such experiment would represent the "ultimate" analysis of an excitation process.

Conclusion:

The measurements of polarization Stokes parameters for the sodium ($3^2P_{3/2,1/2}$ - $3^2S_{1/2}$) and potassium ($5^2P_{2/3,1/2}$ - $4^2S_{1/2}$) resonance lines, employing partially spin-polarized atoms as targets and unpolarized electrons as projectiles, provide valuable information about the interaction channels of the excitation process. For example, the increase of approximately 10% in the values of P_3 (Figs.4(a) and 4(b) for sodium and potassium atoms provide some evidence (Jitschin et al. (1984)) that the exchange interaction contributions in the excitation process are relatively smaller at higher energies while these are comparable to the Coulomb direct interaction channels at lower incident electron energies. Future investigations of resonance radiation using spin polarized electrons as projectiles and polarized alkali atoms as targets are expected to yield the much needed data for formulating adequate theoretical models for electron impact excitation of atoms.

References

Anderson, N. and Bartschat, K. 1984, *Comments At.Mol.Phys,* **29**:157.
Bederson, B., 1969, *Comments At.Mol.Phys.* **1, 71** and **2**:65.
Blum, K. and Kleinpoppen, H. 1975, 1976, *Int.J.Quantum Chem.* **S9, 415** and **510**:231.
Blum, K. and Kleinpoppen, H. 1983, *Adv.At.Mol.Phys.* **19**:188.
Born, M. and Wolf, E. 1970, *Principles of Optics* (Pergamon Press, New York).
Bukhari, M.A.H., Chaudhry M.A., Beyer, H-J., Duncan, A.J., Campbell, D.M. and Kleinpoppen, H., 1995, to be published.
Chaudhry, M.A., Bukhari, M.A.H., Beyer, H-J., Duncan, A.J., Campbell, D.M. and Kleinpoppen, H., *Proc. of XIV Int. Conf. on Atomic Physics, Colorado, August (1994),* 2P-11.
Chen, S.T. and Gallagher, A.C., 1978, *Phys.Rev.* **A17**:551.
Heddle, D. and Gallagher, J.W. 1989, *Rev.Mod.Phys.* **61**:221.
Hils, D., Jitschin, W. and Kleinpoppen, H. 1981, *Appl.Phys.*, **25**:39.
Jitschin, W., Osimitsch, S. Reihl, H., Kleinpoppen, H. and Lutz, H.O. 1984, *J.Phys.B. At.Mol.Phys.* **17**:1899.
Kennedy, J.V., Myerscough, V.P. and McDowell, M.R. 1977, *J.Phys.B.: At.Mol.Phys.* **10**:3759.
Kleinpoppen, H. 1971, *Phys.Rev.* **A3**:2015.
Moisewitch, B.L. and Smith, S.J. 1968, *Rev.Mod.Phys.* **140**:238.
Moores, D.L. and Norcross, D.W. 1972, *J.Phys.B.: At Mol.Phys.* **5**:1482.
Osimitsch, S. 1983, *Diplomarbeit and 1989 PhD Work,* University of Bielefeld, Germany.
Simpson, J.A. and Kuyatt, C.E. 1963, *J. of Res., National Bureau of Standards,* **67C**:279.
Teubner, P.J.O. and Weigold, E., eds. 1992, *Inst. of Phys. Conf. Ser. 122,* Bristol.
Tripathi, A.N. and Mathurt, K.J. 1973, *J.Phys.B: At.Mol.Phys.* **6**:1431.

STUDIES WITH POLARIZED POSITRONS

Gaetana Laricchia

Department of Physics and Astronomy
University College London
Gower Street
London WC1E 6BT

INTRODUCTION

Positrons (e^+), or anti-electrons, can be valuable probes of fundamental interactions between charged particles and atoms, molecules, liquids and solids[1-4]. Positronium (Ps) is the bound state of a e^+ and an electron (e^-) and precision measurements of its energy levels[5], decay rates[6] and modes[7] can provide rigorous tests of bound state QED calculations. Recently, the realization of quasi-monoenergetic beams of such exotic atoms[8] has enabled the controlled scattering of Ps from simple atomic targets[9]. In atomic collisions with positrons, experimental studies have advanced to the study of problems concerned with the near-threshold[10] and differential[11] behaviour of scattering processes as well as beginning to address more dilute targets such as atomic hydrogen[12] and the alkalis[13].

In addition to its intrinsic interest, all of this work is of relevance to astrophysics[14], the semiconductor industry[2-4], medical diagnostics[2], plasma physics[15] and the synthesis of atomic antimatter[16]. Spin-polarized positrons have been used eg to test weak interaction effects, discrete symmetries and derive information about the polarization of the medium in which they annihilate[17,18].

Several recent reviews are available on various experimental aspects of positron physics[19-21]. In this article, the basic elements and methods of positron polarimetry will be outlined and recent progress in this field indicated.

SOURCES OF POLARIZED POSITRONS

The fast positrons and electrons ($\beta^\pm$) emitted in the decay of a radioisotope possess a net helicity as a result of parity non-conservation in the weak interaction[22,23]. The β^+ (β^-), originating from such a process, are spin-polarized along the same (opposite) direction to their velocities ($\underline{v}$), thus exhibiting a positive (negative) helicity ($\pm v/c$) or right (left)- handedness, c being the speed of light. Since the degree of polarization is proportional to the speed of the emitted particles, at the relativistic energies characteristic in these decays, their polarization is almost unity.

At higher energies, polarized positrons and electrons may also result via the emission of synchrotron radiation in a storage ring. At LEP a polarization of 40% is now routinely achieved at around 40GeV[24].

The interactions of slow positrons with matter are relatively insensitive to spin because of (i) the lack of the exchange interaction which, in the electron case, produces large spin-dependent effects and (ii) the repulsive static interaction with the atomic cores which suppresses the positron penetration into the atom. The positron, therefore, is partly screened from the nuclear charge and this results, in comparison to electrons, in smaller spin-orbit effects. This is illustrated in table 1 showing the differences between the spin-up and spin-down phaseshifts calculated[25] for 100eV electrons and positrons colliding elastically with Xe atoms. Also shown in table 1 are calculations[26] for e^+ scattering from mercury which illustrate the growing importance of spin-orbit effects as the increasing incident energy allows the positron to penetrate more deeply into the atom.

Table 1. Differences (in radians) between spin-up and spin-down phaseshifts for various partial waves, l, in the elastic scattering of

$e^\pm$ (100eV) - $_{54}$Xe Stauffer & McEachran[25]			e^+ - $_{80}$Hg Hasenburg[26]			
l	e^-	e^+	l	35eV	150eV	800eV
1	-10^{-1}	10^{-4}	1	10^{-5}	10^{-4}	10^{-3}
2	-10^{-2}	10^{-4}	2	10^{-5}	10^{-4}	10^{-2}
3	-10^{-2}	10^{-4}	3	10^{-5}	10^{-4}	10^{-3}
4	-10^{-4}	10^{-5}	4	10^{-6}	10^{-4}	10^{-3}
5	-10^{-5}	10^{-5}	5	10^{-7}	10^{-5}	10^{-3}

The weak spin dependence in e^+-matter encounters has certain advantages; for example it simplifies the theoretical descriptions of some of these events and, more importantly, it allows the 'moderation' of the fast β^+ to slow e^+ whilst retaining their spin-polarization. Over the past twenty years or so, considerable advances in the available fluxes of monoenergetic positrons have resulted from an increasing understanding of the interactions of β^+ with the materials (moderators) employed to slow them down[20,21]. These advances have resulted[27] in moderator efficiencies approaching 1% and enabled the study of the interactions of monoenergetic e^+ and Ps with atoms and molecules[1] as well as the explorations of surfaces and interfaces[2-4].

If implanted into a solid[20-21], β^+ thermalize (eg via core excitations, the production of plasmons, electron-hole pairs and phonons) within a time ($\sim 10^{-12}$s) which is short in comparison to their lifetime in the medium ($\sim 10^{-10}$s). Consequently, some thermal e^+ might diffuse to a surface of the moderator and, if the material has a negative positron work-function, be ejected into the vacuum. Such e^+, emitted with energies of a few eV, can then be easily confined, accelerated and focussed to form a beam. If non-magnetic materials are used as moderators, the spin polarization is largely (up to 70%) preserved independently of the atomic number of the moderator[28]. The arrangement, shown in figure 1, was designed to optimize the polarization of a e^+ beam[28]. It employed a low-Z material for the source backing in order to minimize β^+ backscattering; an absorber was inserted between the source and the moderator to remove/premoderate the very low/high energy component of the β^+ spectrum and

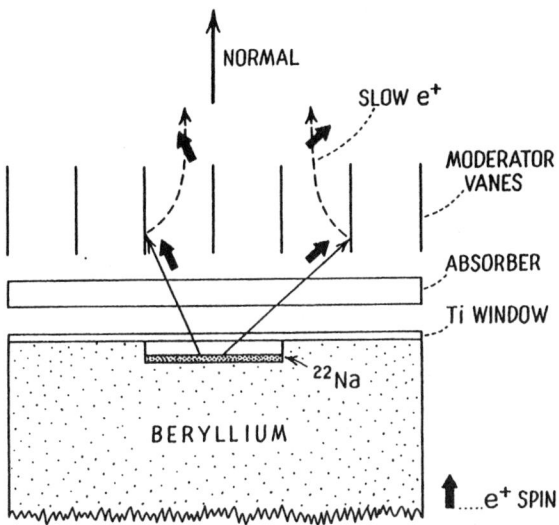

Figure 1. Source-absorber-moderator arrangement for the production of slow polarized positrons[28].

e$^+$ emitted at large angles from the moderator were removed geometrically. In this respect, recently developed thin foil-moderators[29], which operate in a transmission geometry, might be advantageous.

POSITRON POLARIMETRY

Notable exceptions to the spin insensitivity of positron-matter interactions are annihilation and positronium formation. As illustrated in the previous discussion on moderation, a fast positron in a medium will lose most of its energy before annihilating. At the low velocities typical of annihilation events, charge conjugation invariance requires the annihilation of a free e$^+$ with an e$^-$ of opposite spin to result primarily into an even number of photons and in the case of parallel spins into an odd number. The annihilation rate is roughly proportional to α^n where α is the fine structure constant and n the number of photons emitted. The single photon annihilation mode, however, is suppressed, in comparison to the three-photon emission, by the conservation of linear momentum which requires the presence of a recoiling third body. Hence, the annihilation of free e$^+$ in a medium proceeds primarily with e$^-$ of opposite spin via the emission of two photons. This is one of the principles employed in e$^+$ spin polarimetry. Another, based on Ps formation, will be discussed below.

In the ground-state, positronium may be formed in a para-state (p-Ps, $^1S_0$) where the constituent particles have opposite spins or in an ortho-state (o-Ps, $^3S_1$) where the latter has three substates with corresponding magnetic quantum numbers m=0,±1. The same selection rules discussed in the case of free e$^+$ annihilations also apply for Ps. The predominant unperturbed annihilation mode of o-Ps is via 3-γ rays and its mean ground-state lifetime is approximately 142ns, whilst p-Ps annihilates principally into 2-γ rays with a mean lifetime of ~0.125ns. The four ground-state eigenfunctions, Ψ(m), of Ps may be represented as follows:

$$\Psi_o(1) = \uparrow\Uparrow \qquad \Psi_o(0) = 1/\sqrt{2}\ (\uparrow\Downarrow + \downarrow\Uparrow)$$

$$\Psi_o(-1) = \downarrow\Downarrow \qquad \Psi_p(0) = 1/\sqrt{2}\ (\uparrow\Downarrow - \downarrow\Uparrow)$$

where $\uparrow(\downarrow)$ and $\Uparrow(\Downarrow)$ denote the up (down) spin of the e^- and e^+ respectively.

A static magnetic field, B, mixes the two m=0 states, whilst the m=±1 states remain unaffected; the perturbed eigenstates consisting of a linear combination of the two unperturbed m=0 states:-

$$\Psi'_o = (1+y^2)^{-\frac{1}{2}} (\Psi_o + y\Psi_p) \qquad \Psi'_p = (1+y^2)^{-\frac{1}{2}} (\Psi_p - y\Psi_o)$$

with $\qquad y = x\{1+(1+x^2)^{\frac{1}{2}}\}^{-1} \qquad$ and $\qquad x \propto B/W$

W being the hyperfine splitting between the two unperturbed states.

If polarized e^+ are used, the formation of Ps in either of the perturbed m=0 states depends on the orientation of the e^+ spin polarization with respect to the magnetic field, the singlet-like Ps being favoured if the e^+ are polarized parallel to the magnetic field[23]. This can be seen by considering Ps formed by a fully polarized beam of e^+ incident on unpolarized e^- placed in a B-field. The possible combinations of the e^- and e^+ spins may be expressed as:-

$$\Psi_o(-1) = \downarrow\Downarrow \qquad \Psi_1 = \uparrow\Downarrow \qquad (1)$$

if the e^+ spin polarization $\underline{P}$ is antiparallel to $\underline{B}$ and

$$\Psi_o(1) = \uparrow\Uparrow \qquad \Psi_2 = \downarrow\Uparrow \qquad (2)$$

otherwise. Note that Ψ_1 and Ψ_2 are not Ps eigenstates. Expanding them in terms of Ψ'_o and Ψ'_p, at high B-field strengths it is found that

$$\Psi_1 \propto \Psi'_o \qquad \text{and} \qquad \Psi_2 \propto \Psi'_p \qquad (3)$$

implying that the state corresponding to Ψ_1 is longer lived than that of Ψ_2. At B-field strengths ~0.5T, their average lifetimes are sufficiently different to enable the identification of these states by reversing either $\underline{B}$ or $\underline{P}$. Hence, if both $\underline{P}$ and $\underline{B}$ are known, information about the polarization of the electron involved in the annihilation may be deduced by measuring the asymmetry in the formation of triplet-like and singlet-like Ps states.

A number of experimental methods have been employed to study spin-dependent phenomena with positrons[17,18], for example the measurement of the Ps lifetime spectrum, the monitoring of the annihilation characteristics of free e^+ and/or Ps. Figure 2 represents the lifetime spectrum obtained with a positron polarimeter[30] based on Ps formation, in this case, from an evacuated MgO powder placed in a magnetic field of ±0.29T. The spectra (raw data) have been resolved into the unperturbed Ps component (corresponding to $^3S_1$, m=±1) and the two magnetically perturbed components. As expected from equations (1-3), the perturbed component obtained with $\underline{P}.\underline{B}<0$ has more counts at later times than the corresponding spectrum with reversed magnetic field direction.

The average momentum of an annihilating e^+-e^- pair may be extracted by measuring, for example, the Doppler shift of the emitted photons or the Angular Correlation of the Annihilation Radiation (ACAR), the latter effectively corresponding to the measurement of the deviation from collinearity, θ, of the γ-rays in the final state. Generally, if Ps formation may occur in the medium, these latter measurements may manifest (i) a narrow component (ie small θ) due to the decay of slow p-Ps, superimposed onto (ii) a broader component due to the annihilation of thermalized e^+ with (nonthermal) e^- and (iii) a background resulting from annihilations of energetic pairs and 3γ (and higher order) events. From (ii), information may be obtained about the momentum of the e^- with spin opposite to that of the e^+. Additionally, in the presence of a B-field strong enough that the mixed Ps states both annihilate essentially via the 2γ-mode, but sufficiently weak for their lifetimes to be appreciably different, the

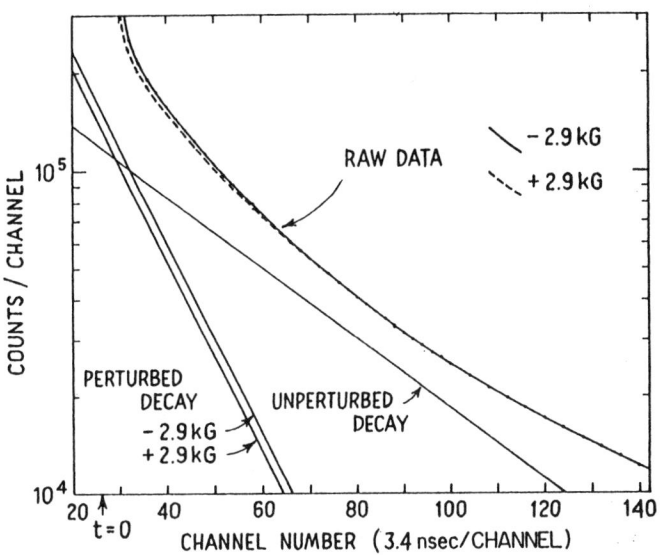

Figure 2. Raw data for Ps decay. Note the change in intensity of the magnetically perturbed component upon reversal of the magnetic field[30].

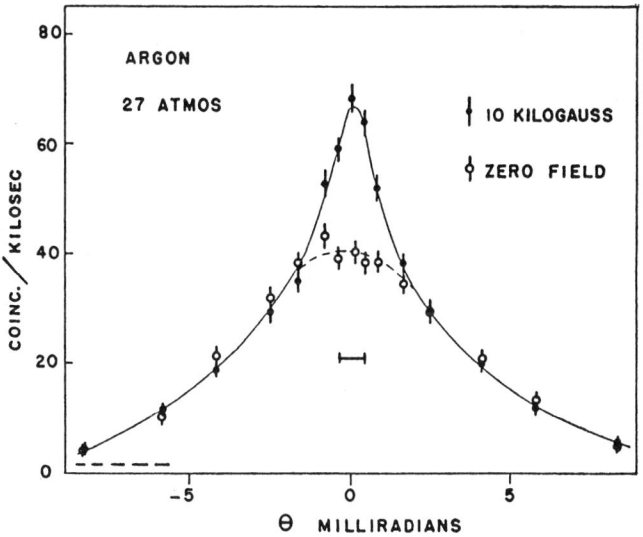

Figure 3. Angular correlation data of positrons emitted from ^{22}Na and annihilating in Argon gas[23].

triplet-like state may be distinguished from the singlet-like state by the considerably lower momentum arising from the longer thermalization time available to the former.

The results of Page and Heinberg[23], who used this latter technique to demonstrate the right-handedness of β^+ from ^{22}Na, are shown in figure 3. By measuring the asymmetry upon reversal of the B-field in the triplet-like Ps formation, they deduced a residual e^+ polarization of 0.4 from Ps formation. A similar ACAR technique, but a different method, was used by Hanna and Preston[31] to establish the polarization of β^+ from ^{64}Cu. They observed changes in the high-momentum tail of component (ii) when the magnetization of a saturated Fe sample was reversed. This was interpreted as arising from e^+ annihilations with e^- from different energy bands under the two conditions of magnetization (the preferential annihilation with the majority-spin e^- occurring when their spin polarization was opposite to that of the e^+).

APPLICATIONS

Bulk Magnetism

Polarized e^+ have been utilized to study the spin-polarized energy bands of ferro- and ferrimagnetic metals and alloys primarily via the ACAR technique[17] which has evolved to the study of eg the electronic structure of half-metallic ferromagnets[32] and three-dimensional electron momentum densities of some of these materials[33]. Also, a Doppler broadening method has been recently used to demonstrate the sensitivity of polarized e^+ to paramagnetic centres in insulators, semiconductors[34].

Surface Magnetism

In a metal, the e^- density is too high for Ps to be formed[35], however, Ps formation may occur in the low-density tail of the e^- distribution at the surface. By measuring the asymmetry in Ps formation from magnetized targets when either the beam polarization or the magnetization is reversed, surface magnetism may be investigated. The method has the advantage of combining spin sensitivity and surface selectivity. Figure 4 shows the spin

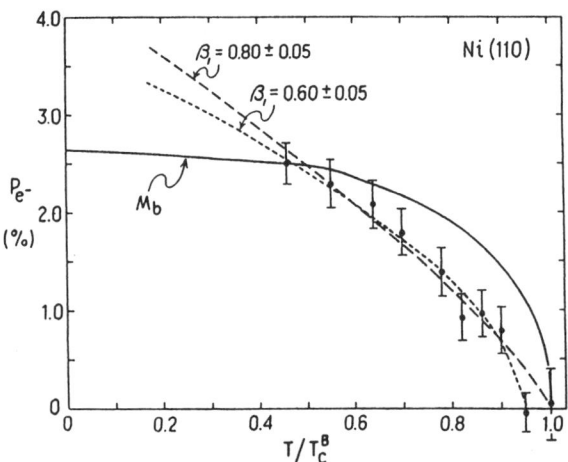

Figure 4. Temperature dependence of the polarization of the electrons captured from a magnetized Ni(110) surface[36].

polarization, P_{e^-}, of the e⁻ captured from a Ni(100) surface, decreasing with temperature until zero is reached at the Curie temperature[36].

Chiral and Other Media

The difference in the population of Ps states formed in a B-field parallel and antiparallel to the β⁺ polarization has been used to observe depolarization effects in some organic compounds, in contrast to what was found in some inorganic materials[37]. Since all materials investigated were diamagnetic, it has been proposed that these effects are the result of a magnetic interaction between the magnetic moment of e⁺ and that of the molecular triplet-states excited via intersystem crossing[37].

The question as to whether e⁻ or e⁺ can distinguish between molecules of opposite chirality continues to generate interest. For a discussion of e⁻ studies, the reader is referred to other articles in this volume. Limits on the asymmetries in Ps formation have been reported for aminoacid enantiomers[38]. Also e⁺ depolarization effects, relative to the external B-field, have been found to be sensitive to particular enantiomers of other chiral media[39].

OUTLOOK

It is anticipated that positrons will continue to aid the understanding of fundamental interactions in nature. Progress in such understanding, and in its application, is supported by concomitant advancements in the production of positron sources, moderators and instrumentation. The quality of a beam may, in this context, be characterized by quantities such as flux and intrinsic polarization. Both of these depend, among other things, on the end-point energy, E_{max}, of the β⁺ spectrum. Calculations[28] have suggested that ¹¹C (E_{max}~1MeV) might provide the highest flux of slow polarized positrons. Its short lifetime ($t_{1/2}$=20min), however, necessitates on-line production. In the light of recent developments in positron beams and sources[1-4], including those from compact cyclotrons[40,41], bright highly polarized beams of e⁺ might be widely available in the not-too-distant future.

It is envisaged that such sources and beams would enable the exploration of the importance of magnetic interactions in e⁺ scattering from paramagnetic molecules and that of exchange and quenching in Ps-atom (molecule) collisions. In the latter respect, a beam of polarized Ps atoms could be produced via neutralization of a polarized e⁺ beam in a gaseous target and the temporal de-selection of the m=0 states. Bright polarized Ps sources would also assist tests of CP, T and CPT invariance[42]. Auger spectroscopy induced by the annihilation of spin polarized positrons has been suggested[43] as a new technique for studying the magnetic properties of surfaces.

Among the more exotic investigations which may result are the observation of the Bose condensation of Ps[44] and the production of polarized antiprotons via the hyperfine interaction in an anti-hydrogen atom[18].

Acknowledgments

Thanks are due to Prof H Kleinpoppen and Dr R Newell for the invitation to write this article; Dr AP Mills and Prof C Wilkin for helpful comments on the script; Dr M Biasini and co-workers at UCL for useful discussions; the Engineering and Physical Sciences Research Council for supporting most of the positron research at UCL and the Nuffield Foundation for the award of a 1994 Science Research Fellowship.

REFERENCES

1. "Positron Interactions with Atoms, Molecules and Clusters" W Raith ed., Baltzer, Basel (1994)
2. "Positron Annihilation" Zc Kajcsos and Cs Szeles ed., Trans Tech, Aedermannsdorf (1992)
3. "Slow Positron Beam Techniques for solids and Surfaces" E Ottewitte and AH Weiss, ed., AIP, New York (1994)

4. "Positron Studies of Solids" A Dupasquier and AP Mills ed, Societa' Italiana di Fisica, Bologna (1994)
5. eg MS Fee et al Phys Rev Lett 70:1397 (1993); D Hagena et al Phys Rev Lett 71:2887 (1993); RS Conti et al Phys Lett A 177:43 (1993)
6. eg JS Nico et al Phys Rev Lett 65:1344 (1990); AH Al-Ramadham and DW Gidley Phys Rev Lett 72:1632 (1994)
7. eg S Orito et al Phys Rev Lett 63:597 (1989); S Adachi et al Phys Rev A 49:3201 (1994)
8. eg G Laricchia and N Zafar "Positron at Metallic Surfaces" A Ishii, ed., Trans Tech, Aedermannsdorf, 347 (1992); G Laricchia (1994) in 4
9. N Zafar, G Laricchia and M Charlton (1994) in 1
10. G Laricchia, J Moxom and M Charlton Phys Rev Lett 70:3229 (1993); J Moxom, G Laricchia and M Charlton J Phys B 26:L367 (1993); G Laricchia and J Moxom Phys Lett A 174:255 (1993); J Moxom et al Phys Rev A in press (1994); P Ashley et al (1994) in 1
11. J Moxom et al J Phys B 23:L613 (1992); A Kover, G Laricchia and M Charlton J Phys B 26:L1 (1993), 27:2409 (1994); L Dou et al Phys Rev Lett 68:2913 (1992); A Schmitt et al Phys Rev A 49:R51 (1994)
12. G Spicher et al Phys Rev Lett 64:1019 (1990); W Sperber et al Phys Rev Lett 68:3690 (1992); GO Jones et al J Phys B 26:L483 (1993); S Zhou et al Phys Rev Lett 72:1443 (1994)
13. eg SP Parikh et al Phys Rev A 47:1535 (1993)
14. eg M Leventhal and BL Brown "Positron (Electron) - Gas Scattering" WE Kauppila, TS Stein and J Wadehra, ed., World Scientific, Singapore 140 (1986)
15. eg RG Greaves, MD Tinkle and CM Surko Phys Plasmas 72:352 (1994)
16. eg "Antihydrogen" J Eades, ed., Baltzer, Basel (1993)
17. eg S Berko "Positron Annihilation" AT Stewart and LO Roellig ed, Academic Press, New York (1967); S Berko and H Pendleton Ann Rev Nucl Part Sci 30:543, 1980
18. eg A Rich Rev Mod Phys 53:127 (1981); Hyp Int 44:125 (1988)
19. WE Kauppila and TS Stein Adv At Mol Phys 26:1 (1990); M Charlton and G Laricchia J Phys B 23:1045 (1990); Comm At Mol Phys 26:253 (1991); G Laricchia and M Charlton (1992) in 2
20. AP Mills Hyp Int 44:107 (1988); (1992) in 2; (1994) in 4
21. PJ Schultz and KG Lynn Rev Mod Phys 60:701 (1988)
22. TD Lee and CN Young Phys Rev 105:1671 (1956); CS Wu et al Phys Rev 105, 1457 (1957)
23. LA Page and M Heinberg Phys Rev 106:1220 (1957)
24. CERN PPE/93-53 and P Clarke (1994) Private Communication
25. AD Stauffer and RP McEachran in "Positron (Electron)-Gas Scattering" WE Kauppila, TS Stein and JM Wadehra, eds, World Scientific, Singapore (1986)
26. KJ Hasenburg Phys B 19:L499 (1986)
27. R Khatri et al Appl Phys Lett 57:2374 (1990)
28. J Van House and PW Zitzewitz Phys Rev A 29:96 (1984)
29. N Zafar et al Appl Phys A 47:409 (1988); E Gramsch, J Trowe and KG Lynn Appl Phys Lett 51:1862 (1987)
30. G Gerber et al Phys Rev D 15:1189 (1977)
31. SS Hanna and RS Preston Phys Rev 106:1363 (1957)
32. KEHM Hanssen et al Phys Rev B 42:1533 (1990)
33. eg H Kondo et al J Phys Cond Matt 4:4595 (1992) and references therein
34. U Lauff et al Phys Lett A 182:165 (1993)
35. A Held and S Kahana Can J Phys 42:1908 (1964)
36. DW Gidley, AR Koymen and T Weston Capehart Phys Rev Lett 49:1779 (1982)
37. A Bisi et al Nuovo Cim 11:635 (1989); 12:831 (1990); G Consolati and F Quasso J Phys B 26:4623 (1993)
38. DW Gidley et al Nature 297:639 (1982)
39. A Bisi, N Gambara and L Zappa Nuovo Cim 14:617 (1992); 15:1315 (1993)
40. KF Canter Private Communication (1994)
41. J Gordon, M Kruip and A Reed Phys World 7:38 (1994)
42. BK Arbic et al Phys Rev A 37:3189 (1988); W Bernreuther et al Z Phys C 41:143 (1988); M Skalsey and J Van House Phys Rev Lett 67:1993 (1991)
43. A Weiss "Positron at metallic Surfaces" A Ishii, ed., Trans Tech, Aedermannsdorf, 317 (1992)
44. PM Platzman and AP Mills Phys Rev B 49:454 (1994)

POLARIZATION EFFECTS IN SIMULTANEOUS ELECTRON PHOTON EXCITATION

W R Newell

Department of Physics and Astronomy
University College London
Gower Street
London WC1E 6BT
United Kingdom

INTRODUCTION

The influences of radiation fields on collision processes and interactions is quite universal since all scattering processes take place in radiation fields even if it is only the thermal background. In the present article we will be concerned with radiation fields produced by lasers and how the radiation and its polarization, through virtual interactions, influences the collision process. Electron-atom scattering in the presence of an external electromagnetic radiation field was initially studied by Goppert-Mayer (1931) in association with her theoretical work on two-photon excitation. However, it is only by combining the more recent developments of laser technology and the expertise of high-resolution electron scattering that an experimental study of the simultaneous interactions of photons and electrons with a discrete atom or molecule can be investigated. Such interactions are of practical importance in the heating of plasmas by radiation and laser-induced gas breakdown phenomena, in addition to being of fundamental interest in the understanding of three-body interactions and the relative coupling between radiation fields and particles.

The effect of direct laser radiation on atoms and molecules has been well reported in the area of multiphoton ionization (see for example Agostini et al 1971) with the demonstration of such phenomena as above threshold ionisation (ATI) and the coulomb explosion (Codling et al 1989). There has however been less emphasis on the role of particle and photon polarization in these experiments. Such phenomena as MPI are laser produced and are distinct from another class of phenomena which are laser assisted phenomena. The distinction is simply that with the absence of photons the laser produced reaction does not exist whereas with no photons present in a laser assisted reaction scattering can still take place between the other two particles. In the later case an example is the free-free scattering of an electron from an atom in the presence of a non resonant laser field.

The transfer of photons to and from electrons while they are undergoing an elastic collision is described by the equation

$$e(E_i) + nh\nu + \text{ATOM} \rightarrow e(E_i \pm mh\nu) + (n \mp m)h\nu + \text{ATOM}$$

In this case the atom is not excited by the projectile nor by the photons but the electron does absorb or emit m photons. In this type of scattering the atom acts only as a third body to conserve the energy and momentum in the reaction. These free-free or Inverse and Stimulated Bremmstrahlung transitions have been reported by several authors (eg Weingartshofer et al 1979) and require high-power laser intensities of $10^8 \rightarrow 10^{12}$ watts/cm² in order to achieve a multi-photon ($m > 1$) transfer to the scattered electrons.

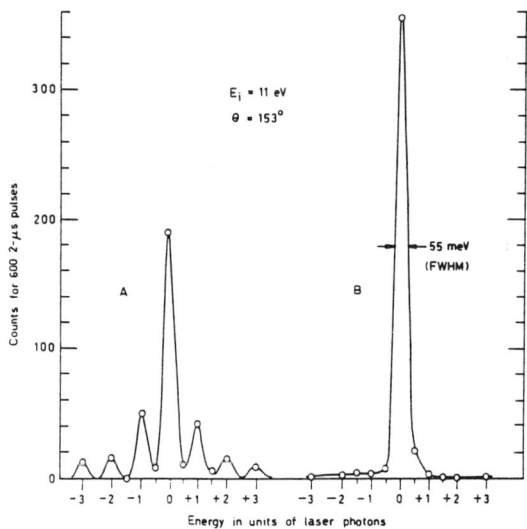

Figure 1 Multiphoton exchange in free-free electron scattering in helium (see text for discussion)

An example of the multiphoton exchange is given in figure 1 for 'elastic' electron scattering from helium in the presence of a CO_2 laser beam. In free-free scattering the atom plays only a passive role but there exists an excitation process in which the incident electron (E_i) combines with a photon ($h\nu$) to cause an excitation of the atom, Simultaneous Electron-Photon Excitation (SEPE). This particular reaction in helium ($1^1S \rightarrow 2^3S$) is described by the equation

$$e(E_i) + nh\nu + \text{He}(1^1S) \rightarrow e(E_j = E_i - \Delta E + h\nu) + (n - 1)h\nu + \text{He}(2^3S)$$

in which the excitation of the 2^3S stationary state in helium, which lies at ΔE above the 1^1S ground state, is accomplished by the absorption of one quantum of radiation $h\nu$ from a laser field combined with a simultaneous inelastic electron scattering in which the electron provides the energy decrement ($\Delta E - h\nu$) required to excite the 2^3S state. Since no stationary state exists between the 1^1S and 2^3S levels this excitation must proceed by a virtual interaction. The simultaneous electron photon excitation cross section of the triplet spin exchanged level in He (2^3S) is given in figure 2 (Mason and Newell 1990) which shows the laser assisted cross section and the direct non laser assisted cross sections for plane polarized laser radiation.

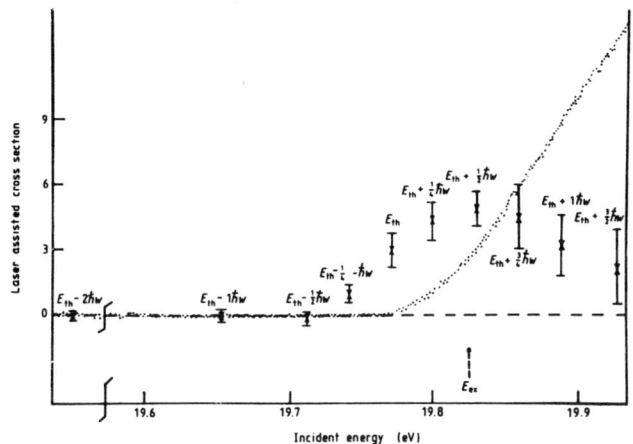

Figure 2 SEPE cross section in helium (•) in c.w. CO_2 laser field and the direct cross section (....) for the 2^3S level. E_{ex} is the excitation threshold of the 2^3S level and the E_{th} is the energy at which the direct signal is first detected.

THE PHYSICAL PROCESS

The general problem of the interaction of an electron, atom and laser field can be written symbolically as

$$|P_f A_f \nu_f\rangle = S|P_i A_i \nu_i\rangle$$

where P_i, A_i and ν_i represent the initial states of the electron, atom and photon, respectively, S is the operator for the reaction and P_f, A_f and ν_f are the final states of the competing particles. However, a real theory of this process must consider all the parameters involved namely the laser polarisation and power, the electron energy and spin, and the nature of the atomic target. In addition, we must ask, is the combined scattering instantaneous or time-evolved and in what sense do these particles interact in pairs, ie, $(e, h\nu)$, $(A, h\nu)$, (e, A) and is there any hierarchy of pair interactions? A rigorous solution should account for all of this; however, even the (e,A) interaction is difficult to treat accurately, especially at the threshold region of the excitation process, and the addition of radiation will certainly increase its complexity.

It will be instructive to outline the derivation of the laser assisted photon exchange cross section for the scattering of electrons in a radiation field with exchange of photons between the scattered electron and the radiation field to demonstrate the dependence on laser polarization.

The electron will have a function form, of a plane wave, outside the laser beam but a modified form inside the laser beam. The pondermotive potential U_p, produced by the electric field in the focused laser, will be neglected in this derivation.

The Schrödinger equation for an electron moving in a linearly polarised laser potential $A(r,t)$ is

$$ih\frac{\partial \chi_q}{\partial t} = \frac{1}{2m}\left[P^2 + \frac{e}{c}A(rt)\right]^2 \chi_q^{(rt)}$$

where

$$A(r,t) = \frac{e}{\omega}E(r)\cos(k.r - \omega t)$$

In the situation where laser wavelength is >> electron wavelength (ie $k << q$), which normally applies, then the amplitude $E(r)$ of the radiation field will change slowly over $\lambda (= 2\pi/q)$ giving (Low, 1958; Kroll and Watson, 1973)

$$\lambda \frac{dE}{dr} << E$$

Then the solution of Schrödinger equation is given by

$$\chi_q = \exp i\left[q.r - \frac{\varepsilon_q t}{h} - \frac{e}{m\omega^2}q.E(r)\sin(\omega t - k.r)\right]$$

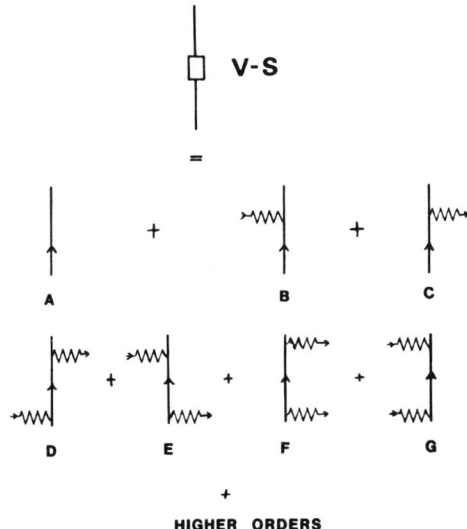

Figure 3 Volkov solution (v-s) of Schrödinger equation for the interaction of EM wave (∿) and electron (↑). The solution consists of an infinite number of amplitudes given in increasing order of photon coupling, i.e., A is zero order, B and C first order, D E F and G second order, etc

This is normally called the Volkov solution and is shown in Feynman form in figure 3 and it adiabatically becomes the free electron wave function $\exp i [q.r - \varepsilon_q t/t]$ (ie plane wave) outside the laser field. The set χ_q of unperturbed electron states can be used in the description of the electron scattering from an atomic potential $V(r)$. Since laser wavelength >> range of $V(r)$ we can make the dipole approximation ie $k.r = 0$ since $k.r$ will vary very little of the range of the potential $V(r)$ and we get $\chi_q = \exp i\, [q.r - \varepsilon_q t/t - q.\alpha_o\, \omega t]$ where $\alpha_o = eE(o)/m\omega^2$ is the amplitude of the simple harmonic motion and $E(o)$ is E at $r = 0$.

This equation for χ_q shows the laser inducing quanta into the free electron wave functions. These induced quanta are virtual and it is only when the electron scatters and therefore accelerates or recoils that the virtual 'off-shell' induced quanta are converted to real 'on-shell' quanta.

Now within the Born approximation the S-scattering matrix for the electron transition from momentum q to q' is written as

$$S_{qq'}^B = -\frac{i}{\hbar}\langle \chi_{q'} V \chi_q \rangle$$

Now using the T-matrix (Newell 1993) the photon exchange cross section is given as

$$\frac{d\sigma^B}{d\Omega}(qq'n) = J_n^2(\Delta q.\alpha_o)\left[\frac{d\sigma^B}{d\Omega}(qq')\right]_{ELASTIC}$$

which is a direct product of a radiation term and a scattering term

$$= \frac{q}{q'} J_n^2(\Delta q.\alpha_o)|f(\theta)|^2_{ELASTIC}$$

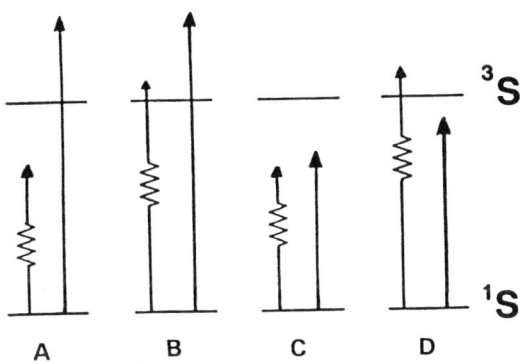

Figure 4 Different combination of electron (↑) and photon energies (⌇) in exciting the (2³S) level of helium

In this case n photons are transferred while the electron momentum changes by $\Delta q (=q'-q)$. The nth order Bessel function J_n describes the radiation term and $f(\theta)$ is the scattering amplitude for elastic scattering.

This equation can be generalised to include inelastic scattering events (Newell 1992 and references therein) and a completely non perturbative solution to these problems has been given by Burke et al (1991) in which the R-matrix method is coupled with Floquet theory.

It is clear from the argument of the Bessel function that there is a critical dependence on the radiation polarization, through α_o, which has the direction of the electric field in the laser beam, and the orientation of the momentum transfer vector Δq which describes the dynamics of the electron scattering. If the laser polarization direction is perpendicular to Δq then $\underline{\Delta q} \cdot \underline{\alpha_0}$ is zero where as it is maximum when $\underline{\Delta q}$ is parallel to $\underline{\alpha_0}$. Additionally in SEPE neither the electron nor the photon(s) separately have sufficient energy to excite the state; it requires the combined energy to cause excitation. Therefore in figure 4 only process 'c' will give rise to simultaneous electron-photon excitation.

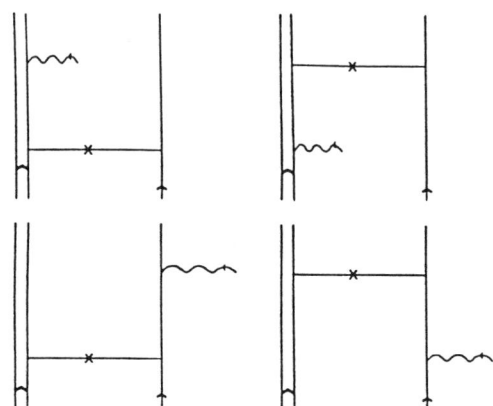

Figure 5 First order Feynman diagrams for SEPE

A simple direct perturbative picutre of the coupling between the radiation field, (considering only n = 1) electron and atom is given by the four Feynman diagrams in figure 5 where each diagram represents one particular scattering amplitude. The single line, ↑, represents the electron and the double line, ⇑, denotes the atom. Time runs vertically in the figure and -x- denotes the electron-atom interaction with ∿ denoting the photon interaction. The role of these four Feynman diagrams in free-free scattering has been studied by Dubois et al (1985) in atomic hydrogen while the SEPE cross-sections in atomic hydrogen and helium have been calculated by Jetzke et al (1984, 1987).

RESULTS

The main experimental requirements in this work is that the energy spread ($\Delta\varepsilon$) of the electron scattering apparatus is much less than the photon energy, $\Delta\varepsilon \ll h\nu$, in order that the photon transfer effects are resolved. The experimental procedures and requirements for different laser wavelengths and for both pulsed and c.w. laser systems have been detailed by several authors (Mason and Newell 1987, 1989; Wallbank et al 1988; Luan et al 1991). In all the experimental work discussed in this paper the signature of the excitation is the production of the metastable 2^3S state. These metastable atoms have sufficient internal energy (19.8 eV) to make them easily detectable using a channel electron multiplier; a schematic diagram of the collision geometry for the three intersecting beams (laser, electron and gas) is given in figure 6. The time-of-flight spectra of the metastable atoms are recorded with and without the laser beam present with the difference in the metastable flux being a monitor of the SEPE cross sections.

Referring to equation for $d\sigma^B/d\Omega$ it is clear that the vector nature of the Bessel function argument is a monitor of the polarization dependence of the SEPE process. Rotating $\underline{\alpha_0}$

about the $\underline{\Delta q}$ direction will selected, in the photon transfer channel, the direction of the momentum transfer vector. Averaging $\underline{\alpha_0}\cdot\underline{\Delta q}$ over all orientations, as is the case for circularly polarised light, should yield a reduction in the SEPE intensity of $2/\pi$.

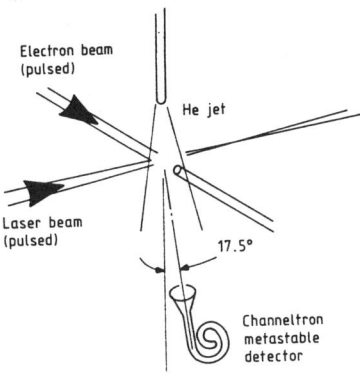

Figure 6 Schematic arrangement of the interaction system; see text for discussion

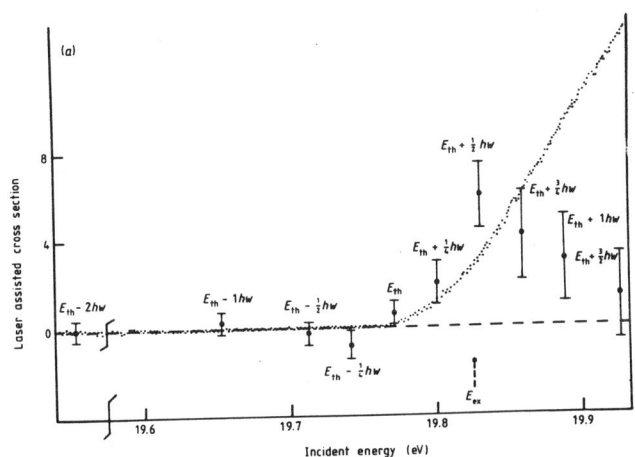

Figure 7 SEPE in Helium(●)using circulry polarized radiation.

Figure 7 shows the SEPE cross section for the $He(1^1S) \rightarrow He(2^3S)$ (Mason and Newell 1990) transition using circularly polarised c.w. CO_2 10.6 μm radiation; compare with figure 2 which shows the SEPE cross section using plane polarised radiation. In both polarization configurations the maximum in the laser-assisted cross section is detected at $E_{th} + \frac{1}{2}\hbar\omega$ at which energy the peak of the electron beam profile is coincident with the spectroscopic excitation threshold, E_{ex}, of the He(2^3S) state (E_{ex} = 19.817 eV). However the signal intensity for circularly polarised light is lower than for the plane polarised light by approximately 0.6. This reduction is in good agreement with the $2/\pi$ reduction predicted above.

At incident energies greater than $E_{th} + \tfrac{1}{2}\hbar\omega$ a similar reduction in the laser-assisted cross-section was observed when using circularly polarised light. However below $E_i = E_{th} + \tfrac{1}{2}\hbar\omega$ (=E_{ex}) there is a distinct change - see figure 7 - in the shape of the observed laser-assisted cross section when circularly polarised radiation is used. Such a difference indicates that the scattering process in the threshold region is influenced by the polarization vector of the laser. The propensity for scattered electrons to move along the α_0 direction thus influencing the momentum transfer vector Δq will be more apparent in the threshold region of the excitation process. The polarisation-dependent effect observed here in the laser-assisted cross section with circularly polarised light demonstrates the dependence of the SEPE process upon the orientation of the linear polarisation of the laser beam with respect to the incident electron beam direction. The dependence of the SEPE in helium excitation using a pulsed CO_2 laser (10^8w/cm^2) as a function of the orientation ϕ of α_0 with Δq has been reported by Wallbank et al (1990).

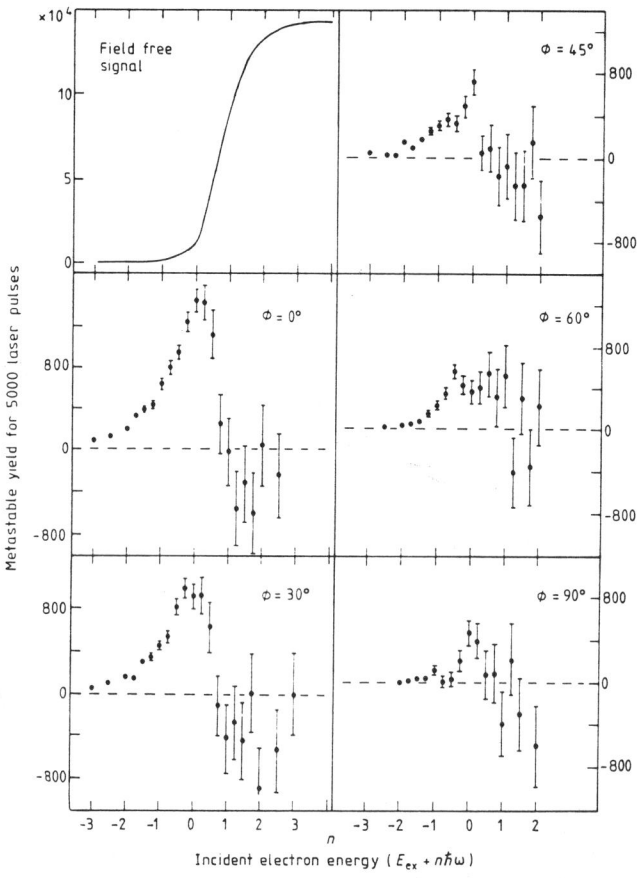

Figure 8 The integrated additional metastable signal as a function of electron energy (in units of quanta from threshold) for various polarisation angles ϕ. The field-free metastable excitation cross section is also shown in this energy region.

The data showing the behaviour of the laser-assisted metastable signal as a function of the angle ϕ between the laser polarisation and the incident electron direction are given in figure 8. The yield of metastable atoms is highest when the laser polarisation is parallel to the incident electron direction ($\phi = 0°$). The signal near the excitation threshold decreases as the angle ϕ is increased until at $\phi = 90°$ approximately only 30% of the signal at $\phi = 0°$ remains.

However even when α_0 and Δq are perpendicular the signal is still finite indicating possibly a failure of the scattering approximations used in the derivation of equation for $d\sigma^B/d\Omega$. Multiphoton ($n > 1$) effects are also evident in the data in figure 8 and similar MP SEPE has been reported by Luan et al (1991) using pulsed YAG radiation.

CONCLUSIONS

The influence of laser polarization in the SEPE process is clearly demonstrated. However there have been no corresponding studies of the nature of the spin polarization of the scattered electrons in such excitations which have a role especially as the scattering is a spin exchange mechanism. Additionally in MPI, especially when intermediate states and level shifts are involved, the spin polarization of the electrons has not been investigated.

ACKNOWLEDGEMENTS

The author would like to acknowledge the facilities in the Department of Pure and Applied Physics, QUB, which were used in the preparation of this paper.

REFERENCES

Agostini P, Antonetti A, Breger P, Crance M, Migus A, Muller H G and Petite G, 1971, J Phys B 22:273.
Burke P G, Francken P and Joachain C J, 1991, J Phys B.
Codling K, Frasinski L J and Hatherly P A, 1989, J Phys B 22:L321.
Dubois A, Maquet A and Jetzke S, 1985, Phys Rev A 34:1888.
Faisal F H M, 1987, Theory of Multiphoton Processes, Plenum Press.
Ferrante G, Leone G and Trombefta F, 1982, J Phys B 15:L475.
Goppert-Mayer M, 1931, Ann Phys Lpg 9:273.
Jetzke S, Board J and Maquet A, 1987, J Phys B 20:2887.
Jetzke S, Faisal F H M, Hippler R and Lutz M S O, 1984, Z Phys A 315:271.
Kroll N M and Watson K M, 1973, Phys Rev A 8:804.
Laun S, Hippler R and Lutz H O, 1991, J Phys B 24: 3241.
Low F E, 1958, Phys Rev 110:974.
Mason N J and Newell W R, 1987, J Phys B 20:L323.
Mason N J and Newell W R, 1989, J Phys B 22:777.
Mason N J and Newell W R, 1990, J Phys B 23:L179.
Mittleman M H, 1982, Theory of Laser Atom Interactions, Plenum Press.
Newell W R, 1992, Comm Atm Mol Phys 28:59.
Newell W R, 1993, Conference Proceedings 13 Atomic Physics, ed W Walther, T W Hansch and B Neizert, AIP (New York), p 540.
Weinsgartshofer A, Clarke E M, Hones J K and Jung C, 1979, Phys Rev 19:2371.
Wallbank B, Holmes J K, LeBlanc L and Weingartshofer A, 1988, Z Phys D 10:467.

POLARIZATION CORRELATION AND COHERENCE LENGTHS OF TWO-PHOTON RADIATION

Alan J Duncan and Zahoor A Sheikh
Atomic Physics Laboratory
University of Stirling
Stirling FK9 4LA
Scotland

INTRODUCTION

There has been much recent interest in the measurement of coherence lengths of photon wavepackets particularly in relation to the correlated two-photon pairs produced in the process of parametric down conversion [1,2]. The coherence length of photons has also been regarded as an important parameter in interpretation of experiments involving cascade photons [3] carried out to test Bell's inequality [4] and the limitations of hidden-variables theories. Conventionally, coherence length is measured by observing the dependence of fringe visibility on path length difference in a beam splitting interferometer but the advent of the study of two-photon pairs has given rise to other interesting ways of carrying out such measurements [1,2]. Here a new method of determining the coherence length, coherence time and bandwidth of the two photons produced in the two-photon decay of metastable atomic deuterium is described. This new method makes use of the subtle interplay between the entangled polarization properties and the spectral properties of the two-photon radiation.

CHARACTERISTICS OF THE TWO-PHOTON DECAY

For metastable atomic deuterium [D(2S)] electric dipole and electric quadrupole transitions from the $2S_{1/2}$ to $1S_{1/2}$ state are forbidden. As a result the D(2S) state has a long lifetime of about 1/8 second and decays primarily by the simultaneous emission of two photons as indicated in Fig 1.

Contributions to the decay of the $2S_{1/2}$ state are also possible by:

(i) single photon magnetic dipole transitions which only become significant, however, when relativistic effects are important, for example in high Z hydrogenic ions. For deuterium itself the lifetime of this type of transition is about 4×10^5 seconds [5] and can be neglected for most purposes;

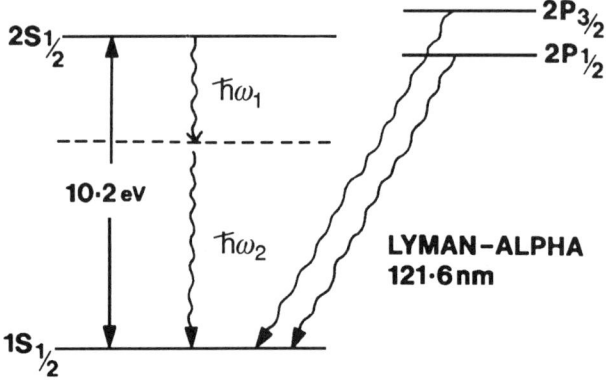

Figure 1. Energy level diagram for atomic deuterium neglecting hyperfine structure (not to scale).

(ii) a cascade, involving the sequential emission of two photons through the $2P_{1/2}$ state which, because of the Lamb shift, lies slightly below the $2S_{1/2}$ state in energy. For deuterium the associated lifetime is about 5×10^9 seconds [6] and hence this process can also be effectively neglected;

(iii) other types of two-photon transition. For example, the $2S_{1/2}$ state could decay by emitting two quadrupole photons [7] but the effect of such processes is negligible for the present purposes;

It is also worth noting that since the electric dipole operator mediating the two-photon decay process is diagonal in the electronic and nuclear spin, fine and hyperfine structure play no part in determining the decay process [8,9] or the properties, particularly the polarization correlation properties, of the two photons emitted.

The theoretically predicted [8,9,10] spectral distribution $A(\omega)$ of the photons emitted in the two-photon decay process is shown in Fig 2, where $A(\omega)d\omega$ represents the probability of detection of one photon of the pair in the range $d\omega$ in the neighbourhood of angular frequency ω. It can be seen from Fig 2 that each photon of the pair can be detected with any frequency (energy) between 1.55×10^{16} rad s^{-1} (10.2 eV) and 0 rad s^{-1} (0 eV) subject to the requirement that the sum of the frequencies (energies) is given by $\omega_o = 1.55 \times 10^{16}$ rad s^{-1} (10.2 eV). The spectral distribution curve has a maximum at a frequency (energy) of 0.775×10^{16} rad s^{-1} (5.1 eV) corresponding to a wavelength of 243 nm. Experimentally this fact is important since it allows measurements of photon characteristics to be made in air using simple optical components.

For photon pairs for which the recoil momentum of the atom is less than the uncertainty in momentum resulting from localisation, the frequency component of the state vector representing the two-photon radiation can be written in the form [11]

$$|\psi> = \int a(\omega)|\omega>_1|\omega_o-\omega>_2 d\omega \qquad (1)$$

where $|\omega>_1$ and $|\omega_o-\omega>_2$ are the state vectors representing photons with the complementary frequencies ω and $\omega_o-\omega$, and $|a(\omega)|^2 = A(\omega)$.

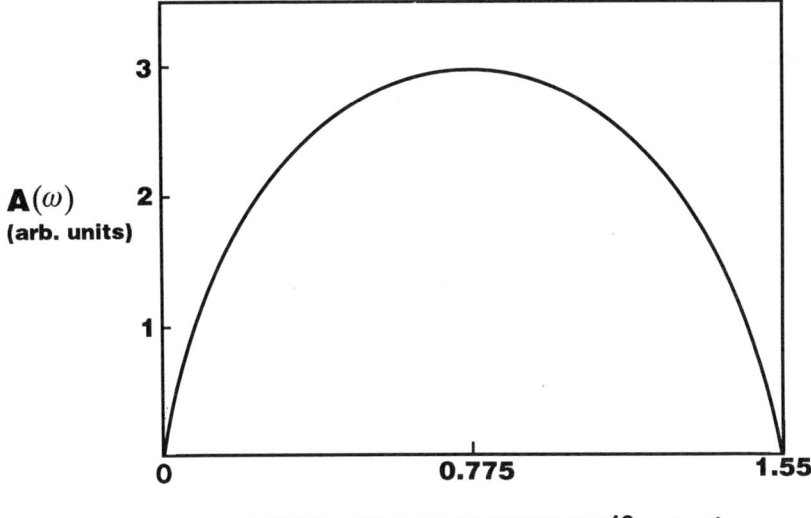

Figure 2. Predicted two-photon spectral distribution for the two-photon decay of metastable atomic deuterium.

The polarization properties of the two photons are also well known [12, 13]. Of particular interest is the situation where the two photons are detected in diametrically opposite directions, say the ±z directions. On the basis of consideration of conservation of angular momentum and parity it can be shown [14] that the polarization component of the two-photon state vector takes the form

$$|\psi> = \frac{1}{\sqrt{2}} (|x>_1|x>_2 + |y>_1|y>_2) \qquad (2)$$

where $|x>_1$ represents a photon on the right-hand-side (+z) say, of the source polarized in the x direction, $|x>_2$ a photon on the left hand side (-z) say, of the source also polarized in the x direction with corresponding definitions for $|y>_1$ and $|y>_2$.

It is interesting to note that both the frequency and polarization components of the state vectors above are in what is now commonly referred to as an "entangled" form. Before any detection event takes place neither the polarization nor frequency (energy) of a single photon is defined or, indeed, can be assigned any meaning. The entangled state vector represents the properties of the photon pair not single photons. However, if, for example, in this case where an arrangement is made to detect photons in diametrically opposite directions, a detection of a photon of particular polarization and frequency (energy) is made on one side of the source then the polarization and frequency (energy) characteristics of the photon on the other side are then known with certainty. In other words, in a sense, the properties of the photon on the right (say) can be considered to be determined by the measurement which is chosen to be made on the left.

PRINCIPLE OF THE METHOD

The basic experimental arrangement for the analysis and detection of the two-photon radiation is shown in Fig 3. The two photons emitted by the source are detected in

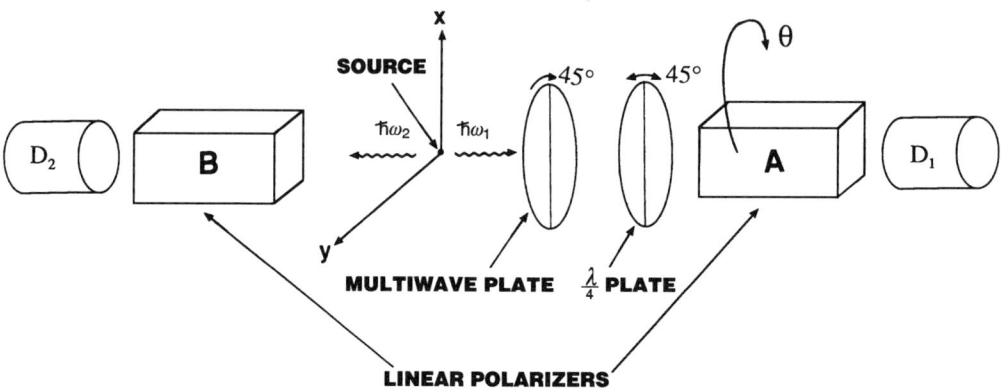

Figure 3. Schematic diagram of the experimental arrangement using two linear polarizers A and B and two detectors D_1 and D_2. The transmission axis of polarizer A is rotated through angles θ with respect to the x axis while that of polarizer B is fixed parallel to the x axis. The fast axis of the multiwave plate is set at 45° to the x axis while the fast axis of the quarter-wave plate is set at ±45° relative to the x axis to analyse circular polarization.

coincidence by the detectors D_1 and D_2. A linear polarizer with its transmission axis orientated in the x direction is placed on the left hand side of the source ensuring that upon detection of a photon on the left the complementary photon on the right, before being detected itself, can to all intents and purposes be regarded as also polarized in the x direction. In the absence of any wavelength filter on the left the frequency of the photon on the right is indeterminate and the radiation on the right can be considered either to consist of a sequence of single frequency photons with a spectral distribution determined by the spectrum of the source or, alternatively and equivalently, as a sequence of minimum wavepackets with the spectrum and coherence length of each wavepacket determined by the spectral characteristics of the two-photon source.

If this radiation on the right is now passed through a uniaxial birefringent multiwave plate with its axis at an angle of 45° to the x axis as shown in Fig 3, its state of polarization will be changed and it will also be depolarized to an extent depending on the thickness d of the multiwave plate. The state of polarization of the radiation emerging from the multiwave plate may be monitored by measuring the Stokes parameters P_1, P_2 and P_3 of the radiation on the right detected in coincidence with the radiation on the left. The Stokes parameters are defined as [15],

$$P_1 = \frac{I(0°) - I(90°)}{I(0°) - I(90°)}, \quad P_2 = \frac{I(45°) - I(-45°)}{I(45°) - I(-45°)}, \quad P_3 = \frac{I(RHC) - I(LHC)}{I(RHC) - I(LHC)} \qquad (3)$$

where $I(0°)$ is the strength of the coincidence signal when the transmission axis of polarizer A is set at an angle θ = 0° to the x axis with corresponding definitions for $I(90°)$, $I(45°)$ and $I(-45°)$. Similarly $I(RHC)$ and $I(LHC)$ refer to the strength of the coincidence signal with the achromatic quarter-wave plate shown in Fig 3 in place orientated with its axis at ±45° to the x axis so as to detect, respectively, right-handed-circularly (RHC) polarized light or left-hand-circularly (LHC) polarized light.

If the retardation of the multiwave plate of thickness d is ϕ then

$$\phi = \frac{(n_e - n_o)\omega d}{c} \quad (4)$$

where n_e and n_o are respectively the extraordinary and ordinary refractive indices of the material of the multiwave plate. It is easy to show [16] then that for monochromatic radiation incident upon the multiwave plate the Stokes parameters of the emerging radiation are given by,

$$P_1 = \cos\phi, \quad P_2 = 0, \quad P_3 = -\sin\phi. \quad (5)$$

However, in the present case, averaged over the spectral distribution $A(\omega)$, the expected values for P_1, P_2 and P_3 of the emerging radiation are,

$$P_1 = \int_0^{\omega_o} \cos\phi \, A(\omega) d\omega \,\Big/\, \int_0^{\omega_o} A(\omega) d\omega$$

$$P_2 = 0 \quad (6)$$

$$P_3 = -\int_0^{\omega_o} \sin\phi \, A(\omega) d\omega \,\Big/\, \int_0^{\omega_o} A(\omega) d\omega$$

Examination of the above expressions reveals that if the birefringence $(n_e - n_o)$ were frequency independent P_1 and P_3 would be precisely the Fourier cosine and Fourier sine transforms of the spectral distribution $A(\omega)$ with the quantity $(n_e - n_o)d/c$ acting as the "time" variable.

It is also interesting to note that in the wavepacket picture with the individual photon represented by a wavepacket in the form

$$f(t) \cos \omega_c t \quad (7)$$

where ω_c is the centre angular frequency of the wavepacket and $f(t)$ represents its envelope, it can be shown that the total polarization

$$P = \sqrt{P_1^2 + P_2^2 + P_3^2}$$

is given by

$$P = \frac{\int f(t) f(t - \delta/c) dt}{\int |f(t)|^2 dt} \quad (8)$$

where

$$\delta = (n_e - n_o)_{\omega_c} d - \omega_c \left[\frac{\partial(n_e - n_o)}{\partial \omega}\right]_{\omega_c} d \quad (9)$$

is the separation of the orthogonally polarized wavepacket envelopes produced by passage through the multiwave plate. For quartz it can be shown from knowledge of its birefringence [17] that $\delta = 1.677 \times 10^{-2} d$.

The total polarization P, therefore, represents in this approximation the autocorrelation of the wavepacket envelope, the Fourier transform of which gives the power spectrum $A(\omega) = |a(\omega)|^2$ of the source. More simply, the value of δ which first results in reduction of the total polarization to zero gives an estimate of the coherence length and coherence time of single photons and the bandwidth of the two-photon source.

APPARATUS

The metastable hydrogen source and the general experimental procedures have been described in detail elsewhere [12]. In summary, a 1-keV beam of metastable atomic deuterium, of density about $10^4 cm^{-3}$, is produced by charge exchange, in caesium vapour, of deuterons extracted from a radio-frequency ion source. Deuterium is used rather than hydrogen since a more stable beam with less collision induced noise is produced in this way. After exiting from the charge exchange cell the beam passes through collimating apertures and then the observation region into the monitor region at the end of the apparatus where it is quenched in an electric field, the resulting Lyman-α signal being used to normalize measurements taken over a long period. The two-photon radiation emitted by a small portion of the beam, after passing through 5 mm thick fused silica vacuum windows one on either side of the beam, is collected and collimated by two 50 mm diameter fused silica lenses each with a focal length of 43 nm at a wavelength of 243 nm placed diametrically on either side of the beam about 50 mm from its centre. The lenses produce an image of the source on the two photomultiplier cathodes each of which is 53 cm from the centre of the beam. The pulses from the photomultipliers are fed to a conventional coincidence circuit [18] to produce a time correlated spectrum which is displayed on a multichannel analyser.

The linear polarizers consist of twelve 2 mm thick fused silica plates optically polished flat to 2λ at 243 nm and set nearly at Brewster's angle. The transmission efficiencies M and m for light polarized parallel to and perpendicular to the transmission axis of the polarizers are $M = 0.936 \pm 0.008$ and $m = 0.032 \pm 0.001$ at a wavelength of 243 nm. The efficiencies M and m have a weak wavelength dependence. More details of the characteristics of the linear polarizers are given elsewhere [19].

The achromatic quarter-wave plates (Halle-Nachfl) consist of a combination of two double-plates, one of crystal quartz and one of magnesium fluoride. The plates of diameter 19.5 mm are cut parallel to the optic axis and polished flat to an accuracy of $\lambda/10$. The retardation produced by these plates shows some dependence on wavelength which has been described previously [19]. It is, of course, necessary to use such achromatic plates because of the continuum nature of the radiation from the source which emits at all wavelengths between 121.6 nm and infinity. However, in practice, because of absorption in oxygen there is an effective short wavelength cut-off at about 185 nm which, in turn, implies a long wavelength cut-off at the complementary wavelength of 355 nm. Hence only photons with wavelengths in the range 185 nm to 355 nm can contribute to the coincidence signal.

The "multiwave" plates themselves, placed one at a time on the right hand side of the source, are, in practice, chosen to be a series of zero-order half-wave plates at wavelengths of 200 nm, 243 nm, 300 nm and 486 nm. These plates consist of two flat pieces of crystal quartz of slightly different thicknesses cut parallel to the optic axis and placed in contact with their optic axes perpendicular to give "effective" thicknesses d = 7.69 µm, 10.84 µm, 14.56 µm and 26.27 µm deduced from the known birefringence properties of quartz [17]. Additional effective thicknesses of d = 3.15 µm and 18.53 µm are obtained by placing the 200 nm and 243 nm plates in series with their optic axes respectively at right angles and parallel. An effective thickness of d = 37.11 µm is obtained by placing the 243 nm and 486 nm plates in series with their axes parallel.

EXPERIMENTAL RESULTS AND COMPARISON WITH THEORY

In carrying out the experiment, the Stokes parameters P_1, P_2 and P_3 are measured for each multiwave plate and combination of multiwave plates described above. From considerations of symmetry $P_2 = 0$, and it was confirmed that, within the limits of experimental error, this was true for a representative sample of the multiwave plates. In practice the measured values P_1, P_2 and P_3 must be corrected for the imperfection of polarizer A by dividing the measured values by the "polarization" $\Pi = (M-m)/(M+m)$ of polarizer A to obtain the "true" values P_1', P_2' and P_3' for the Stokes parameters of the radiation emerging from the multiwave plate. Using the values for M and m quoted in the previous section $\Pi = 0.934$. The total polarization P is then defined by

$$P = \sqrt{P_1'^2 + P_2'^2 + P_3'^2}. \tag{10}$$

The measured values of P_1' and P_3' and the resultant value of P are shown in Fig 4.

Calculation of the theoretical prediction is based on equation 6 modified to take account of the limitations of the experiment. Since the radiation on the right cannot be considered to be in a state of pure linear polarization before passing through the multiwave plate because of the imperfection of polarizer B on the left, for monochromatic radiation the Stokes parameters become

$$P_1 = \Pi\cos\phi, \quad P_2 = 0, \quad P_3 = -\Pi\sin\phi \tag{11}$$

where Π in this case is the polarization of polarizer B (which is identical to polarizer A). Averaged over the spectral distribution, the theoretically predicted Stokes parameters P_1', P_2', P_3' are hence given by the modified equations

$$P_1' = \int_{\omega_1}^{\omega_2} \Pi\cos\phi\, A(\omega)d\omega \Big/ \int_{\omega_1}^{\omega_2} A(\omega)d\omega$$

$$P_2' = 0 \tag{12}$$

$$P_3' = -\int_{\omega_1}^{\omega_2} \Pi\sin\phi\, A(\omega)d\omega \Big/ \int_{\omega_1}^{\omega_2} A(\omega)d\omega$$

with the integrations extending over the frequency range from $\omega_1 = 5.31 \times 10^{15}$ rad s^{-1} (355 nm) to $\omega_2 = 1.02 \times 10^{16}$ rad s^{-1} (185 nm). Using these equations the values for P_1', P_2' and P_3', and hence also the total polarization P, can be found by numerical integration from a knowledge of $A(\omega)$ and the frequency dependence of M, m and ϕ. The results of these calculations are shown as the solid lines in Fig 4.

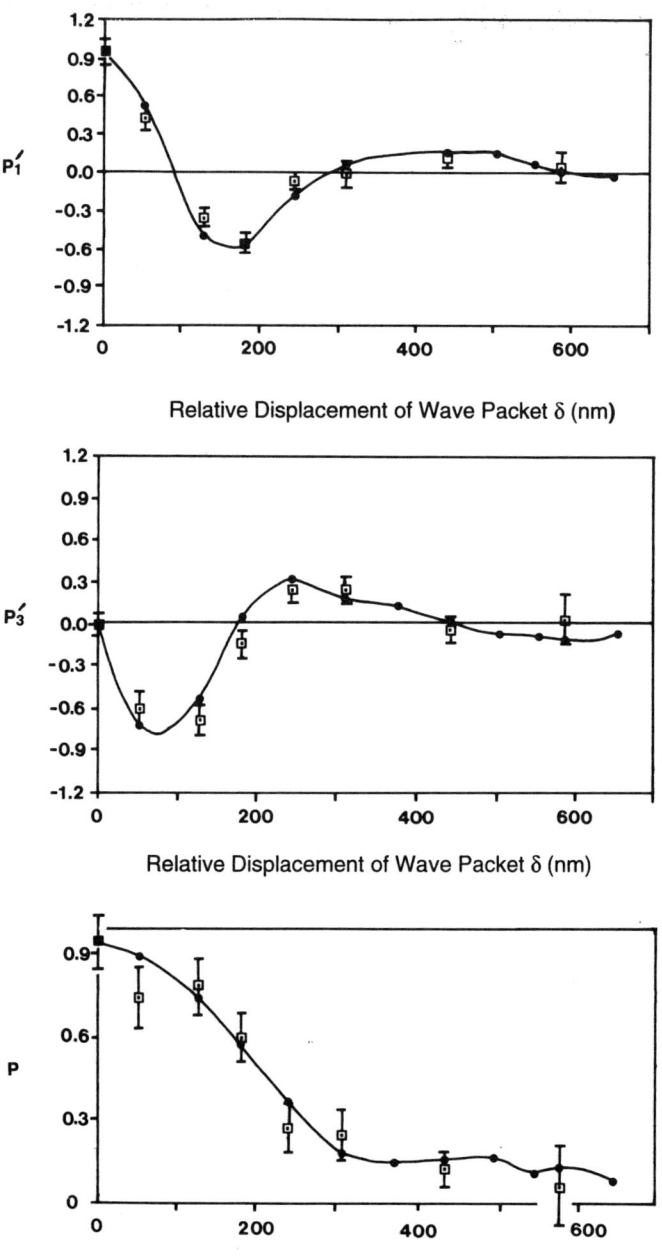

Figure 4. Variation of the Stokes parameters P_1' and P_3' and the total polarization P, with the separation δ between the orthogonally polarized wavepacket envelopes produced by passage through the multiwave plate. The solid lines represent the theoretical predictions.

CONCLUSIONS

It is clear from examination of Fig 4 that there is very good agreement between the theoretical predictions and the experimental results. Although it must always be remembered that the experiment measures the properties of the two-photon pair detected in coincidence, it is useful and interesting to consider the experiment as an investigation of the properties of the single photon on the right detected in coincidence with the complementary photon on the left. On this basis, for the photon on the right the multiwave plate plays the role of the beam splitting interferometer in a Fourier transform spectrometer. The results for P_1', P_3' and P shown in Fig 4 are clearly closely related to the Fourier cosine, sine and autocorrelation of the "top-hat" spectral distribution resulting from that shown in Fig 2 with sharp cut-offs at frequencies of 5.31×10^{15} rad s^{-1} and 1.02×10^{16} rad s^{-1}. Unfortunately, the strong frequency dependence of the birefringence ($n_e - n_o$) prevents a straightforward deduction of the spectral distribution from the experimental results by inverse Fourier transformation.

However, adopting the photon wavepacket description it is expected that the total polarization P will drop to a low value first when the orthogonally polarized ordinary and extraordinary wavepackets no longer overlap significantly after passage through the multiwave plate. Hence, it can be deduced from the results that the photon wavepacket regarded as a minimum packet is approximately of length 350 nm. In other words the coherence length of the wavepacket is to be considered approximately to be 350 nm, the corresponding coherence time 1.2 fs and the bandwidth 0.83×10^{15} Hz. It is worth noting that the coherence length of 350 nm represents approximately only one and a half wavelengths at the theoretically predicted 243 nm centre wavelength of the two-photon radiation.

Finally, it is interesting to observe how the subtle interplay between the polarization and spectral properties of the two-photon radiation has been used to allow a measurement to be made of the coherence length and spectral width. The method clearly allows the interpretation that a single photon of a two-photon pair is to be considered to have a very short coherence length determined by the lifetime of the virtual intermediate state of the decay process rather than the very long lifetime of 1/8 second of the metastable atomic state itself. The very long coherence length associated with the lifetime of 1/8 second of the metastable state is, however, undoubtedly the coherence length appropriate in considerations relating to fourth order interference experiments [11].

ACKNOWLEDGEMENTS

The authors wish to thank Dr H J Beyer, Dr M A Chandhry, Dr H Hamdy and Professor H Kleinpoppen for useful discussions. One of us (ZAS) wishes to thank B Z University, Multan, for its support.

REFERENCES

1. C.K. Hong, Z.Y. Ou and L. Mandel, Phys.Rev.Lett. 59, 2044 (1987).

2. J.G. Rarity and P.R. Tapster, Phys.Rev. A41, 5139 (1990).

3. A. Aspect, P. Grangier and G. Roger, Phys.Rev.Lett. <u>47</u>, 460 (1981).

4. J.S. Bell, Physics <u>1</u>, 195 (1964).

5. F.A. Parpia and W.R. Johnson, Phys.Rev. <u>A26</u>, 1142 (1982).

6. J. Shapiro and G. Breit, Phys.Rev. <u>113</u>, 179 (1959).

7. C.K. Au, Phys.Rev. <u>A14</u>, 531 (1976).

8. M. Göppert-Mayer, Ann.Phys. <u>9</u>, 273 (1931).

9. G. Breit and E. Teller, Astrophys.J. <u>91</u>, 215 (1940).

10. L. Spitzer and J.L. Greenstein, Astrophys.J. <u>114</u>, 407 (1951).

11. Z.Y. Ou and L. Mandel, Phys.Rev.Lett. <u>61</u>, 54 (1988).

12. W. Perrie, A.J. Duncan, H.J. Beyer and H. Kleinpoppen, Phys.Rev.Lett. <u>54</u>, 1790, 2647E (1985).

13. T. Haji-Hassan, A.J. Duncan, W. Perrie, H. Kleinpoppen and E. Merzbacher, Phys.Rev.Lett. <u>62</u>, 237 (1989).

14. A.J. Duncan, in *Progress in Atomic Spectroscopy, part D* (edited by H.J. Beyer and H. Kleinpoppen), pp 447-505, Plenum Press, New York (1987).

15. M. Born and E. Wolf, *Principles of Optics*, Pergammon Press, Oxford (1975).

16. Z.A. Sheikh, PhD Thesis, University of Stirling, Stirling (1993).

17. G. Kaye and T Laby, *Tables of Physical and Chemical Constants*, Longmans, London (1968).

18. D. O'Connell, K.J. Kollath, A.J. Duncan and H. Kleinpoppen, J.Phys.B: At.Mol.Opt.Phys. <u>8</u>, L214 (1975).

19. T. Haji-Hassan, A.J. Duncan, W. Perrie, H. Kleinpoppen and E. Merzbacher, J.Phys.B: At.Mol.Opt.Phys., <u>24</u>, 5035 (1991).

20. J.G. Rarity, P.R. Tapster, E. Jakeman, T. Larchuk, R.A. Campos, M.C. Teich and B.E.A. Saleh, Phys.Rev.Lett. <u>64</u>, 1348 (1990).

SOME THEORETICAL AND COMPUTATIONAL RESULTS FOR ELECTRON MOLECULE SCATTERING.

D.G.Thompson
Department of Applied Mathematics and Theoretical Physics
Queen's University of Belfast
Belfast BT7 1NN
N.Ireland

1 Introduction

This paper is concerned with two theoretical projects being carried out at Queen's University and the University of Münster in Germany.

In the first project we consider the scattering of transversely polarised electrons by oxygen molecules and calculate the amount of depolarisation. In such an open shell molecule the depolarisation can be due to exchange and spin-orbit effects. However oxygen is a light molecule and the latter is expected to be negligible and is not considered in this calculation.

The second project, on the other hand, is concerned with the scattering of longitudinally polarised electrons by closed shell molecules. Here the exchange effect is zero and we must introduce a spin-orbit potential. We look for chiral effects in chiral and achiral oriented molecules, i.e effects which depend on the "handedness" of the system. The properties of chiral molecules, in particular the scattering of polarised light by such molecules, have been known for many years but similar investigations for electron scattering are much more recent.

2 Electron scattering by O_2

2.1 Scope of the calculations

The theoretical work was given impetus by the experiments of Hegemann et al (1991) on the elastic scattering of electrons by O_2, NO, which is another open shell molecule, and Na. While the change in polarisation was large as a function of scattering angle for Na the experimentalists were surprised that the variation for O_2 and NO was very small. Since both atom and molecules are open shell systems it was argued that we should expect similar behaviour in the depolarisation experiments.

The experimental result for elastic scattering has been confirmed (da Paixão et al, (1992); Fullerton et al, (1994)); the calculations have been extended to electronic excitation where considerable variation has been found (Fullerton et al, (1994)). Calculations have also been carried out for oriented O_2 where large variations have been found for particular orientations, even for elastic scattering (Nordbeck et al,(1994)).

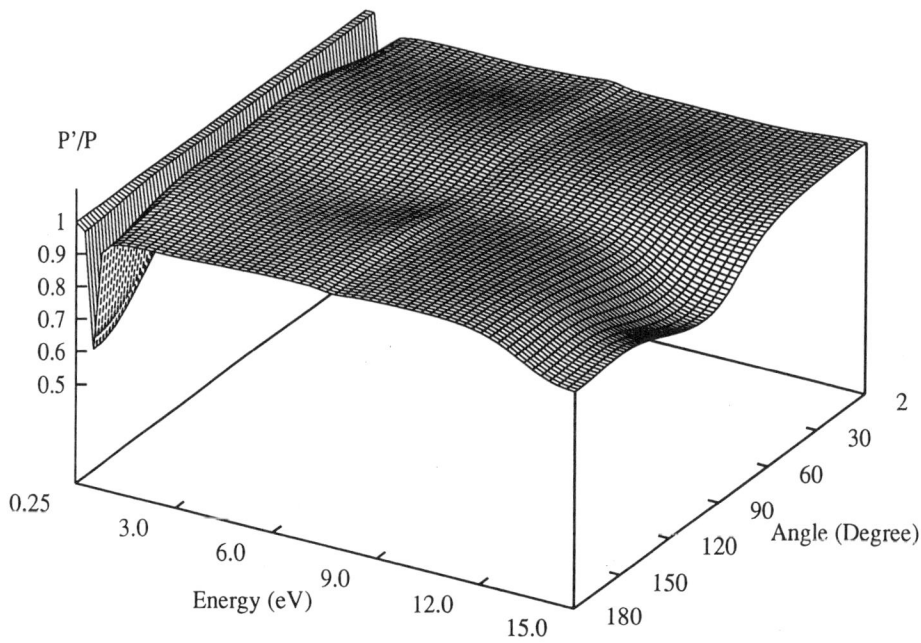

Figure 1 Polarisation fraction, P'/P, for X $^3\Sigma_g^- \to$ X $^3\Sigma_g^-$. (From Fullerton et al, (1994)).

2.2 Summary of the theory

Define a space fixed coordinate frame (X,Y,Z) and a molecular frame (x,y,z); (x,y,z) can be obtained from (X,Y,Z) by rotation through Euler angles $\Omega \equiv (\alpha, \beta, \gamma)$

We scatter electrons from a fixed molecule whose internuclear axis is along z. The electronic state i of the molecule is defined by λ_i, the component of molecular angular momentum along z, by S_i the total molecular spin, and by M_{S_i} the component of S_i along Z (we note that the spin-orbit interaction can be neglected and orbital and spin parts can be dealt with separately).

The electron is incident along Z and scattered in the XZ plane; let its spin state be m_s with respect to Z and its component of orbital angular momentum be m_l with respect to z.

Calculations of a T matrix have been carried out by Noble and Burke (1992) for several scattering states (ΛS) where $\Lambda = \lambda_i + m_{l_i}$ and S is the total spin ($S_i \pm 1/2$); for given S the scattering equations are independent of $M_S = M_{S_i} + m_{s_i}$.

The scattering amplitude for transitions from state I to state J with the electron scattered along **r** is

$$f(\lambda_J, S_J, M_{S_J}, m_{s_J}; \lambda_I, S_I, M_{S_I}, m_{s_I}) = \sum_S (S_I M_{S_I} 1/2 m_{s_I} | S M_S) F^{(S)}$$
$$(S_J M_{S_J} 1/2 m_{s_J} | S M_S)$$

where the $(a\alpha b\beta|c\gamma)$ are Clebsch-Gordon coefficients, and

$$F^{(S)} = i(\pi/k_I k_J)^{1/2} \sum_\Lambda \sum_{l_I} \sum_{l_J} \sum_{m_{l_J}} i^{l_I - l_J} (2l_I + 1)^{1/2}$$
$$D^{l_I *}_{0, \Lambda - \lambda_I}(\Omega) \, D^{l_J}_{m_{l_J}, \Lambda - \lambda_J}(\Omega) \, T^{\Lambda S}_{l_I, \Lambda - \lambda_I; l_J, \Lambda - \lambda_J} \, Y_{l_J m_{l_J}}(\hat{r})$$

(See Fullerton et al (1994) for further details)

An O_2 state is either a triplet or a singlet; for the triplets S can be 1/2 and 3/2, for the singlets only 1/2.

We consider the electron beam to have initial polarisation P_Y perpendicular to the scattering plane. The polarisation of the beam after scattering, P'_Y, by an ensemble of unpolarised target molecules can be obtained using the density matrix analysis of Bartschat and Madison (1988).

Neglecting the spin-orbit interaction we obtain, for triplet to triplet transitions

$$P'_Y / P_Y = 1 - (8/27) \frac{|F^{(1/2)} - F^{(3/2)}|^2}{\sigma}$$

where

$$\sigma = (1/3)(|F^{(1/2)}|^2 + 2|F^{(3/2)}|^2)$$

Since the spin-flip cross section is

$$w_{SF} = (1/2)(\sigma(\uparrow\downarrow) + \sigma(\downarrow\uparrow))$$
$$= (4/27)|F^{(1/2)} - F^{(3/2)}|^2$$

we have

$$P'_Y / P_Y = (1 - 2w_{SF}/\sigma) \quad (1)$$

Thus we see that the depolarisation fraction, P'_Y / P_Y is close to one if the spin-flip cross section is small compared to the total differential cross section.

Table 1 O_2 states used in calculation of Noble and Burke (1992)

State	calculated excitation energy (ev)	Experimental excitation energy (ev)
X $^3\Sigma_g^-$	0.00	0.00
a $^1\Delta_g$	0.93	0.98
b $^1\Sigma_g^+$	1.47	1.65
c $^1\Sigma_u^-$	5.49	6.12
C $^3\Delta_u$	5.68	6.27
A $^3\Sigma_u^+$	5.81	6.47
B $^3\Sigma_u^-$	10.86	9.25
$^1\Delta_u$	13.16	11.8
$^1\Delta_u^+$	14.67	13.25

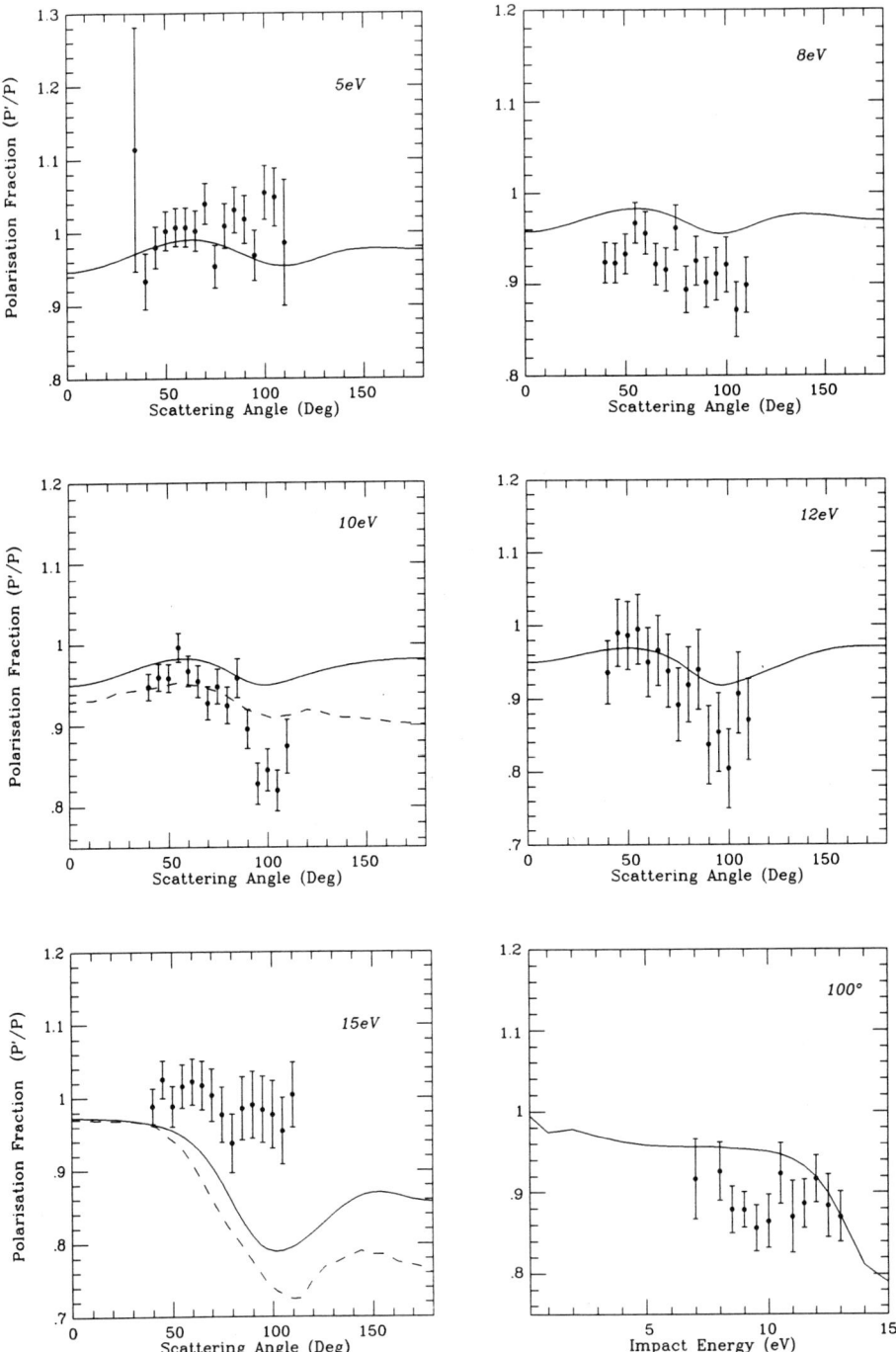

Figure 2 Theoretical and experimental values of P'/P for $X^3\Sigma_g^- \to X^3\Sigma_g^-$ in O_2: (a) 5 ev, (b) 8 ev, (c) 10 ev, (d) 12 ev, (e) 15 ev, (f) 100 degrees; full curve, theoretical values of Fullerton et al (1994); broken curve, theoretical values of da Paixão et al (1992); •, experimental values: (c) and (e), Hegemann et al (1991); (a), (b) and (d), Hegemann (1993); (f), Schroll (1993).

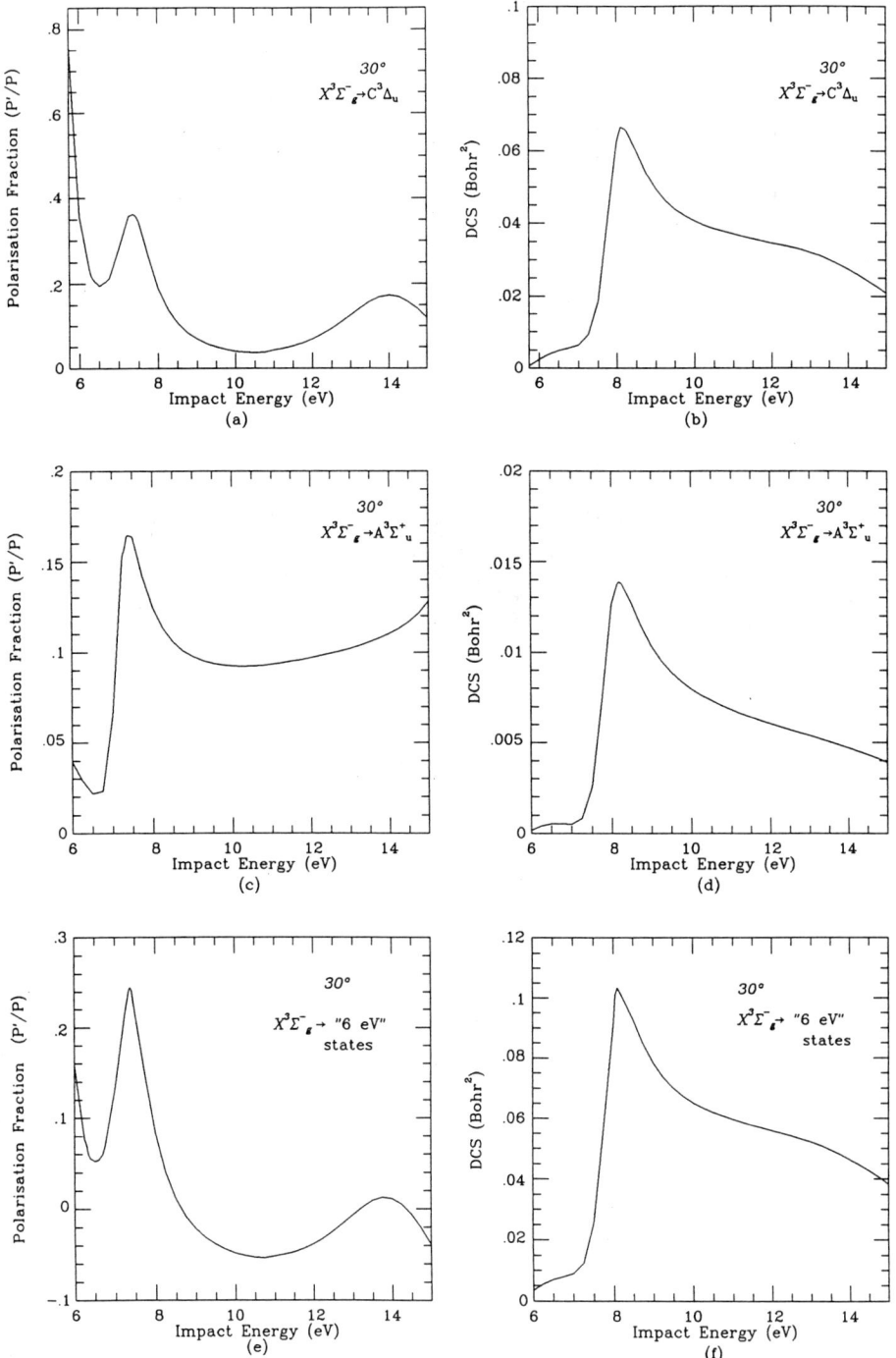

Figure 3 Polarisation fractions and differential cross sections for the 6 ev group of states of O_2 at 30 degrees: (a) and (b), $X^3\Sigma_g^- \to C^3\Delta_u$; (c) and (d), $X^3\Sigma_g^- \to A^3\Sigma_u^+$; (e) and (f), $X^3\Sigma_g^- \to$ 6 ev group of states. (from Fullerton et al (1994))

2.3 The scattering calculation

The depolarisation calculations of Fullerton et al (1994) have been made using the T matrix elements of Noble and Burke (1992) who carried out a 9 state R matrix calculation; the states and their energies are given in Table 1. The calculation gives three negative ion resonance states, $^2\Pi_g$ at 0.700 ev , width 0.026 ev , $^2\Pi_u$ at 7.845 ev , width 1.027 ev , and $^4\Sigma_u$ at 14.17 ev , width 3.39 ev . The position of the $^4\Sigma_u$ resonance is certainly too high since the calculation includes only 30 percent of the long range polarisability. There is also some discussion about the position of the $^2\Pi_u$ resonance; recent experimental differential cross sections (Middleton et al, (1994)) for excitation of the $a^1\Delta_g$ and $b^1\Sigma_g^+$ states suggest the resonance state may be at a higher energy. Further theoretical work needs to be done on this problem.

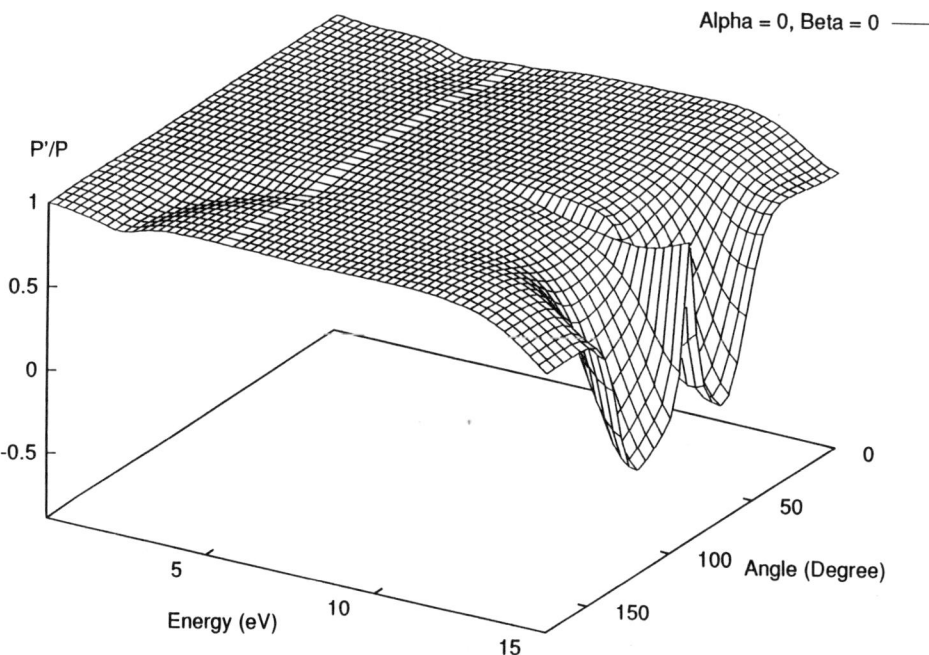

Figure 4 P'/P for $X^3\Sigma_g^- \to X^3\Sigma_g^-$ for O_2 oriented with $\alpha = 0$ and $\beta = 0$. (from Nordbeck et al, (1994))

2.4 Results for elastic scattering (random orientations)

The general structure of P'/P for $X\,^3\Sigma_g^- \to X\,^3\Sigma_g^-$ is seen clearly in Figure 1. There is structure at very low energies due to the $^2\Pi_g$ resonance but in general the behaviour is quite flat and close to 1. The effect of the $^2\Pi_u$ resonance is only seen as a very small 'crease'. There is more structure at the larger energies, probably due to the $^4\Sigma_u^-$ resonance.

More detail is given in figure 2, together with experiment and the theoretical results of da Paixão et al (1992), which is only a 3 state approximation. The experiment shows a slight dip at 100 degrees for 10 and 12 ev, otherwise it is very close to 1. The present theoretical results also show a dip but at too high an energy. However, as we have noted, improvements to the scattering calculation may lower the position of the $^4\Sigma_u^-$ resonance and lead to even better agreement with experiment.

2.5 Results for excitation to the "6 ev states" (random orientations)

There is a group of 3 states with energy approximately 6 ev above the ground state: $^3\Delta_u, ^3\Sigma_u^+$ and $^1\Sigma_u^-$. The polarisation analysis for the ground state to "6 ev" triplet transitions is as for elastic scattering, but for the triplet to singlet transition we obtain $P'/P = -1/3$.

The theoretical results for P'/P and σ for the triplet to triplet transitions at 30 degrees, see figures 3a-3d, show considerable structure around 8 ev due to the $^2\Pi_u$ resonance. Unfortunately there are no experimental results for these transitions, but it is likely that when they do become available they will be an average over all three transitions. For this reason it is interesting to note in figures 3e and 3f theoretical results for P'/P and σ averaged over the three states. 3e is similar in shape to the $^3\Delta_u$ but lower in size; the effect of the $^3\Sigma_u^+$ is small because of its small cross section.

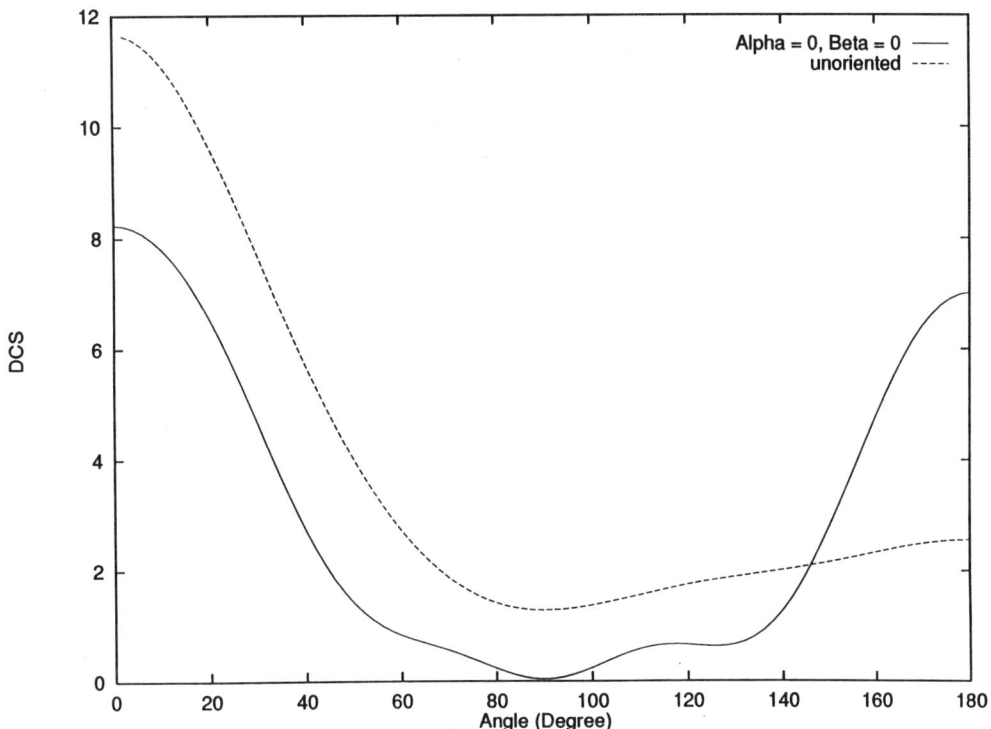

Figure 5 Calculated differential cross section at 15ev for $X^3\Sigma_g^- \rightarrow X^3\Sigma_g^-$. The dotted line is for randomly oriented O_2 and the continuous curve for O_2 oriented with Euler angles $\alpha = 0$ and $\beta = 0$. (from Nordbeck et al, (1994))

2.6 Results for elastic scattering - oriented molecules

In their paper using the Schwinger Variational method da Paixão et al (1992) gave a few results for oriented O_2 which showed large variations in P'/P. This has been investigated further in Nordbeck et al (1994).

We present their results for the orientation given by the Euler angles $\alpha = 0, \beta = 0$, (i.e, the molecule is oriented along the laboratory Z axis). There is little variation in P'/P except at the higher energies around 90 degrees. For this orientation we can show that only Σ scattering symmetries contribute to the scattering amplitude, i.e there are no contributions from the Π_g and Π_u resonances, as can be clearly seen in figure 4. Furthermore at 90 degrees the Σ_u contribution is zero, leaving the Σ_g contribution which is very small at 15ev. We note from equation 1 that when $F^{(1/2)}$ and $F^{(3/2)}$ are very small then both w_{SF} and σ are very small; small relative changes in the amplitudes can lead to large variations in P'/P. This is illustrated in figures 5 and 6. The differential cross section, see figure 5, is very small at 90 degrees and leads to large variations in P'/P as in figure 6.

Nordbeck et al (1994) conclude that it is likely that large variations are possible only near the minima of the differential cross sections; not many collisions are involved in this region and the contribution to the randomly oriented P'/P results is small.

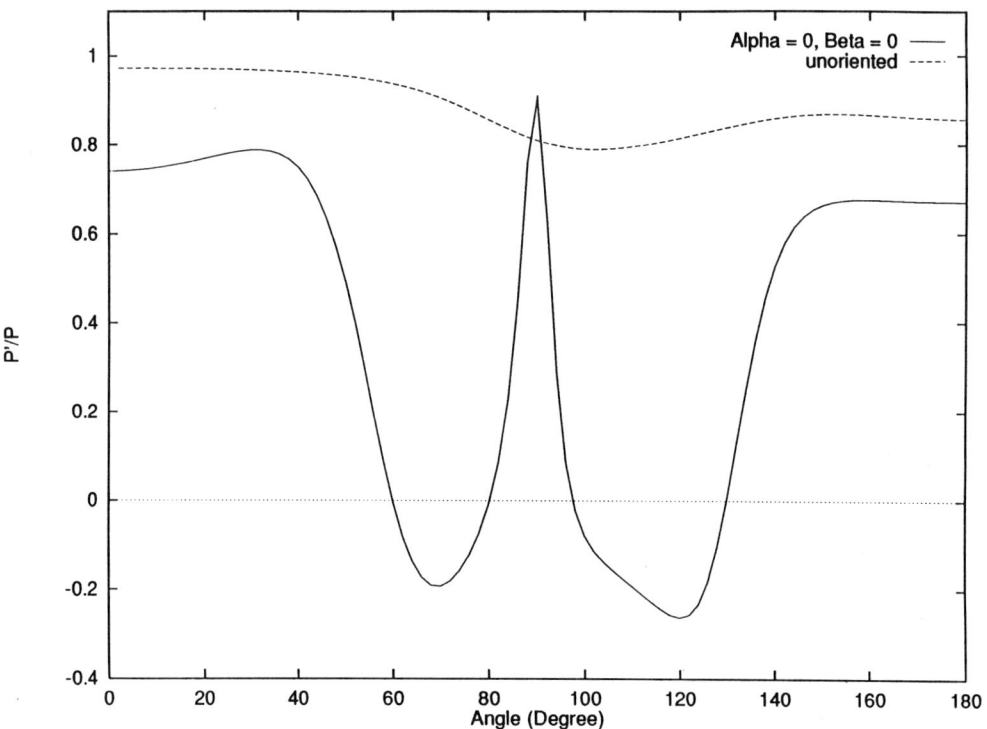

Figure 6 P'/P for $X^3\Sigma_g^- \to X^3\Sigma_g^-$. Calculated values at 15ev; the dotted curve is for randomly oriented O_2 and the continuous curve for O_2 oriented with $\alpha = 0$ and $\beta = 0$. (from Nordbeck et al, (1994))

3 Chiral effects in electron scattering

3.1 Electron optic dichroism

Chirality is usually thought of as being concerned with the "handedness" of systems, whether they are distinguishable from their mirror image, or equivalently, from their image under the spatial inversion operation - we will see below how this concept has to be extended.

In this discussion we consider only the elastic scattering of electrons by spinless molecules. If the initial spin state is $\chi(m_0)$ and the final spin state is $\chi(m_1)$ the scattering is characterised by four scattering amplitudes $f(m_1, m_0)$ which together give the scattering matrix

$$\begin{bmatrix} f(1/2,1/2) & f(1/2,-1/2) \\ f(-1/2,1/2) & f(-1/2,-1/2) \end{bmatrix}$$

For electron scattering by **atoms** (through angles $\theta\phi$) this matrix has the form (cf Kessler, (1985))

$$\begin{bmatrix} f & -ge^{-i\phi} \\ ge^{i\phi} & f \end{bmatrix}$$

i.e. there are only two independent amplitudes. We find that if the initial beam is unpolarised the final beam is polarised perpendicular to the scattering plane; if the beam is initially longitudinally polarised the differential cross section is independent of the polarisation.

But what would we expect if we replace the target atoms by chiral molecules, or indeed by any molecule? Farago (1980) coined the term "electron optic dichroism" in analogy to light scattering by chiral molecules. The properties are as follows:

(a) Compare the attenuations of a beam of longitudinally polarised electrons, polarisation $\pm P$. There should be a difference for chiral molecules. This was confirmed by Campbell and Farago (1985) for camphor.

(b) An initially unpolarised beam should attain longitudinal polarisation. An experiment by Beerlage et al (1981), again in camphor, was unable to confirm this.

(We note that there is also a chiral property analogous to optical activity but we will not discuss it in this paper.)

3.2 Summary of symmetry properties of the scattering matrix

(Fuller details of sections 3.2, 3.3 and 3.4 can be found in Johnston et al (1993)).

We consider the elements of the scattering matrix to be the matrix representation of an operator M_{op}:

$$f(m_1, m_0) = \langle m_1 | M_{op} | m_0 \rangle$$

We assume that the interaction between the electron and the molecule is invariant under spatial inversion and time reversal, and rotations of the whole system. M_{op} must have the same properties, i.e it is a proper scalar.

We set up a coordinate system:

$$\begin{aligned} \mathbf{n_1} &= (\mathbf{k_1} + \mathbf{k_0})/|\mathbf{k_1} + \mathbf{k_0}| \\ \mathbf{n_2} &= (\mathbf{k_1} - \mathbf{k_0})/|\mathbf{k_1} - \mathbf{k_0}| \\ \mathbf{n_3} &= \mathbf{n_1} \times \mathbf{n_2} \end{aligned}$$

where the electron has initial wave vector $\mathbf{k_0}$ and final wave vector $\mathbf{k_1}$.

We express M_{op} in terms of the Pauli matrices σ:

$$M_{op} = 1 g_0 + \sum_{i=1}^{3} g_i \mathbf{n_i} \cdot \sigma_i$$

where the g_i depend on the electron energy and the orientation of the molecule. With the assumption that M_{op} is a proper scalar we can calculate the invariance properties of the g_i functions under spatial inversion ($\mathbf{r} \to -\mathbf{r}$) and time reversal ($\mathbf{k_0} \to -\mathbf{k_1}$ and $\mathbf{k_1} \to -\mathbf{k_0}$). Results are given in Table 2. We note that g_1 is a time-even pseudo scalar and g_2 is a time-odd pseudoscalar.

Table 2. Transformation properties of the g_i functions. (from Johnston et al, (1993))

	Spatial inversion	Time reversal
g_0	+	+
g_1	−	+
g_2	−	−
g_3	+	+

3.3 Time-even and time-odd chirality

If we define a laboratory fixed coordinate system (XYZ) so that k_0 is along Z and (XZ) is the scattering plane, then we can show that (Johnston et al, (1993))

$$\begin{aligned}
2g_0 &= f(1/2, 1/2) + f(1/2, 1/2) \\
2g_1 &= (f(1/2, 1/2) - f(-1/2, -1/2))\cos\theta/2 \\
     &\quad + (f(1/2, -1/2) + f(-1/2, 1/2))\sin\theta/2 \\
2g_2 &= -(f(1/2, 1/2) - f(-1/2, -1/2))\sin\theta/2 \\
     &\quad + (f(1/2, -1/2) + f(-1/2, 1/2))\cos\theta/2 \\
2g_3 &= i(f(1/2, -1/2) - f(-1/2, 1/2))
\end{aligned}$$

If we scatter an electron beam with longitudinal polarisation ($\pm P$) and measure the intensities of the final beam, I(P), we find that the difference in attenuation

$$A = (I(P) - I(-P))/(I(P) + I(-P))$$

is

$$A = (C_e + C_o)/\sum_{i=1}^{3}|g_i|^2 \qquad (2)$$

where

$$\begin{aligned}
C_e &= 2[Re(g_0 g_1^*)\cos\theta/2 - Im(g_3 g_1^*)\sin\theta/2] \\
C_o &= 2[Re(g_0 g_2^*)\sin\theta/2 - Im(g_2 g_3^*)\cos\theta/2]
\end{aligned}$$

We also find that the if the initial beam is unpolarised the induced polarisation in the k_1 direction is

$$P = (C_e - C_o)/\sum_{i=1}^{3}|g_i|^2 \qquad (3)$$

The chiral effects, depending on g_1 and g_2, are a combination of a time-even and a time-odd pseudo scalar. Johnston et al (1993) introduce the terms time-even and time-odd chirality.

We note that for electron scattering by atoms $g_1 = 0 = g_2$. (c.f the expression for the scattering matrix)

3.4 Conditions for g_1 and g_2 to be zero.

Since the crucial terms for chiral effects are g_1 and g_2 it is important to obtain the conditions for them to be non-zero. Johnston et al (1993) found that:

(i) The molecules should be chiral or the scattering system should be chiral. By "system" we mean molecule + electron trajectory; a system is chiral if it is distinguishable from its image under spatial inversion. As an example of a chiral system consider scattering by a diatomic molecule lying along the Y axis with electrons incident along the Z axis and scattered in the XZ plane. Carrying through the spatial inversion operations, X→−X, Y→−Y and Z→−Z, it is easy to see that the transformed system is distinguishable from the original system, even under any natural rotation, provided the molecule is hetero-nuclear and not homo-nuclear. As another example consider scattering by water molecules; the system is chiral unless the scattering plane coincides with one the molecular symmetry planes.

(ii) However (i) is not a sufficient condition. It is found that we have to look at the invariance properties of the system under time-reversal.

For g_1 to be non-zero the system must be distinguishable from its image under the combined operations of spatial inversion and time-reversal.

For g_2 to be non-zero the system must be distinguishable from its image under time-reversal.

The example given in (i) for the diatomic molecule is actually of time-odd chirality, i.e $g_1 = 0$ but $g_2 \neq 0$.

We note further that for randomly oriented systems the terms involving g_2 average to zero - i.e time -even chirality only - and for randomly oriented systens involving achiral molecules terms involving g_1 also average to zero.

Table 3 A, (equation 2) and P, (equation 3) for oriented HCl ($\alpha, \beta, \gamma = \pi/2, \pi/2, 0$) and oriented H_2S ($\alpha, \beta, \gamma = \pi/4, \pi/4, 0$) at 5 ev.

Angle	P(HCl)	P(H_2S)	A(H_2S)
30	.0018	-.0001	.0006
60	.0032	-.0009	.0017
90	.0037	-.0022	.0028
110		.0121	-.0124
120	.0036	.0003	-.0014
150	.0030	-.0006	.0002

3.5 The size of the effects

Estimates of the size of these effects have been made for a model chiral molecule using perturbative techniques (Blum and Thompson, (1989); Blum et al, (1991); Fandreyer et al, (1991); Johnston et al, (1993)). Significant effects were obtained particularly near sharp resonances and minima in the differential cross-section.

Calculations have been carried out recently for oriented H_2O, H_2S and HCl, (Greer, (1994); Greer and Thompson, (1995) ; Smith and Thompson,(1995)) solving the full static exchange equation for the molecule and also including a spin-orbit term for the interaction of the incident electron with the heavy atom. Preliminary results for A (equation 2), and P (equation 3) are shown in Table 3 for H_2S and HCl. The H_2S moleule is oriented with respect to a laboratory frame of axes through Euler angles

$(\alpha\beta\gamma) \equiv (\pi/4, \pi/4, 0)$, and the HCl through $(\alpha\beta\gamma) \equiv (\pi/2, \pi/2, 0)$ which correspond to the time-odd case discussed above. (Note that $P = -A$ for time-odd chirality).

We note that the size of the effect is small but not insignificant. As for the model molecule calculation the values can become much larger near cross section minima. It is probable that values for oriented chiral molecules are several orders of magnitude greater than for randomly oriented ensembles.

Acknowledements

The author is very pleased to acknowledge the work done by colleagues on the projects discussed in this review. Belfast: Prof. P.G.Burke, Miss C.M. Fullerton, Dr R.A.Greer and Mr I.M.Smith; Münster: Prof. K.Blum and Mr G Wöste; Belfast and Münster: Dr. C.Johnston; Daresbury Laboratory: Dr C.J.Noble. This work was supported by the Science and Engineering Research Council and by NATO Collaborative Research Grant CRG930056.

References

Bartschat,K., and Madison,D.H.,1988,J.Phys.B:At.Mol.Opt.Phys.,21,2621
Beerlage,M.J.M.,Farago,P.S., and Van der Wiel,M.J.,1981 J.Phys.B:At.Mol.Phys.,14,3245
Blum,K. and Thompson,D.G.,1989,J.Phys.B:At.Mol.Opt.Phys.,22,1823
Blum,K.,Fandreyer,R., and Thompson,D., 1990, J.Phys.B:At.Mol.Opt.Phys.,23,1519
Campbell,D.M., and Farago,P.S.,1985,Nature,318,52
da Paixão,F.J.,Lima,M.A.P., and McKoy,V.,1992, Phys.Rev.Lett.,68,1698
Fandreyer,R.,Thompson,D., and Blum,K.,1990, J.Phys.B:At.Mol.Opt.Phys.,23,3031
Farago,P.S.,1980,J.Phys.B:At.Mol.Phys.,13,L567
Fullerton,C.M.,Wöste,G.,Thompson,D.G.,Blum,K., and Noble,C.J., 1994, J.Phys.B: At.Mol.Opt.Phys., 27,185
Greer,R.A.,1994, Phd thesis,Queen's University of Belfast
Greer,R.A., and Thompson,D.G.,1995,J.Phys.B:At.Mol.Opt.Phys.,submitted
Hegemann,T.,1993,Doctoral Thesis,Universität Münster
Hegemann,T.,Oberste-Vorth,M.,Vogts,R., and Hanne,G.F.,1991, Phys.Rev.Lett.,66,2968
Johnston,C.,Blum,K., and Thompson,D.,1993,J.Phys.B:At.Mol.Opt.Phys.,26,965
Kessler,J.,1985,Polarised Electrons (Berlin-Springer)
Middleton,A.G.,Noble,C.J.,Anderson,M.W.B.,Teubner,P.J.O., Wöste,G., Burke,P.G., Brunger,M.J., Fullerton,C., and Blum,K., 1994,J.Phys.B:At.Mol.Opt.Phys,to be published
Noble,C.J., and Burke,P.G.,1992,Phys.Rev.Lett.,68,2011
Nordbeck,R-P.,Fullerton,C.M.,Wöste,G.,Thompson,D.G., and Blum,K., 1994, J.Phys.B: At.Mol.Opt.Phys.,27,5375
Schroll,S.,1993,Diplom Thesis,Universität Münster
Smith,I.M., and Thompson,D.G.,1995,to be published

SPIN DEPENDENT ELECTRON SCATTERING FROM ORIENTED MOLECULES: AN EXPERIMENTAL APPRAISAL

N J Mason

Department of Physics and Astronomy
University College London
Gower Street
LONDON WC1E 6BT

INTRODUCTION

The development of electron-photon coincidence techniques and polarised electron scattering experiments has led to a major improvement in our understanding of electron-atom scattering interactions (see A. Crowe and W Raith elsewhere in this volume). It is now possible to perform so called "perfect scattering experiments", in which the maximum information on the electron-atom scattering dynamics is measured, namely the magnitudes and relative phases of the scattering amplitudes. Comparison of these experimental results and rapidly developing theory (using the most powerful modern super-computers) is encouraging, so that it is perhaps now possible to state that our understanding of this most fundamental of collision processes is now largely complete and in the future will only require "fine tuning".

Corresponding studies of electron-molecule scattering are less advanced. The proficiency of internal modes of excitation (eg rotation and vibration) and the multi-centred nature of the molecular target make it difficult to parameterise both theoretical and experimental treatments of the electron-molecule scattering problem. At present we do not have an exhaustive set of experimental data on the excitation cross section of any molecule. Different theoretical treatments are conflictory in their predictions and often disagree with experimental results. Indeed, unlike the electron-atom scattering interaction where helium is the agreed standard for testing both theory and experiment, there is as yet no agreement on a standard electron-molecule scattering interaction against which new theory and experiment can be judged.

However the interaction of electrons with molecules is important in all areas of science from aeronomy (auroral and ionospheric phenomena) to cellular biology. For example, one of the most intriguing problems in chemical evolution is the origin of optical asymmetry in biomolecules, whereby amino acids in the natural proteins belong overwhelmingly to the L series of optical isomers whereas natural sugars are made up almost exclusively of D optical isomers. In contrast, if optically active molecules are synthesised in the laboratory, right-handed and left-handed molecules are produced with equal probability. Therefore the study of the electron interaction with optically active molecules has grown in interest in recent years, simulated in part by the Vester-Ulbricht hypothesis on the origin of the chirality of biological molecules, and by the work of Farago (1980, 1981) who predicted interesting new spin dependent effects due to the "parity violation" originating from the chiral structure of the target.

Several experimental studies have been attempted to measure the geometrical and dynamical relations between spin polarisation phenomena and the dissymmetry of the target. However these experiments have been inconclusive. The pioneering experiment of Campbell and Farago (1987) reporting electron optical dichroism with polarised electron scattering from D(+) and L(-) camphor molecules has not been able to be repeated, despite the efforts of research groups in Munster, Germany and Rice, USA. Analagous positron impact experiments (Bisi et al (1992)) have been equally inconclusive despite the additional sensitivity available from positronium yield studies, which in part negate the problems of low detection efficiency prevalent in Mott detection of polarised electrons.

Theoretical analysis of such effects were pioneered by Rich et al (1982) and developed by Hayashi (1985, 1988) who, using multiple scattering theory, showed that all such polarisation phenomena are enhanced if resonances are present. However, it was only with the work of Blum and Thompson (1989 and this volume) and Kohl and Shipsey (1992) that a quantitative investigation of spin dependent electron-molecule scattering began. Blum and Thompson using a trial molecule Bi H_3, (since it is not possible for modern theory to tackle a molecule as large as camphor) whose chirality is generated by non-equal bond lengths, predicted the strongest asymmetry in polarised electron scattering to occur at both resonance energies and in the minimum of differential cross sections, but with magnitudes still significantly less than those predicted by the experiment of Campbell and Farago.

Non-Chiral (achiral) molecules may also show dichroism if they are fixed in space, i.e. if they are spatially oriented. Circular dichroism in the angular distribution (CDAD) of photoelectrons has been demonstrated in a series experiments by Schonhense and co-workers and recently the first experiments on electron scattering from oriented molecules have been performed. It is therefore now possible to discuss new experiments that demonstrate chirality in non-chiral but oriented molecules. In this review these early experiments will be discussed, the techniques for producing oriented molecules in the gaseous, liquid and solid state will be described and the feasibility of new spin polarised electron experiments analyzed.

CIRCULAR DICHROISM IN ORIENTED MOLECULES

Schonhense and co-workers have demonstrated that circular dichroism in the angular distribution (CDAD) of photoelectrons is an effective tool for studying oriented molecules. CDAD is the difference between photo-electron currents ejected at a definite angle by left and right circularly polarised light. If achiral molecules are neither oriented or aligned, CDAD is zero. However, should the molecules be spatially oriented, Cherepkov (1982, 1983) predicted that CDAD asymmetries are of the same order of magnitude as the differential cross section itself and hence would be easily observable in a photo-emission experiment. The optical activity of unoriented chiral molecules is due to the dissymmetry of their structure and is described by the electric dipole-magnetic dipole interference terms consequently the measured asymmetries are weak. The optical activity of oriented molecules is associated with the dissymmetry of the geometry of the experiment and occurs in the electric dipole approximation and hence the asymmetries are that much stronger. Accordingly, using the terminology proposed by Barron (1986) the optical activity of oriented molecules is a "false chirality".

Two types of molecular orientation may be considered. The first is the spatial orientation of molecules in a gas beam due to the orientation of dipole moments in the molecules. The second, known as rotational orientation, is the orientation of the angular momentum vectors of the molecules. Both types of alignment have been verified experimentally.

Gas phase studies

Polar molecules cooled in a supersonic expansion into their lowest vibrational ground state and only a few rotational states, can be oriented by means of the interaction of their dipole moments with an electrostatic hexapole lens and orientation field plates (the linear Stark effect). The hexapole acts as a state selector focusing those molecules in states exhibiting a positive Stark effect while all other molecules are either not influenced or are dispersed (P R Brooks, (1976)). An electric guiding field, which becomes steadily weaker downstream, but never zero, can then transfer the oriented molecules from the hexapole region to a collision region. (See figure 1).

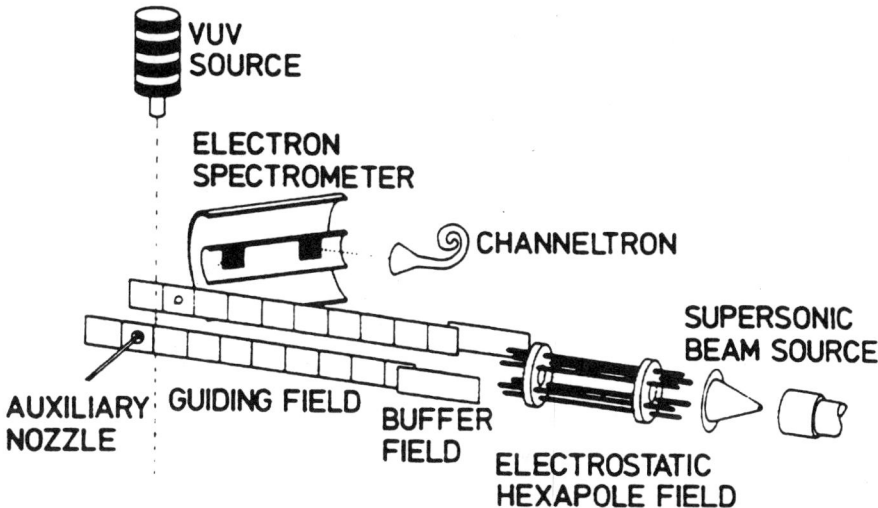

Figure 1. Schematic diagram of the hexapole apparatus used by Kaesdorf et al (1985)

The "orientation" of the molecules is then defined as a measure of the molecular dipole moment μ with respect to the electric field E, angle θ. Defining K and M as the quantum numbers of the total angular momentum J projected onto the molecular symmetry axis and on the electric field direction respectively, symmetric top molecules in specific rotational states J, M, K> with $<\cos\theta> = KM/J(J+1) \neq 0$ are then oriented and the rotational probability distribution $P_{JKM}(\cos\theta)$ can be calculated quantum mechanically. The force exerted on the molecules in the homogeneous electric field of the hexapole F = - grad W_{stark} (where W_{stark} is the interaction energy (= - μ E cos $<\theta>$)) separates molecules with different orientation of their axis relative to the local electric field direction spatially such that for a hexapole voltage U_o only molecules with a specified degree of local orientation can pass the hexapole exit aperture.

The first experimental demonstration for CDAD using gas phase oriented molecules was performed by Kaesdorf et al (1985). Diatomic closed-shell molecules with permanent electric dipole movements rotate perpendicularly to μ and so require enormous electric fields (~ 10^7 V/cm) to overcome the rotational inertia of the molecules. Kaesdorf et al, therefore, chose the symmetric top molecule CH_3I whose rotational states have momentum J, with a component parallel to μ, thus lower electric fields may be used in the hexapole to orient the molecules. Oriented CH_3I molecules were photoionised by VUV radiation and the asymmetry for photoelectrons emitted parallel or anti-parallel to the molecular axis measured. If the methyl group is directed towards the spectrometer, a photoelectron current I^+ is detected, if the iodine atom is directed towards the spectrometer, a current I^- is detected. Figure 2 shows the asymmetries I^-/I^+ measured by Kaesdorf et al for the $^2E_{1/2}$, $^2E_{3/2}$ ionic states resulting from ionisation of the lone-pair-type orbital 2e, strongly localised on the iodine atom. For both HeI radiation (21.22eV) and NeI radiation (16.67 and 16.85eV) a small but finite asymmetry is observed, in contrast unoriented molecules (φ) show no asymmetry.

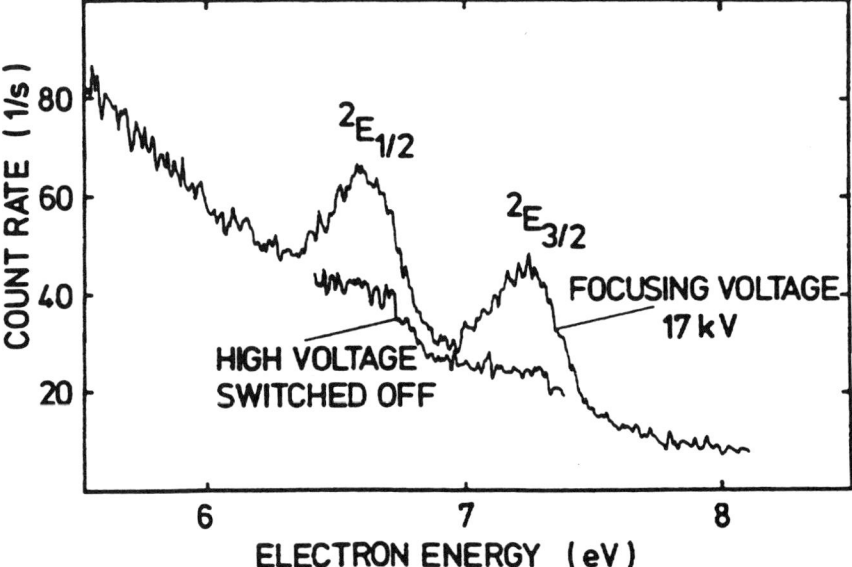

Figure 2. Photoelectron spectrum of oriented and focused CH_3I molecules by use of NeI VUV radiation (Kaesdorf et al (1985)

Circular Dichroism in photo-emission from adsorbates

The orientation of molecules adsorbed on surfaces has been well documented. (See Plummer and Eberhard (1982), Palmer and Rous (1992)). Physisorbed molecules in coupling to the substrate experience both an attractive Van der Waals force and a repulsive force arising from the shorter range overlap of the charge densities of the molecule and the surface, such that if the attractive force is dominant the physisorbed molecules may align themselves perpendicular to the surface, whereas if the repulsive forces are greater the molecules will be parallel to the surface. Spatial orientation within the adsorbate follows.

Similarly, chemisorbed molecules on crystal surfaces may also be fixed in space, a chemical bond being formed by charge transfer between the adsorbed molecule and the surface. Carbon monoxide adsorbed on various transition metal surfaces is perhaps the best-characterised molecular chemisorption system. It is predominantly found in an upright orientation with the carbon end pointing towards the surface such that the rotational degree of freedom is quenched.

In contrast to gas phase studies, on surfaces a high target density may be obtained, resulting in huge photoelectron intensities, hence allowing CDAD studies to be made in the

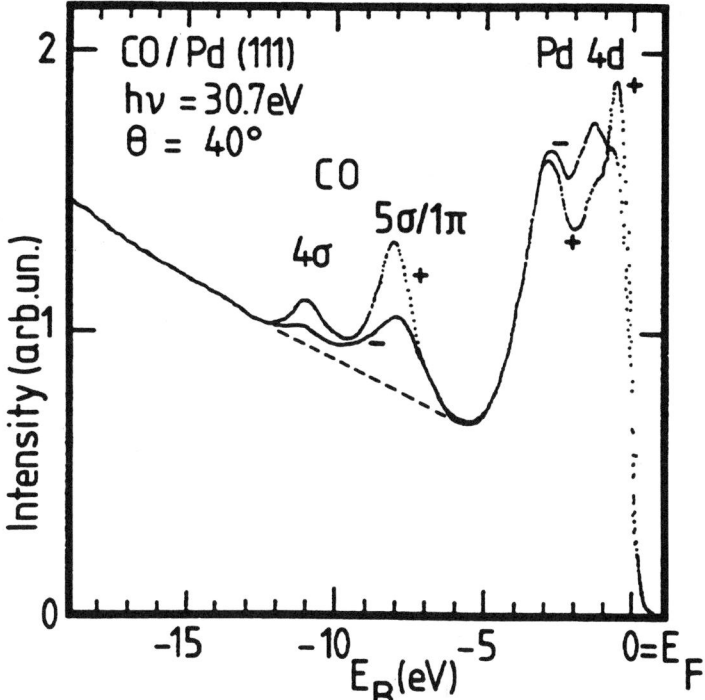

Figure 3. Typical photoelectron spectra of CO on Pd(111) illustrating CDAD. + and - denote right- and left-handed circularly polarized light, respectively. The binding-energy scale refers to the Fermi energy E_f of the substrate crystal. [Westphal et al (1989)]

solid phase with greater statistical accuracy. However, a disadvantage of this method is that the mutual interaction of the molecules and their bonding to the surface must be taken into account.

Figure 3 shows a photoemission spectrum for both right circularly polarised light (+) and left circularly polarised light (-) from the BESSY synchrotron. [Westphal et al (1989)]. Huge CDAD asymmetries are observed. Introducing an asymmetry function ACDAD as

$$ACDAD = \frac{I^+(\Theta) - I^-(\Theta)}{I^+(\Theta) + I^-(\Theta)}$$

allows the asymmetry to be mapped as a function of emission angle (figure 4). The measured asymmetries reach high values of 80% and are in good agreement with numerical calculation confirming that CDAD occurs within the pure electric-dipole approximation.

In a series of experiments Westphal et al (1991) demonstrated CDAD for NO, benzene and CH_3I on Pd (111). CDAD provided information on both the geometry of the adsorbate (molecular orientation angles with respect to the surface), electronic structure and photoemission dynamics (transition matrix elements and phase-shift differences). The possible extension of CDAD to magnetic systems, where it is analogous to the magnetic circular dichroism in X-ray photoabsorption [see Van der Laan this volume], and to the study of alignment of molecules in liquid surfaces remains an open question since there is presently no theoretical treatment of CDAD in these cases. Nevertheless, CDAD is already becoming established as a tool for the characterization and monitoring of surface layers and in the future, may become a standard technique, albeit requiring a synchrotron facility!

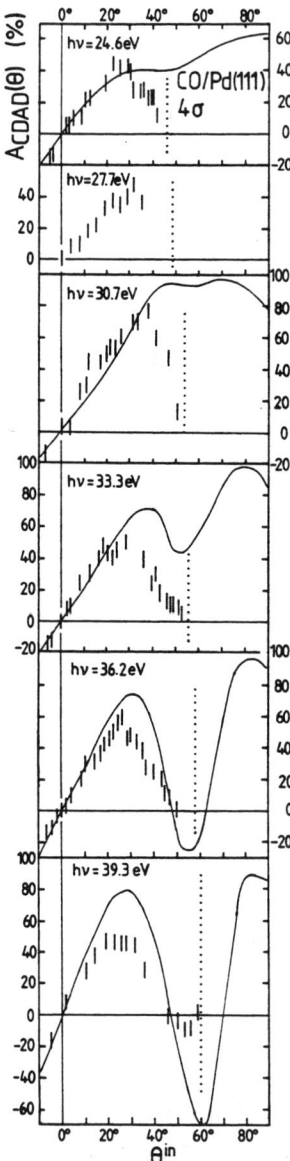

Figure 4. Circular dichroism in the angular distribution (CDAD) for CO on Pd(111) at selected photon energies experiment solid line theory. [Westphal et al (1989)]

ELASTIC ELECTRON SCATTERING FROM ORIENTED MOLECULES

A strong orientational dependence in the differential cross sections of elastic scattering from molecules assumed to be fixed in space was first predicted by the calculations of Fink and co-workers (1989 a, b) and the first experimental confirmation of these effects has recently been reported by Volkmer et al (1992, 1993). Hence a study of electron dichroism from oriented molecules is now possible.

Fink et al considered electron scattering from CH_3I with the electron wave front and molecular bond either parallel or perpendicular to one another and predicted that the scattering cross sections were sufficiently different that an apparatus could be built to determine the orientation of the molecules by an electron scattering technique. It was necessary to introduce an additional scattering angle to define the orientation of the electric field direction with respect to the scattering plane described by the polar angle θ_s and the azimuthal angle χ_s (see figure 5). ϑ is the usual scattering angle between the incident and outgoing electron momenta K_o and K. The elastic differential cross section $(d\sigma/d\Omega)_{aligned}$ at high incident energies and large scattering angles is then found using the independent atom model (IAM). With the CH_3I molecule reduced to CI for convenience (the three hydrogen atoms scatter only weakly compared to the carbon and iodine).

$$\left(\frac{d\sigma}{d\Omega}\right)_{aligned} = \left(\frac{d\sigma}{d\Omega}\right)_{random} + M_{JKM}(\theta_s, \vartheta\chi_s)$$

Where M_{JKM} is an interference term containing additional contributions originating from the molecular orientation and $(d\sigma/d\Omega)_{random}$ is the electron collision cross section for randomly oriented molecules. [Volkmer et al (1992)].

In any experiment, an ensemble of quantum states of oriented molecules is produced in the scattering region, if the individual states [denoted by $i = J, K, M>$] are populated with a relative population fraction ω_i, the intensity, (I), of electrons scattered into a solid angle $\Delta\Omega$ is

$$I(\theta_s, \vartheta, \chi_s) = \sum_i W_i I_i + I_B$$
$$= I_{RB} + I_o N_t \sum_i W_i M_i \Delta\Omega$$

where I_o is the current density of the electron beam, N_t is the number of target molecules, I_B is the background scattering intensity due to the residual gas and I_{RB} the sum of I_B and orientation-independent part of the contribution.

To determine the intensity contribution from randomly oriented target molecules $I_R(\theta_s, \vartheta) = C(\theta_s, \vartheta)$ the background from residual gas scattering is measured and subtracted from $I_{RB}(\theta_s, \vartheta)$. A direct measure of the orientational molecular interference contribution, M_i, is then found by determining the relative change of

$$\Delta\sigma(\theta_s, v) = \frac{I(\theta_s, v) - I_{RB}(\theta_s, \vartheta)}{I_R(\theta_s, v)}$$
$$= \sum_I W_i M_i(\theta_s, \vartheta)\left(\frac{d\sigma}{d\Omega}\right)_{random}$$

$\Delta\sigma$ is independent of computational models, background intensity, the incident electron beam current, the solid angle of the detector and the apparatus asymmetry.

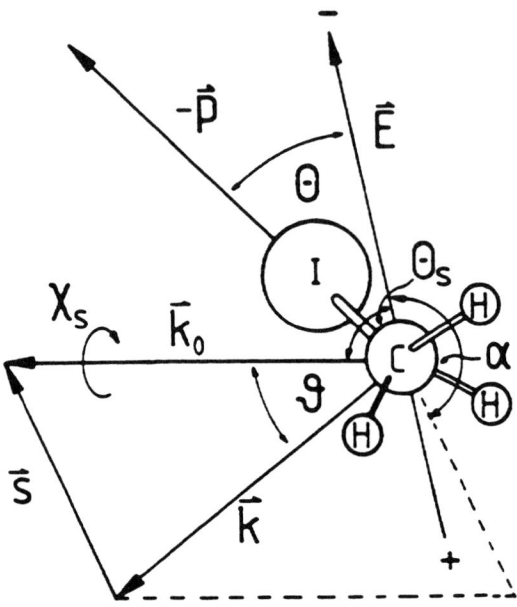

Figure 5. Schematic of the scattering geometry illustrating the relative orientations of the vectors **E, p, K₀, k and s**. Volkmer et al (1992)

Volkmer et al (1992) measured $\Delta\sigma$ for CH_3I with the CH_3 group pointing preferentially towards the incident electrons ($\theta_s = 0°$) and away ($\theta_s = 180°$). Using the design of Kaesdorf (figure one) a CW supersonic molecular beam with translational energy of ~100 meV passed through a hexapole assembly comprising six rods charged to $\pm U_o$ KV. A central beam stop prevented molecules with $<\cos\theta_i> = 0°$ from entering the interaction region while focusing molecules exhibiting a positive Stark effect $<\cos\theta_i> < 0$ into a guiding field for transport to the scattering region. The electron beam (700 - 1000 keV, 10 µA) crossed the molecular beam perpendicularly while being parallel to the orientation field. The detector was composed of a retarding field analyzer and a channeltron, and could be rotated about the scattering plane by means of a stepper motor. The molecular interference term could then be monitored as a function of the transferred momentum S (for scattering angles 4 - 15°). By turning off the guiding fields the orientation could be turned off without effecting the electron scattering and I_{RB} (θ_s, ϑ) determined. By removing the hexapole voltage any inherent apparatus asymmetry could be investigated.

Figure 6 shows the results of Volkmer et al (1992). When U_o = OKV $\Delta\sigma(\theta_s)$ is zero, within error, as is to be expected if there is no significant apparatus asymmetries. At Uo = ± 7 KV significant variations were found for oriented CH_3I molecules pointing toward ($\theta_s = 0°$) or away ($\theta_s = 180°$) from the incident electron beam.

Using a rotational temperature of T = 70K an average degree of orientation $<\cos\theta>$ of 0.25 was predicted by Volkmer et al. The interference terms $\overline{M}(\theta_g, v) = \sum_i W_i M_i$ were then determined using the independent atom model of Fink et al and compared with experiment. Agreement was found to be good with both theory and experiment predicting a larger maximum for $\theta_s = 0°$ than for $\theta_s = 180°$, but theory predicts a deeper minimum at small momentum transfer, and the zero crossings of $\Delta\sigma$ are predicted to lie at larger angles than experimentally observed.

Since this experiment measures the molecular axis both parallel and anti parallel to the electron beam $\overline{M}$ can be divided into two parts; the pure orientation part $\overline{M}_o(\vec{S})$ and

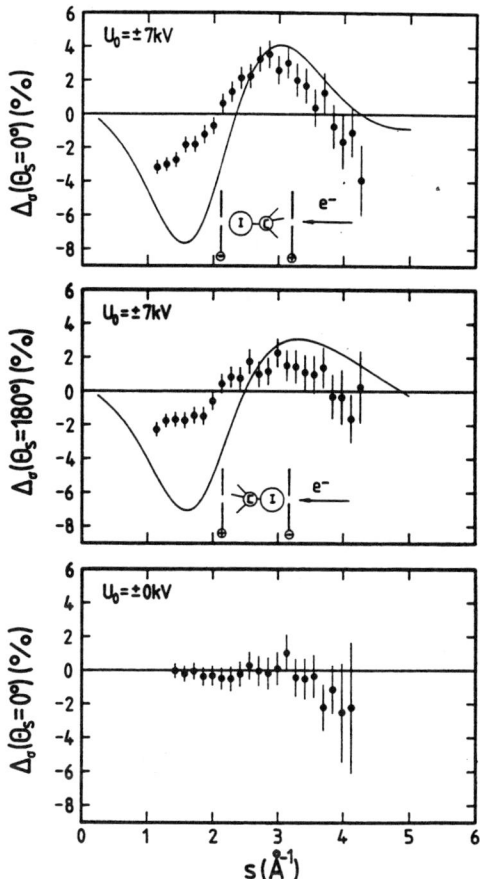

Figure 6. Deviation of the elastic differential cross section for oriented molecules from that of unoriented molecules obtained using Eq. (1) at 1 keV electgron energy (hexapole voltage U_o as indicated). The data points ($\pm$ 1σ statistical error) are for the field orientation angles $\Theta_{s'} = 180°$ (center), and for molecular beam switched off (bottom). The full curves represent the theoretical result for $T_{rot} = 70$ K. Volkmer et al (1992)

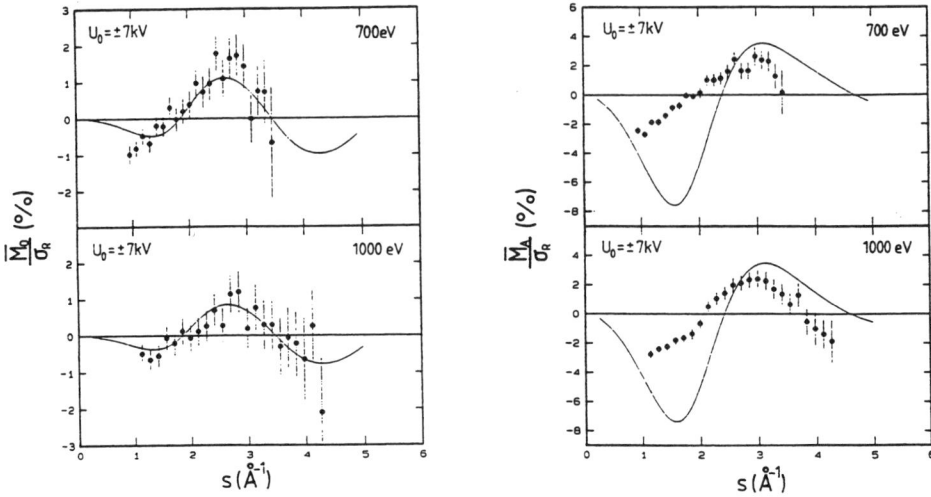

Figure 7. (a) Relative orientation part (b) relative alignment part both normalised to unoriented elastic differential cross section at 700 ml 1000 eV. Solid line theoretical calculation. Volkmer (1993)

alignment part $M_A(\vec{S})$ (see Figure 7). As the incident electron energy is increased the absolute value of the orientation part $\overline{M}_o$ decreases, but the alignment part $\overline{M}_A$ hardly changes. However, when the hexapole field is reduced ($U_o = \pm 4$ KV) $\overline{M}_o$ increases whilst $\overline{M}_A$ decreases. Kohl and Shipsey's theory predicts the variation of $\overline{M}_o$, but not that of $\overline{M}_A$. However, due to the limited knowledge of the rotational state distribution within the CH_3I beam it is too early to fault the theory. Several improvements to the experiment are needed to test theory more vigorously and experiments should be performed at lower incident energies where the cross sections are larger and may offset the low target densities available after selection of orientation (10^8-10^9 molecules per cm^3). Experiments at larger scattering angles may reveal additional information on the spin flip scattering amplitude, but at larger scattering angles the cross sections are smaller (often by several orders of magnitude)

Nevertheless, these early theoretical and experimental studies have demonstrated that electron scattering is a viable tool for molecular orientational studies in the gas phase and that a dichroism may be measured from achiral molecules.

SPIN DEPENDENT ELECTRON SCATTERING FROM ORIENTED MOLECULES

Having discussed the existing studies of optical dichroism for oriented molecules in both gas and solid phases of matter and having considered elastic scattering from oriented molecules, it is timely to consider spin polarised electron scattering from oriented molecules.

In 1985 Campbell and Farago reported an experiment suggesting that spin polarised electrons when scattered from chiral molecules can exhibit electron optic dichroism, in that there is an asymmetry in the scattering of longitudinally polarised electrons from D+ and L- isomers of Camphor, such a dichroism may also be observed if the target is achiral but oriented.

Blum and Thompson (1989) derived a general theory for spin dependent elastic forward scattering of electrons from oriented molecules and presented some numerical estimates for the predicted asymmetries. Using an initially unpolarised electron beam, they showed that both longitudinal and transverse spin polarisation were small but finite in contrast to randomly oriented molecules which show essentially no polarisation. Similarly depolarisation of an elastically scattered beam of transverse polarised electrons incident upon an oriented molecular target is expected to be significant (Blum et al (1990)) whereas the depolarisation (P^1) of polarised (P) electrons scattered from a (random) gaseous target shows no depolarisation [Hegemann et al (1991) and Figure 8].

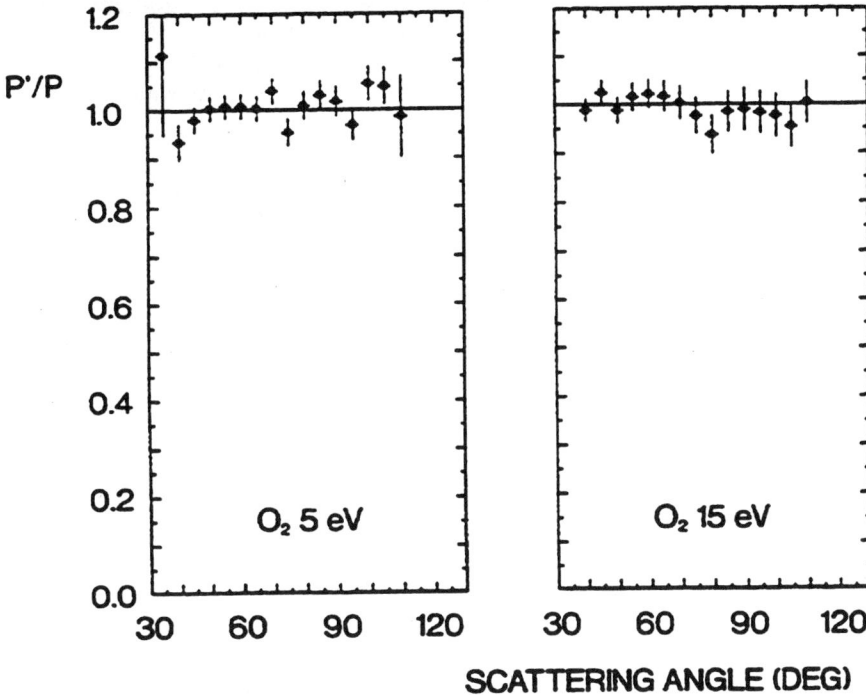

Figure 8. Experimental results of P^1/P plotted against scattering angle for elastic collisions of polarized electrons from 0_2 molecules at 5 and 15 eV.

Furthermore, such depolarisation is expected to increase dramatically when the molecule is excited or the electron forms a resonant state with the target. Asymmetries of $\geq 3 \times 10^{-3}$ have recently been predicted for oriented HCl molecules (see Thompson, this volume) which are, in fact, similar to those reported by Campbell and Farago (1987) for longitudinally polarised electrons scattering from gas phase D^+ camphor ($A = 23 \pm 11 \times 10^{-4}$) or L^- camphor ($A = -50 \pm 17 \times 10^{-4}$).

It is, therefore, timely to perform experiments on the scattering of <u>polarised electrons</u> from oriented molecules. At the present time there are three possible experimental arrangements that could be adopted:
(i)   Spin polarised electron scattering from gas phase oriented molecules
(ii)   Spin polarised electron scattering from adsorbates
and a very recent development
(iii) Spin polarised electron scattering from liquid surfaces.
       Each of these will be discussed in turn.

(i)   Spin polarised electron scattering from gas phase oriented **molecules**:
Following the theoretical suggestions of Blum, Thompson and co-workers, an initially unpolarised beam of electrons scattered from a beam of oriented molecules may gain longitudinal or transverse spin polarisation from the chirality of the molecular system. Alternatively, a beam of polarised electrons (strength P) scattered from a beam oriented

molecules may suffer depolarisation in the scattering event and P^1/P monitored (where P^1 is final polarisation of scattered electron beam). Essentially such an experiment would be a repeat of that of Hegemann et al (1991) in which they sought to determine spin exchange phenomena in gas phase (randomly oriented) O_2 and NO using transverse and longitudinal polarised electrons.

A second series of experiments might involve a <u>fixed polarisation</u> of incident electrons scattering from an oriented molecule beam whose orientation can be changed within the hexapole field. This is essentially a repeat of the classic Campbell and Farago experiment in which oriented molecules replace the camphor isomers, with removal of the hexapole fields allowing any apparatus asymmetries to be allowed for.

However, the number density of the oriented gas beam, produced by a hexapole field, is low ($N \sim 10^{13}$ molecules/cm^3) and coupled with the low detection efficiency of Mott detectors ($< 10^{-3}$), such experiments will require careful design to obtain the maximum transmission efficiency in the electron optics.

(ii) Spin polarised electron scattering from **adsorbates**:

In comparison to gas phase studies, the number density of oriented molecules on a surface may be two orders of magnitude greater. However, once the adsorbate and substrate are chosen the chirality of the system can only be changed by alternating between longitudinal and transverse polarisation of the incident electron beam.

The preparation of surfaces of aligned molecules upon a metallic substrate is now routinely achieved by the condensed matter community (see Palmer and Rous (1992)). Molecular oxygen physisorbed on graphite and Pt(111) has been shown to be oriented parallel and perpendicular to the surface respectively. Using such a target geometry and replacing an unpolarised electron beam with a GaAs derived electron beam, either the scattered current may be monitored by a Mott polarimeter for any depolarisation of the incident polarised electron beam, or the scattered current may be monitored at a fixed angle as a function of electron polarisation.

However, such an experiment is not easily performed and is expensive, requiring UHV to prepare both the surface sample and the polarised electron beam, LEED and Auger analysis to characterise the surface and construction of a Mott detector for polarisation analysis, which may be difficult given their large size should it be necessary to incorporate the detector within the same vacuum chamber as the target.

Some simplification may be possible should organic targets be used rather than gaseous adsorbates. It is well established that certain organic molecules preferentially orientate themselves upon surface substrates when forming a thin film, without the need for UHV preparatory techniques.

Monolayers of hexadecanethiol and octane on a Pt (111) substrate provide thin film of <u>standing</u> and <u>lying</u> alkanes respectively. Such self assembly is based upon the spontaneous adsorption of the film components directly onto substrate and may be performed by emersion in solution and vapour deposition. Such films are both reliable and robust and therefore are perhaps more suitable for spin polarised experiments. (Bruckner et al (1994), Strong and Whitesides (1988).

(iii) Electron scattering from liquid **surfaces**:

The study of liquid surfaces by electron spectroscopy remains in its infancy even though the microscopic understanding of these systems is pivotal to our understanding of many chemical and biological processes.

The orientation of surface active molecules in the presence of fluid flow has recently been established using both ultraviolet photoelectron spectroscopy (UPS) and metastable impact electron spectroscopy (MIES) [see Morgner et al (1993) for a review]. In a binary

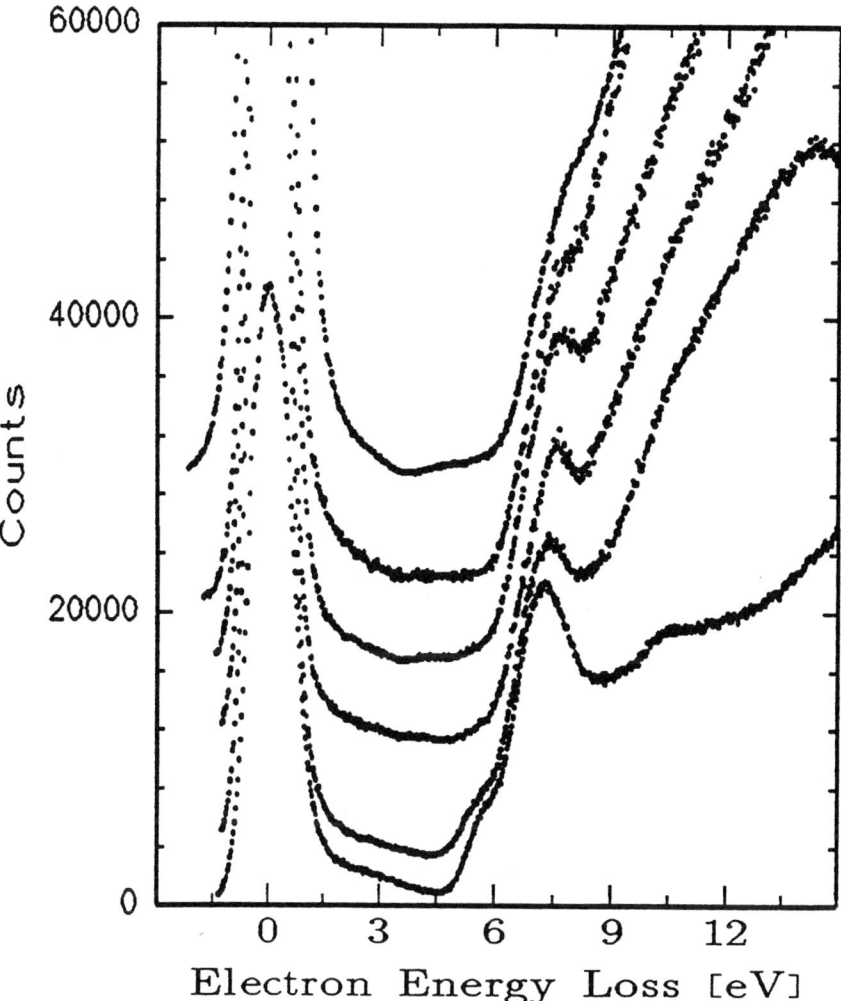

Figure 9. Electron energy loss spectra of a liquid surface with a primary electron energy of 20 eV. Spectrum at the bottom: pure formamide. The upper spectra are taken with increasing amounts of dissolved sodium oleate (a)0.55, (b)1.1, (c)1.6, (d)4.1, (e)11 in units of 10^{-3} mol l. Morgner et al (1994)

mixture of sodium oleate (~1 millimole per litre) and formamide within a liquid jet, the sodium oleate (NaOl) is found to form a layer above the formamide within a few milliseconds. Up to the completion of a closed surface layer the NaOl molecules lie flat on the surface, but after the layer is completed the molecules begin to reorientate themselves into an upright position, which in turn allows more NaOl molecules to be accommodated in the surface.

Recently the first low energy electron energy loss spectroscopy (EELS) measurements of liquid surfaces have been performed to study the temporal evolution of surface composition in binary liquid mixtures (Morgner, 1993). Figure 9(a) shows an EELS spectrum obtained from the surface of pure formamide with electrons of 7 eV primary energy. The peak at ~7.5 eV corresponds to the excitation of a Π state from a lower Π orbital, a so called $\Pi \rightarrow \Pi^*$ transition, a weaker triplet is observed as a shoulder at an energy loss of ~5.5 eV. As small amounts of NaOl are added to the formamide these energy loss features gradually vanish as the NaOl segregates from the bulk and covers the surface [Figure 9b) - e)]. Thus EELS has confirmed earlier UPS and MIES data and in the future may be used to characterise liquid surfaces as it has adsorbate films.

The discovery of positional orientation of surface active molecules in liquid beams introduces the possibility of studying oriented molecules in a new condensed matter environment. Liquid beam studies offer the advantages of higher number density compared to gas phase studies and (in contrast with adsorbates) two (or more) spatial orientations of the molecules within the jet. Spin polarised electron beams scattered from different positions along the length of the jet will encounter different molecular orientations and thence different "chirality". It may then be possible to study "dynamical orientation" with the polarised electrons monitoring the rate of change orientation by detecting change in asymmetries along the liquid jet.

However, the high vapour pressures of many liquids will require maximum differential pumping between both the electron source and Mott detector and the liquid beam. Scattering from the vapour above the surface must also be corrected for and any apparatus asymmetries analyzed.

Nevertheless, studies of aligned molecules on liquid surfaces may prove to be the best experimental arrangement to adopt for studies of electron dichroism.

CONCLUSION In this chapter recent experiments on oriented molecules in both gaseous and condensed phases of matter have been described and the possibility of performing spin polarised electron scattering experiments from oriented molecules have been analyzed. The development of experimental techniques to prepare well characterised targets of oriented molecules and high brightness, highly polarised electron sources make it possible to contemplate a new series of experiments to study the chirality of molecular systems.

If successful, such research may allow the development of new methods to both characterise and manipulate oriented molecules in any of the three phases of matter.

ACKNOWLEDGEMENTS

The author wishes to acknowledge receipt of a Royal Society University Research Fellowship. He also wishes to thank Professor Hans Kleinpoppen for the invitation to write this contribution and for his excellent organisation of the workshops on "Polarised Electron/Polarised Photon Physics".

REFERENCES

L.D. Barron (1986) Chem.Phys.Lett 123 423
A. Bisi, N. Gambara and L. Zappa (1992) IlNuovo Cinerto 14**D** 617
K. Blum, R. Fandreyer and D. Thompson (1990) J.Phys.B.**23** 1519
K. Blum and D Thompson (1989) J.Phys.B**22** 1823
P.R. Brooks (1976) Science **193** 11
M. Bruckner, B. Heinz, H. Morgner (1994) Molecular Phys. (to be published)
D.M. Campbell and P.S. Farago (1987) J.Phys.B**20** 5133
D.M. Campbell and P.S. Farago (1985) Nature **318** 52
N.A. Cherepkov (1982) Chem.Phys.Lett.**87** 344
N.A. Cherepkov (1983) Adv.At.Mol.Phys.**19** 395
P.S. Farago (1980) J.Phys.B**13** L567
P.S. Farago (1981) J.Phys.B**14** L743
M. Fink, A.W. Ross and R.J. Fink (1989)(a) Z.Phys.D.**11** 231 and A. Mihill and M. Fink (1989)(b) Z.Phys.D**14** 77
S. Hayashi (1985) J.Phys.B**18** 1229
S. Hayashi (1988) J.Phys.B**21** 1037
T. Hegemann, M. Oberste-vorth, R. Vogts and G.F. Hanne (1991) Phys.Rev.Lett.**66** 2968
S. Kaesdorf, G. Schönhense and U. Heinzmann (1985) Phys.Rev.Lett.**54** 885
D.A. Kohl and E.J. Shipsey (1992) Z.Phys.D**24** 33
H. Morgner (1994) Low energy electrons for the investigation of liquid surfaces in NATO Advanced Study Institute Patras (1993)
R. Palmer and P.J. Rous (1992) Rev.Mod.Phys.**64** 383
B.W. Plummer and W. Eberhard (1982) Adv.Chem.Phys.**49** 533
A. Rich, J. Van House and R. Hegstrom (1982) Phys.Rev.Lett.**48** 1341
L. Strong and G.M. Whitesides (1988) Langmair **4** 546
M. Volkmer, C.H. Meier, J. Lieschike, A. Mihill and N. Bowering (1993) ICPEAC Aarhus, Denmark abstracts 263
M. Volkmer, C.H. Meier, A. Mihill, M. Fink and N. Bowering (1992) Phys.Rev.Lett.**68** 2289
C. Westphal, J. Bansmann, M. Getzlaff and G. Schönhense (1989) Phys.Rev.Lett **63** 151
C. Westphal, J. Bansmann, M. Getzloff, G. Schönhense, N.A. Cherepkov, M. Brainstein, V.McKoy and R.L. Dubs (1991) Surf.Sci.**253** 205

PHOTOIONISATION AND FLUORESCENCE OF CALCIUM AND STRONTIUM IN THEIR P - D GIANT RESONANCE REGIONS

J B West
Daresbury Laboratory
Warrington WA4 4AD
UK

INTRODUCTION

It is well known that the inner shell photoionisation spectra of the alkaline earth atoms contain prominent autoionising structure, in particular the so-called "giant" resonances corresponding to the excitation of an outer p electron to an empty d-orbital. Partial collapse of the d-orbital wave function is responsible for the large overlap between the p and d wavefunctions and as a result these resonances have large peak cross sections, $\sim 10^{-15}$ cm^2. Through configuration interaction these resonances are also quite complex, as can be seen from the absorption spectrum taken by Mansfield and Newsom (1977) for calcium, in which the 3p - 3d resonance is composed of several discrete features; similarly for strontium, also measured by Mansfield and Newsom (1981), where the structure in the 4p - 4d resonance is much more obvious. This report summarises the measurements made at the Daresbury SRS using electron spectroscopy and fluorescence spectroscopy in these resonance regions, where the aim was to measure the branching ratios for populating the excited states of the ion, and the accompanying photoelectron angular distributions. In this way further understanding of the complex nature of these reonances is to be gained, and in addition these studies, when combined with the measurement of the polarisation of the fluorescence resulting from the residual ion as it decays, thereby determining its alignment, form part of a complete photoionisation experiment. For both atoms, the process being studied is the following, the example for strontium being given at the 490Å 4p - 4d resonance:

Sr ($4p^6 5s^2$ $^1S_0$) + hv (490Å) → Sr* ($4p^5 4d 5s^2$ ^{1,3}P).

This Sr* state can then decay by autoionisation, and it does this mainly to the ground state of

Sr⁺, but there is also a large cross section for decay to an excited state of Sr⁺ as follows:

$$Sr^* (4p^54d5s^{2\ 1,3}P) \to Sr^+ (5s\ ^2S_{1/2}) + e^-,\ or \to Sr^+ (5p\ ^2P_{1/2,3/2}) + e^-$$

The excited Sr⁺ 5p state will then decay by fluorescence to the ground state

$$Sr^+ (5p\ ^2P_{1/2,3/2}) \to Sr^+ (5s\ ^2S_{1/2}) + h\nu\ (4078\text{Å}, \cancel{4215\text{Å}})$$

where the 4215Å component corresponds to decay of the $^2P_{1/2}$ level and is removed by filtering. The usefulness of the fluorescence polarisation measurement does depend on the transition above being pure, ie the excited 5p level in Sr, or the corresponding 4p level in calcium, being populated only via the mechanism shown above and not, for example, by cascade processes. The complementary electron spectrometry measurements, to detect the electron indicated in the above processes, are required to investigate this, and also to provide the information required for the complete experiment. These form the basis of this report.

EXPERIMENTAL METHOD

The experimental geometry for this work is shown in figure 1, with the photon beam incident in the y-direction. The photoelectron angular distribution measurements were made with respect to the primary E-vector component in the x-z plane, where the angle θ is the angle between the ejected photoelectron and the E-vector. The fluorescence polarisation measurement was made for photons emitted in the x-direction, in a separate experiment. The experimental apparatus for both measurements is shown schematically in figure 2, where the incident photon beam is perpendicular to the plane of the figure. Photons in the range 20 - 30 eV, with a resolution of ~50 meV and a peak flux ~10¹¹ photons/sec, were provided by a toroidal grating monochromator fitted to beamline 3 at the SRS (for details of this instrument, see West and Padmore, 1987) in the case of the calcium measurements, and by a normal incidence monochromator (see Holland et al 1989) for the strontium measurements. A capillary light guide was used to bring the photons close to the interaction region from the exit slit of the monochromator. The oven shown in figure 2 is a compact design published by Ross (1994), and gave a vapour density of ~10⁻³ torr in the interaction region. The electron spectrometer, a 150° hemispherical sector of 90 mm mean radius, was fitted with a special shield at its entrance, to prevent its entrance slit assembly being exposed to the metal vapour when moving between the θ=0° and 90° positions. Particularly for strontium, this was necessary to preserve the efficiency and resolution performance of the analyser.

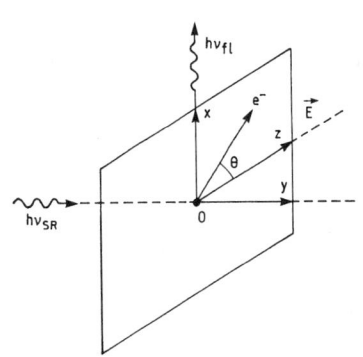

Figure 1   The experimental geometry

The angular distribution of the photoelectrons was calculated from the two measurements at the above angles, using the well known expression for the differential cross section

$$\frac{d\sigma}{d\theta} = \frac{\sigma}{4\pi}\left[1 + \frac{\beta}{4}(3p\cos2\theta + 1)\right] \qquad (1)$$

where β is the angular distribution parameter and σ is the total cross section for the photoionisation channel being measured; for a derivation of this expression see West (1991). The polarisation p of the incident light was measured in a separate experiment by using helium for which $\beta = 2$. The electron spectrometer was calibrated for variations in its efficiency with electron energy by making a series of measurements with argon gas and comparing the data to the known cross section of argon tabulated by Marr and West (1976); also, its angular symmetry was determined by checking the β-parameters measured for argon with those published by Holland et al (1982).

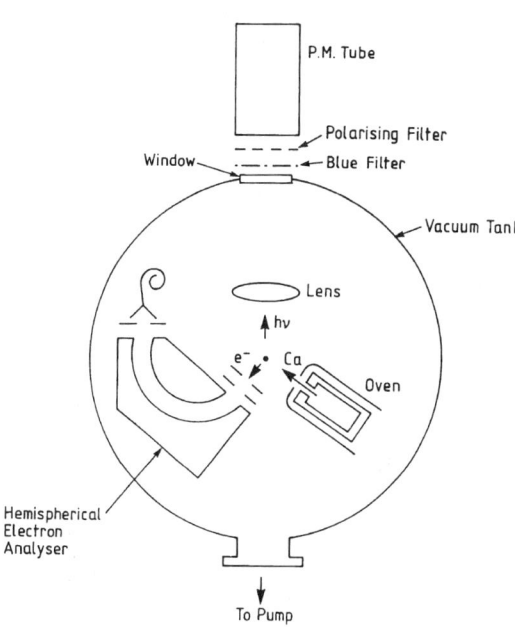

Figure 2   The experimental equipment

The fluorescence data were taken independently of the photoelectron measurements, and to measure the polarisation two measurements were needed, one with the transmission axis of the polarisation analyser shown in figure 2 parallel to the incoming light beam, the other with this axis perpendicular to it. The blue filter shown was chosen to exclude the $^2P_{1/2}$ component of the fluorescence, which is unpolarised and would therefore cause the value of the alignment parameter A_{20} to be too low. The alignment parameter is related to the fluorescence polarisation P_x by the expression (see, for example, Berezhko and Kabachnik, 1977

$$P_x = \frac{3\alpha_2 A_{20}}{\alpha_2 A_{20} - 2} \qquad (2)$$

where the coefficient α_2 is determined by the transition being measured; in this case, for the $^2P_{3/2} \rightarrow {}^2S_{1/2}$ transition $\alpha_2 = 0.5$.

RESULTS

Strontium

In figure 3 the results taken by Hamdy et al (1991) of the total fluorescence intensity for strontium in the region of the 4p -4d resonance are shown. For comparison the singly charged ion yield measurements from Nagata et al (1986), and the electron spectrometry measurements

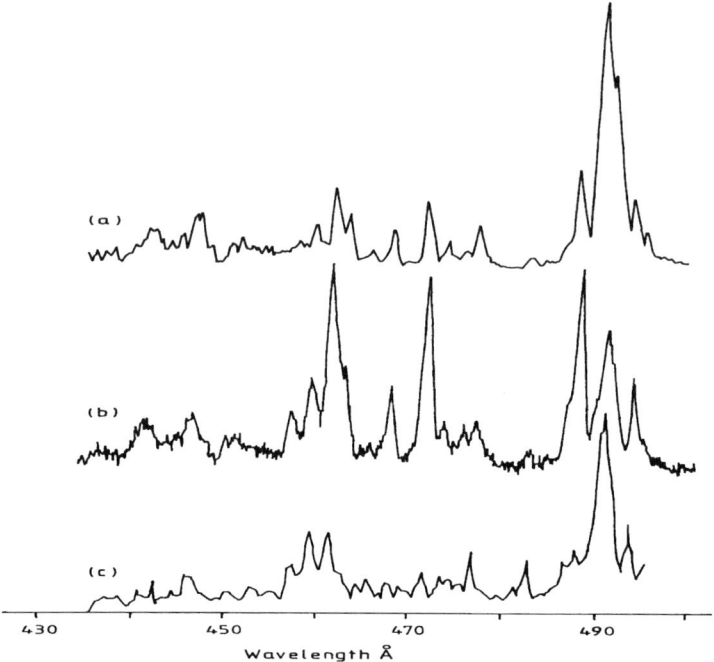

Figure 3 The singly charged ion yield (upper curve), fluorescence yield (middle curve) and 5p partial cross section (bottom curve) for Sr$^+$

of Yagishita et al (1988), who obtained the partial cross section for populating the 5p level by measuring the intensity of the electrons which leave the Sr$^+$ ion in the 5p excited state, are also shown. It is clear that this resonance is composed of at least three components, which were not assigned by Mansfield and Newsom (1981) and indicate that this resonance is indeed complex. The fluorescence spectrum is in effect part of the singly charged ion yield, and could be expected to resemble the fluorescence spectrum; the 5p partial cross section measurements should be identical to it. However, it can be seen clearly from figure 3 that although the peak positions for the three sets of data agree quite well, their relative intensities do not. This is almost certainly due to the presence of cascade processes, the 5p level being populated not just through the mechanism outlined in the introduction, but by higher satellite states being excited and decaying to this level. It is interesting to note that Hamdy et al found that the peak at 487.9Å (25.41 eV) was completely unpolarised, which would be the result of a contribution from cascades. A photoelectron spectroscopy experiment was therefore required to resolve this issue. The data shown in figure 4 were taken at the central peak of the 4p - 4d resonance at 490.8Å (25.26 eV), in the upper part of the figure, and in the lower part is shown the electron spectrum at the 25.41 eV peak. Here the difference is clear; at 25.26 eV there are only three electron peaks, corresponding to leaving Sr$^+$ in its ground state, 4d or 5p excited levels, whereas at 25.41 eV many other levels are also populated. This same result was confirmed by the much more detailed experiment of Jimenéz-Mier et al (1993). It seemed reasonable to conclude therefore that the polarisation measurement made by Hamdy et al at 25.26 eV was free from cascades, since the excited 4d level cannot decay by fluorescence to the ground state of the ion. This was an important finding, since it meant that this measurement could be used to determine

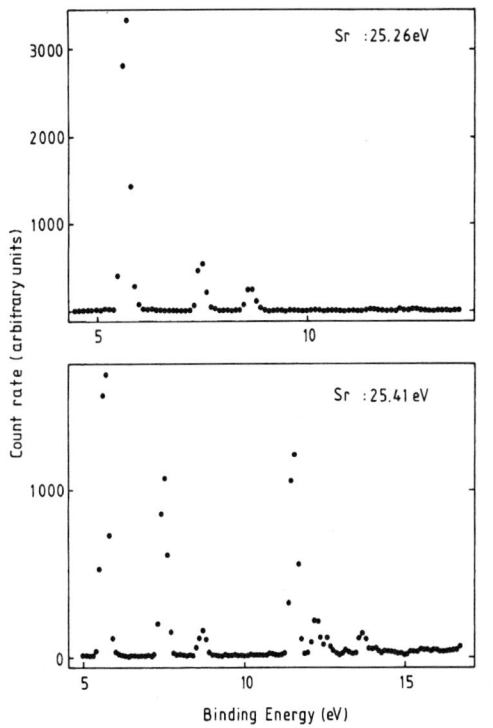

the alignment of the residual ion, using the equation for P_x above, and with the additional information from the angular distribution measurement determine the photoionisation parameters. This was what Ueda et al (1993a) proceeded to do.

In figure 5 are shown the relative partial cross section for populating the Sr^+ 5p level, and the angular distributions for the corresponding ejected electrons. The numbering of the resonance features follows that of Mansfield and Newsom (1981), and most of these features are evident in both the partial cross section and angular distribution parameter. The second strongest peak in the lower half of figure 4 corresponds to excitation to the Sr^+ 6s level; in fact the branching ratio for population of this level rises to 27% at the 25.41 eV resonance, compared to 4% for the 5p level. This highlights the different nature of the

Figure 4 Photoelectron spectra at the photon energies shown

component parts of the complex 4p - 4d resonance, because it indicates that the 25.41 eV resonance may have an uncollapsed Rydberg character, thereby allowing substantial overlap with the Sr^+ 6s wavefunction and in distinct contrast to the main resonance at 25.26 eV. It is notable also that the 5p β-parameter shows a large excursion here towards β = 0, indicating that it is dominated by a p → s transition, and in general reflecting large phase differences between outgoing s and d waves.

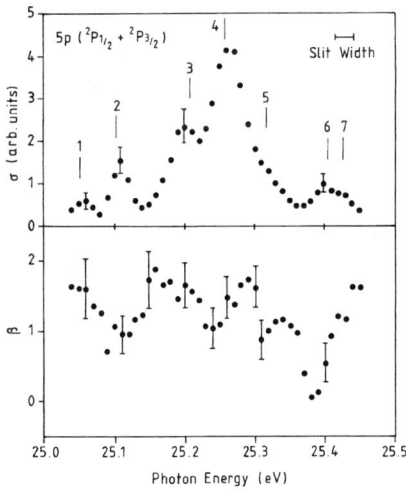

Figure 5 Data for the Sr^+ 5p level

In order to extract the photoionisation parameters, in this case the relative phase and the ratio of the dipole matrix elements corresponding to the s and d outgoing waves, it is necessary to assume LS coupling. Otherwise the $d_{3/2}$ and $d_{5/2}$ waves must be separated and from the two measurements, the alignment parameter and angular distribution parameter, there is insufficient information to do this. Ueda et al (1993a) found that in the region of the resonance at 25.26 eV, the angular distribution parameter for the 5s electrons, shown in figure 6 together with the relative partial cross section for leaving the Sr^+ ion in its ground state, is, within experimental error, ~2. This means that here the two outgoing waves $p_{1/2}$ and $p_{3/2}$ are

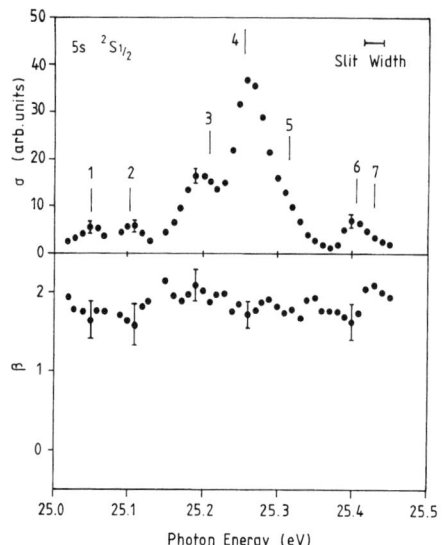
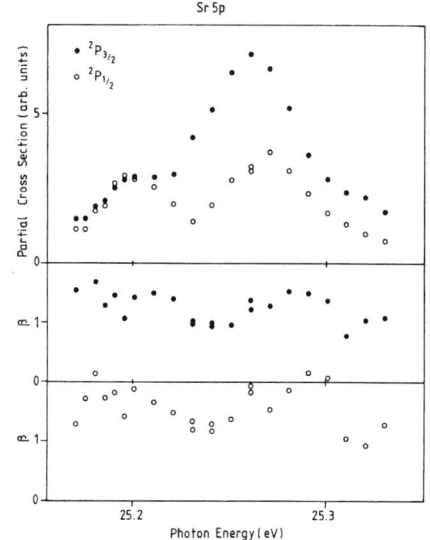

Figure 6 Data for the Sr⁺ 5s electrons

Figure 7 The Sr⁺ 5p spin-orbit components

not separable and the LS coupling approximation appears to be valid. This is certainly not the case at other resonances in the 4p - 4d complex, where the β-parameter falls to ~1.6 in some regions, but it does seem that, surprisingly for a large atom such as strontium, LS coupling can be assumed at the 25.26 eV resonance. In further support of this, in figure 7 are shown the relative intensities and β-parameters for the spin-orbit split components of the 5p doublet, where it is evident that although the ratio $^2P_{3/2}$:$^2P_{1/2}$ varies substantially over the whole region, at the peak of the 25.26 resonance it is 2:1, again as would be expected if LS coupling were valid.

Within the assumption of LS coupling, the alignment parameter A_{20} and β can be defined in terms of the dipole moments D_s and D_d, for the outgoing s and d waves respectively, and the phase difference D between them as follows:

$$A_{20} = -\frac{|D_s|^2 + |D_d|^2/10}{|D_s|^2 + |D_d|^2} \qquad (3)$$

and

$$\beta = \frac{|D_d|^2 - 2\sqrt{2}\,|D_s||D_d|\cos\Delta}{|D_s|^2 + |D_d|^2} \qquad (4)$$

From these equations it is possible to calculate, from the β measurement of the $^2P_{3/2}$ component and the value of A_{20} calculated from equation (2) above, the ratio $|D_s|^2/|D_d|^2$ and the phase difference Δ. Ueda et al (1993a) found the following values:

$$|D_s|^2/|D_d|^2 = 0.157 \pm 0.024 \text{ and } |\Delta| = 114 \pm 30°.$$

Calcium

The fluorescence data in the region of the 3p - 3d resonance, from Hamdy et al, are shown in figure 8, and are compared with the singly charged ion spectra taken by Sato et al (1985). Although it was no longer possible in this experiment to resolve clearly the individual

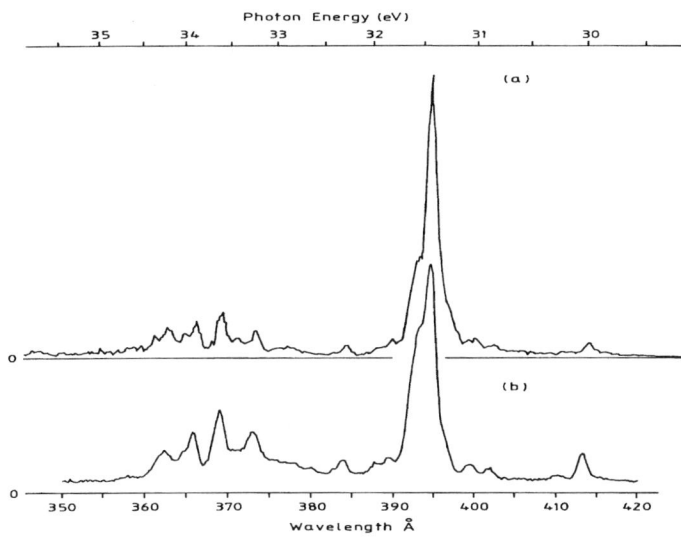

Figure 8 Comparison of the singly charged ion data (upper curve) and fluorescence data (lower curve)

components in the 3p - 3d resonance, differences are evident between the ion and fluorescence spectra. Because of this inability to resolve these peaks it was impossible to apply the same kind of analysis as used above for strontium to the calcium case, even though the assumption of LS coupling would be more reasonable for calcium. Ueda et al (1993b) made electron spectrometry measurements on calcium, and figure 9 shows the data for the cross section for populating the Ca⁺ 4p level and the angular distribution for the corresponding electrons; as before, the numbering scheme follows that of Mansfield and Newsom (1977). Two main structures in the resonance are now clearly resolved, and these features show completely different behaviour for the β-parameter. By taking wide range photoelectron spectra, Ueda et al (1993b) showed that in the calcium case there was substantial intensity in higher Rydberg levels; their data are reproduced in the following table in the form of the percentage population of the levels shown at the two resonance energies.

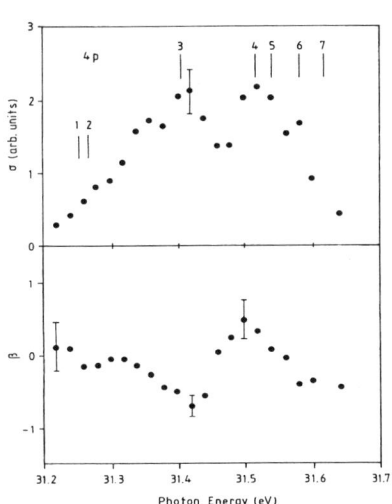

Figure 9 Data for the Ca⁺ 4p level

Energy (eV)	4s	3d	4p	5s	4d	5p	nl	3p^{-1}4s3d
31.41	51	3.7	3.8	1.6	7.4	2.7	13	16
31.52	28	5.0	10	6.3	12	5.8	19	14

It is clear from this that cascades are a serious problem for calcium; for the 31.52 eV resonance over half the intensity is in the population of the higher states, many of which will decay to the 4p level. Using equation (4), all that can be said is that the phase difference term $\cos \Delta$ is positive in the region of the main resonance at 31.41 eV, since β is negative there.

The solution to this difficulty is to make measurements in coincidence, ie to detect the fluorescent photon in coincidence with the photoelectron ejected leaving the calcium ion in the 4p level. In this way it will be possible to exclude cascades from the data, but the analysis becomes more complicated because, by the very nature of the experiment, the fluorescence measurement is no longer being averaged over all possible photoelectron ejection angles. Furthermore, the incident radiation is elliptically polarised, ie it is a sum of linear and circular polarisations, and the angular correlation between the electron and photon is a function of both of these, as shown by theoretically by Klar (1980). It is therefore necessary to measure the circular polarisation component of the incoming light, in general a rather difficult measurement, and yet one of crucial importance as shown by Kämmerling and Schmidt (1991) in their electron angular correlation experiment for double ionisation of xenon. Berezhko et al (1978) have shown that the polarisation of the fluorescent radiation is given by

$$P_x = \frac{\alpha_2 (3A_{20} + \sqrt{6} A_{22})}{\alpha_2 (A_{20} - \sqrt{6} A_{22})} \tag{5}$$

where now both the coefficients A_{22} and A_{20} contain terms which, when applied to the geometry of this experiment, include the linear and circular polarisations of the incident light (Kabachnik, 1994). Lacking an independent measurement of the circular polarisation, the aim of the experiment is to make a series of electron/fluorescent photon coincidence measurements over a range of $\theta > 180°$, where the polarisation of the fluorescence is determined. It should then be possible to extract the circular polarisation of the light as well as the photoionisation parameters by careful fitting of the coincidence data to the theoretical expressions for P_x and β. The first results of such measurements are the subject of a forthcoming publication (Beyer et al, 1994).

CONCLUSION

The branching ratio and photoelectron angular distribution measurements summarised here on the photoionisation of calcium and strontium in their p - d giant resonance regions show high sensitivity to resonant structure and provide a means of investigating and determining the nature of the presently unassigned components within the p - d complex. The measurements of the angular correlation between the photejected electron and the fluorescent photon which results from the decay of the excited are also sensitive to resonant behaviour, and such measurements are now under way. It is also the aim to determine the photoionisation parameters from these experiments, since, using also the electron angular distribution measurement from the non-coincidence experiment, the ratios of the dipole amplitudes, and their relative phases can be calculated. With a comprehensive set of coincidence measurements it should be possible to obtain enough information to calculate these parameters for the case where LS coupling is no longer valid, for example for the subsidiary resonances in the region of the 4p - 4d main resonance in strontium.

ACKNOWLEDGMENTS

It is a pleasure to acknowledge my collaborators in the above experiments, H Kleinpoppen and H-J Beyer, Stirling University, UK; K J Ross, Southampton University, UK; K Ueda, Research Institute for Scientific Measurements, Sendai, Japan; and H Hamdy, Beni-Suef University, Cairo, Egypt. We are grateful for the support provided by the Science and Engineering Research Council to run these experiments, and K Ueda thanks the British Council for financial support in the form of a travel grant.

REFERENCES

Berezhko E G and Kabachnik N M, *J. Phys. B: At. Mol. Phys.* **10**, 2467 (1977)

Beyer H-J, West J B, Ross K J, Ueda K, Kabachnik N, Hamdy H and Kleinpoppen H, to be published

Hamdy H, Beyer H-J, West J B, and Kleinpoppen H, *J. Phys. B: At. Mol. Opt. Phys.* **24**, 4957 (1991).

Holland D M P, Parr A C, Ederer D L, Dehmer J L and West J B *Nucl. Instrum. Methods* **195**, 331 (1982)

Holland D M P, West J B, MacDowell A A, Munro I H and Beckett A G *Nucl. Instrum. Methods* B **44**, 233 (1989)

Jiménez-Mier J, Caldwell C D, Flemming M G, Whitfield S B, and van der Meulen P *Phys. Rev.* A (1993)

Kabachnik N M, 1994 (private communication)

Kämmerling B and Schmidt V, *Phys. Rev. Letts.* **67**, 1848 (1991); see also comment by W R Johnson and K T Cheng, and reply by Kämmerling and Schmidt, *Phys. Rev. Letts.* **69**, 1144 (1992)

Klar H, *J. Phys. B:At. Mol. Phys.* **13**, 2037 (1980)

Mansfield M W D and Newsom G H, *Proc. R. Soc.* A **357**, 77 (1977)

Mansfield M W D and Newsom G H, *Proc. R. Soc.* A **377**, 431 (1981)

Nagata T, West J B, Hayaishii T, Itikawa Y, Itoh Y, Koizumi T, Murakami J, Sato Y, Shibata H, Yagishita A and Yoshino M, *J. Phys. B:At. Mol. Phys.* **19**, 1281 (1986)

Ross K J 1994 *Vacuum* (in press)

Sato Y, Hayaishi T, Itikawa Y, Itoh Y, Koizumi T, Murakami J, Nagata T, Sasaki T, Sonntag B, Yagishita A and Yoshino M, *J. Phys. B: At. Mol. Phys.* **18**, 225 (1985)

Ueda K, West J B, Ross K J, Hamdy H, Beyer H J and Kleinpoppen H, *J. Phys. B: At. Mol. Opt. Phys.* 26, L347 (1993a)

Ueda K, West J B, Ross K J, Hamdy H, Beyer H J and Kleinpoppen H, *Phys Rev A* **48**, R863 (1993b)

West J B and Padmore H A, *Handbook on Synchrotron Radiation* Vol. II ed. G V Marr (North Holland, Amsterdam, 1987) p21

West J B, in *Vacuum Ultraviolet Photoionization and Photodissociation of Molecules and Clusters*, ed C-Y Ng (World Scientific, Singapore, 1991)

FARADAY ROTATION IN THE UV/VUV AND HIGH FIELD MAGNETO-OPTICS

J.P. Connerade

The Blackett Laboratory Imperial College
London SW7 2BZ U.K.

INTRODUCTION

Magneto-optics has a long history which predates quantum mechanics by many years. The crucial discoveries were made by Faraday[1] in 1846 Kerr[2] in 1875 and Zeeman[3] in 1896. Electro-optics came later, when lo Surdo[4] and Stark[5] discovered what is now called the Stark effect, around 1912. Quantum mechanics as such only appeared in 1927.

Notwithstanding this long history, magneto-optics has not received all the attention it deserves at the hands of atomic physicists. Some of the most interesting developments (order-to-chaos transitions, quasi-Landau resonances, etc) were discovered comparatively recently and are still subjects of current research. Even in relation to the Zeeman effect, undergraduate textbooks do not contain much more (indeed, sometimes less) than can be found in Zeeman's celebrated book[6].

In recent times, an explosive growth in experimental and theoretical studies of the high field Zeeman effect[7,8,9] was triggered by a famous experiment[10] in which the transition from coulombic to Landau symmetry was discovered. More recently still, efforts have been made made to extend Faraday rotation studies beyond the visible range of the spectrum into the ultraviolet and vacuum ultraviolet, where synchrotron radiation and laser sources (both naturally polarised) are available. First studies of the Zeeman and Faraday effects in the vacuum ultraviolet have been reported[11].

The present paper contains a brief description of magneto-optical experiments presently in hand at Imperial College.

FARADAY ROTATION WITH SYNCHROTRON RADIATION

Our studies of Faraday rotation began somewhat accidentally while attempting experiments on the high Field Zeeman effect around 1982, using the 500 MeV syn-

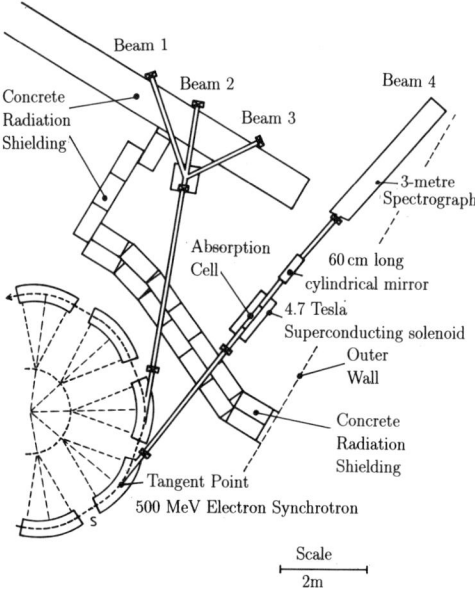

Figure 1: The experimental layout used to observe Faraday rotation spectra using a synchrotron radiation source

chrotron source at the Physikalisches Institut in Bonn. The layout is illustrated in Fig. 1.

What we had not bargained for at the time was the sensitivity of the high dispersion grating we were using to the polarisation of the incident light. As a result, our system was turned into the classical crossed polariser arrangement for the observation of the Faraday effect, with the role of the polariser played by the synchrotron itself and that of the analyser by the grating. Thus, we had an experiment with a new purpose and the following additional features with respect to earlier work: (i) a UV/VUV capability (ii) high spectral resolution and (iii) high magnetic fields.

Our observations with this arrangement therefore broke some new ground. First, by virtue of the high field, we saw rotation angles of many π radians, and could study them as a function of wavelength. Intensity maxima and minima provided an absolute calibration of the rotation angle without any need to rotate the analyser or, indeed, to move any optical component (an important consideration when extending observations into the VUV range).

Secondly, by studying the nature of the resulting patterns we were able to demonstrate the importance of including in the analysis *both* the effects of birefringence leading to pure magneto-optical rotation (MOR) and those of magnetic circular dichroism (MCD) in which one of the two circularly polarised components of the radiation is selectively absorbed. Although previous workers had attempted to separate them experimentally, we found that, by using a more complete theory in which they are *both* included and computed point by joint across the whole profile[12], very accurate relative f-values of atomic transitions can be deduced from quite simple experiments in which the magnetic field is held at a constant value. (We call this approach the Magneto-Optical Vernier (MOV) or balance of MCD against MOR) to distinguish it from pure MOR or MCD methods.

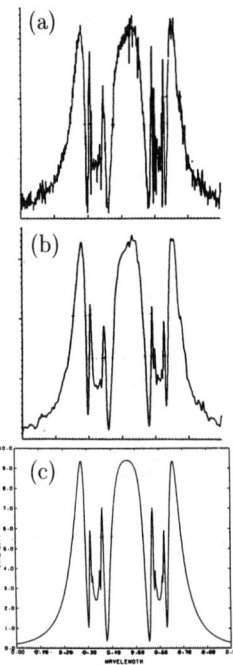

Figure 2: A typical magneto-optical rotation pattern, as observed in a pulsed laser experiment (a) the raw data (b) the same, smoothed by 3-5-3 averaging and (c) as computed by the MOV theoretical approach

Experience shows that the highest Rydberg series members that can be studied by the MOV technique using synchrotron radiation lie around n 28 for alkaline-earth spectra, the limitation being largely due to the spectral resolution available by classical spectroscopy. Alternative techniques, such as the hook method[13] do not extend much above $n \sim 15$.

EXTENSION TO PULSED LASER SOURCES

By exploiting the superior spectral resolution of the tunable dye laser, measurements have been extended up to $n \sim 46$ in alkaline - earth spectra.

One simple way to improve the spectral resolution is to move away from synchrotron radiation and use frequency doubled or frequency mixed laser sources. This step has the additional benefit of allowing greater control over the polarisation properties of the radiation. On one hand, the laser provides more intensity, so that additional optical elements can be used. On the other, polarising optics can be inserted at the fundamental laser wavelengths prior to upconversion, which is both simpler and more efficient than using polarisers in the VUV range.

Since laser sources for the UV/VUV are pulsed, it is appropriate also to combine them with pulsed magnets in the way we describe below. This has the advantage that one can do away with cryogenic technology and the associated complexities.

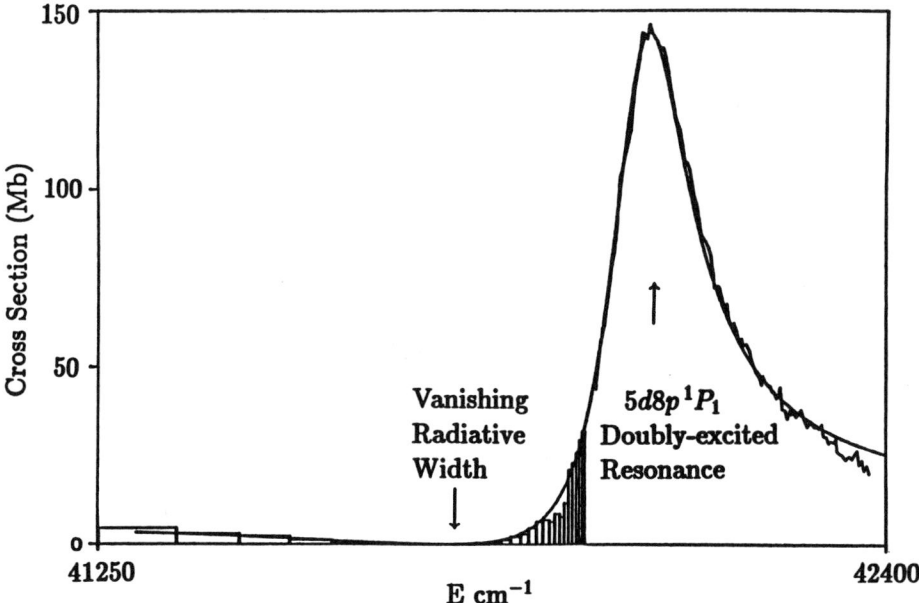

Figure 3: The $\frac{df}{dE}$ plot of the principal series of barium, showing the pronounced intensity minimum in the series and subsequent rise in relative intensity which are due to perturbation by the $5d8p\,^1P_1$ doubly-excited perturber

Of course, the purpose of the experiment is to provide a constant magnetic field, with no electric field of comparable influence. A pulsed B field will produce an electric field, but if the circuit is designed for characteristic times in the ms range, then over the duration of a laser pulse, which typically lasts 10^{-8}s or so, the change in magnetic field strength is very small and the electric field can be neglected.

In Fig. 2, we show a typical magneto-optical pattern recorded in a pulsed experiment together with a computed theoretical profile used for its interpretation. A disadvantage of the laser-based method as opposed to synchrotron radiation is that the laser source must be scanned point by point over the profiles, and that only one profile can be studied at a time. This is very time consuming. The speed of the scan is actually limited by energy dissipation in the pulsed magnet. A full description of the pulsed method, including the timing and charging circuits and the 'crowbar' required to limit the energy dumped into the coil has already been published[14]. In practice, pulsed field magneto-optics has significantly extended the range available to Zeeman and Faraday spectroscopy.

In a later section, we show how new effects such as inter-ℓ mixing and the perturbation of the apparent f-value by the applied magnetic field come into play at high principal quantum number n. Such effects had never been studied before by MOR spectroscopy.

Some of the most interesting interchannel interaction effects occur at high n. In Fig. 3, we show an example in which the $5d8p\,^1P_1$ doubly-excited interloper in the Ba I spectrum perturbs the principal series around n = 22 - 23, resulting in a pronounced minimum. This is a very remarkable effect: because of this perturbation, several

Rydberg states are stabilised against the emission of radiation, i.e. their lifetimes are increased (vanishing radiative width).

An unsuccessful search for this effect by the hook method was reported by Parkinson et al[15]. It resulted in the two points shown as crosses in Fig. 3. The other data are from experiments at Imperial College.

FROM THE UV TO THE VUV

Although mainly of technical interest, it is worth placing on record that we have extended Faraday and Zeeman Spectroscopies into the VUV[16]. A number of fundamental and important spectra such as that of the hydrogen and helium principal series still lie beyond the spectral range presently accessible. At time of writing, studies of VUV spectra in our experiments are still limited by the transmission properties of the standard materials, but polarisers exist in which three or four reflections are employed, and therefore it is possible, at least in principle, to extend observations well below the present limit of about 1500 Å. Another significant issue is the development of suitable radiation sources. Synchrotron radiation of course, is polarised at all wavelengths, but conventional dispersive instrumentation looses spectral resolution roughly in proportion to λ the wavelength. Thus, the combination is unsatisfactory for MOR at small λ. One would prefer to extend laser sources to shorter wavelengths.

The best method for this purpose is the exploitation of sum - difference nonlinear mixing schemes[17] which are, however, limited to about 1200 Å cut-off. We are currently developing a sum-sum four-wave mixing sheme, which will open up the range down to 720 Å.

THE FARADAY EFFECT IN THE CONTINUUM: AUTOIONISING RESONANCES

Because experiments to date have mainly been confined to the energy range below the first ionisation thresholds of free atoms, not much attention has been given to Faraday rotation in the continuum, and until recently there was no theory applicable above threshold.

In principle, there is very little rotation in a pure continuum as one moves away from the threshold. This arises because the absorption spectrum is nearly flat, so that the Zeeman splitting reveals almost no difference between the refractive indices for the two senses of circular polarization. There does exist a $1/\nu$ dependence (ν being the wavenumber) in the rotation through the Faraday rotation formula, which results in a universal (but very small) rotation. The effect is largest near an ionisation threshold, where the limit structure also has some effect.

A much more significant cause of Faraday rotation is the presence of fairly sharp autoionising structure in the continuum. The theory of this rotation is readily written out once it is realised that the standard expression (Fano formula) for an autoionising line can be decomposed into the sum of a Lorentzian and a dispersion profile, for both of which analytic expressions can be obtained giving the variation in refractive index.

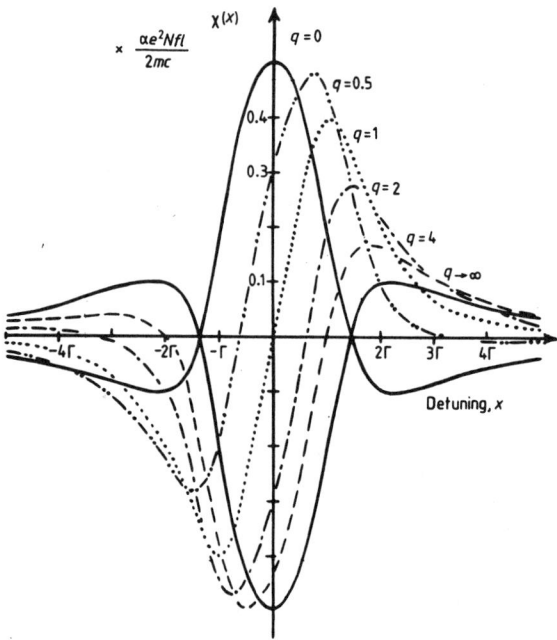

Figure 4: rotation angle as a function of frequency for a Beutler-Fano profile

The result is conveniently plotted as a family of curves, which show the rotation angle and the rotation per unit absorption depth for different profile asymmetries (see Figs 4 and 5).

We have observed Faraday rotation patterns for autoionising lines, and an example is shown in Fig. 6. Note the asymmetry of the whole pattern. This is, we believe, the first observation of a strong rotation effect in the ionisation continuum.

HIGH-FIELD OR ℓ-MIXED FARADAY EFFECT

One of the advantages of MOR as opposed to MCD is its freedom from opacity effects, which arises simply from the fact that the rotation angle is cumulative down the vapour column. This advantage is largely preserved in the MOV method, but only if the outer reaches of the pattern are well developed: as the f-value of the transition decreases (increasing principal quantum number n), one is tempted to wind up the field strength for a given integrated column density, so as to preserve the accuracy and sensitivity of the technique.

Unfortunately, this cannot be continued indefinitely, because of the eventual intrusion of the quadratic Zeeman effect, through the combination of high magnetic field strengths B and high principal quantum numbers n. The quadratic effect[10] scales as $B^2 n^4$, and results in a redistribution of oscillator strengths between transitions to states of different ℓ, which become mixed by the strong field, and therefore accessible as satellites of the main Rydberg line. This leads to an apparent loss of

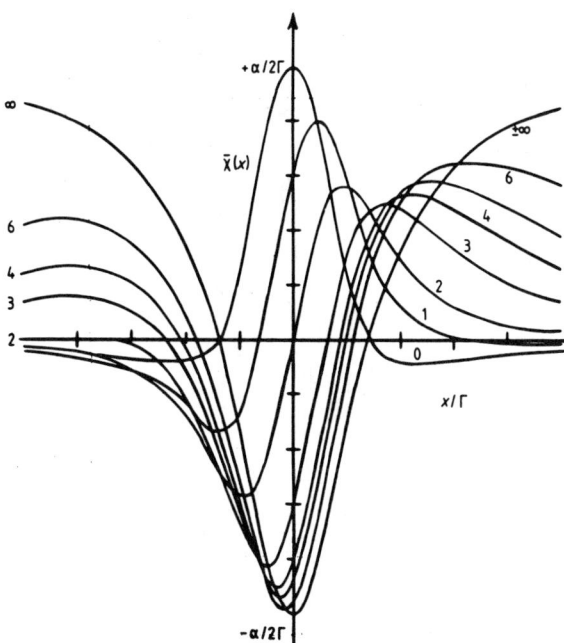

Figure 5: rotation angle per unit absorption depth for a Beutler-Fano profile.

rotation of the main line, and hence to a sudden decline in the measured oscillator strength, apparent in Fig. 7.

Of course, if the redistribution of oscillator strength is understood well enough, then it can be corrected for, and calculations along these lines have been reported[18] which demonstrate the influence of ℓ-mixing on the refractive index of an atomic vapour in a strong magnetic field (see Fig. 8). Although the effect should be discernible at laser resolution, no direct observation of such structure has yet been reported. There is a need for experiments at higher spectral resolution than presently attempted.

The use of the MOV method to measure f-values of transitions to very high Rydberg states must ultimately be limited by the quadratic Zeeman effect: although corrections can be applied at the onset of ℓ-mixing, the spectra grow rapidly in complexity with increasing energy or fields, and one enters the n-mixing range, where it becomes impossible to disentangle the excitations and assign individual f-values to the transitions in the presence of the external field. Clearly, this is the price one has to pay for using a method in which an external influence is applied to the atom. Fortunately, the influence of the quadratic effect is usually extremely small, and indeed much smaller than the effect of collisions. In fact, it has proved possible, by studying MOV patterns in the presence of a foreign gas, to study the broadening of transitions to high Rydberg states resulting from collisions[19] in a range of pressures not readily accessible by other methods.

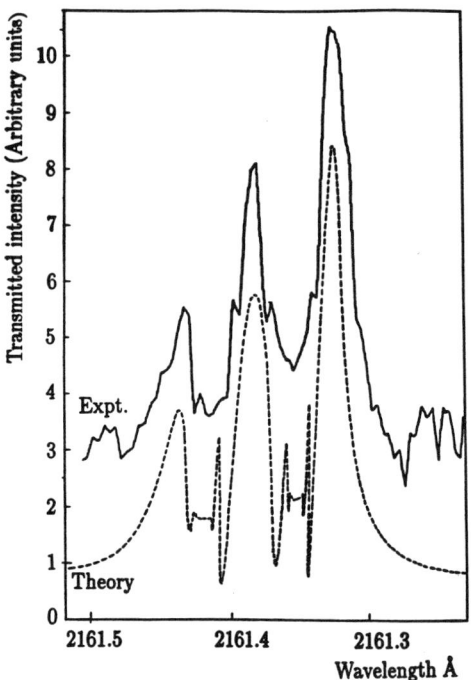

Figure 6: Observed Faraday rotation spectrum for a Beutler-Fano resonance. The calculated profile is the dashed curve. Note the pronounced asymmetry of the pattern.

STUDY OF THE QUADRATIC ZEEMAN EFFECT IN ATOMIC BEAMS

In addition to Faraday rotation experiments, our group at Imperial College has also been studying the quadratic Zeeman effect, and in particular the relative intensities of the ℓ-mixed satellite lines, for many-electron atoms in atomic beams. These experiments have been reported in more detail in a previous paper[20], but are related and deserve mention here. The motivation for this work was the following. Calculations[21] have shown that the intensities of the transitions observed[10] do not quite match theory and, in view of the detection method used in previous experiments, which imply the possibility of saturation as well as opacity effects, it seemed appropriate to repeat the study using atomic beams. We were also interested in performing a beam experiment under a geometry somewhat different from those adopted previously, so as to be able to separate the σ^+ from the σ^- spectra, and we therefore performed an experiment in which the laser beam and the atomic beam were crossed inside a magnet, with the laser beam travelling parallel to the field lines. Since the atom involved (barium) is rather heavy, the motional Stark field remained small. These studies have revealed an interesting new form of dichroism, in that the σ^+ and σ^- structures were found not to be identical in the ℓ-mixing region (see Fig. 9) despite the general assumptions in the literature that they are the same for alkaline-earth spectra, which are dominated by singlet states, and there-

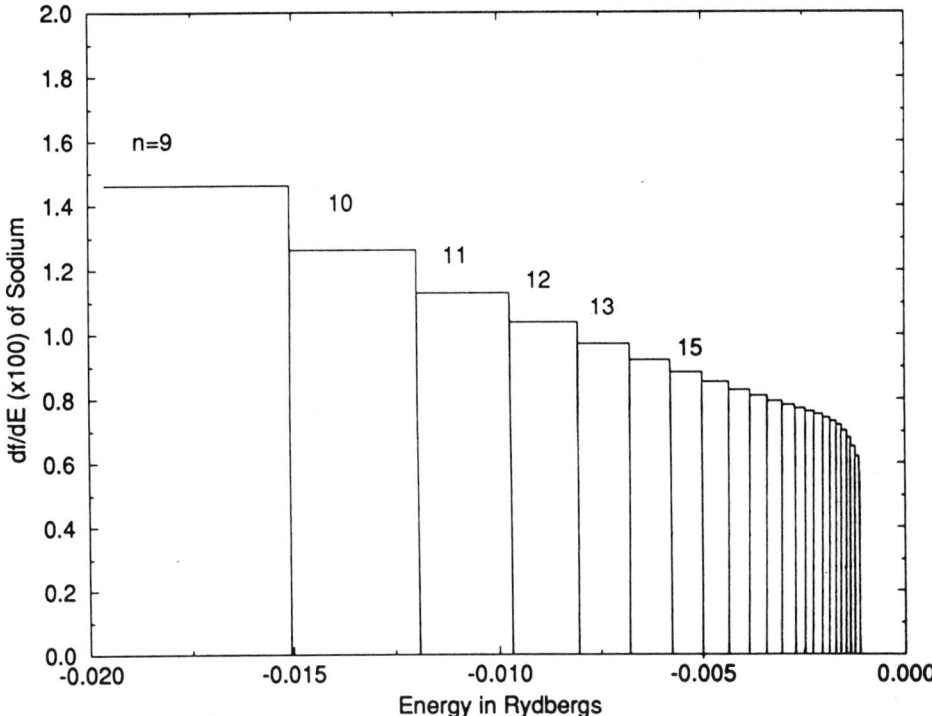

Figure 7: The $\frac{df}{dE}$ curve for the high Rydberg members of the sodium principal series, as measured by magneto-optical spectroscopy. Note the sudden drop in apparent f-value at high Rydberg members, which can be corrected for as described in the text.

fore possess no Paschen-Back effect. The explanation seems to be connected with the many-electron nature of the barium spectrum, and with the presence of nearby doubly-excited states: one way to ascertain this is to study other similar spectra, where the doubly excited transitions occur at different energies relative to the Rydberg spectrum. We have now ascertained that even the earlier spectra[10] did in fact contain such an effect, although it was not suspected at the time. We are continuing our experimental investigations, but it is also clear that more theoretical work is required.

CONCLUSION

In this brief review of the activities at Imperial College in the area of magneto-optics, it is of course not possible to cover many other interesting developments which make this a dynamic and expanding field. However, I hope to have given some flavour of one corner of magneto-optics which may serve as an introduction to any interested reader. The subject of magneto-optics as a whole is a vast one, ranging from applied work (eg designing Faraday isolators for high power lasers) to highly fundamental physics (eg use of the Faraday effect in parity-violation experiments, as practised in the Clarendon Laboratory at Oxford). Whatever one's purpose, the Faraday effect

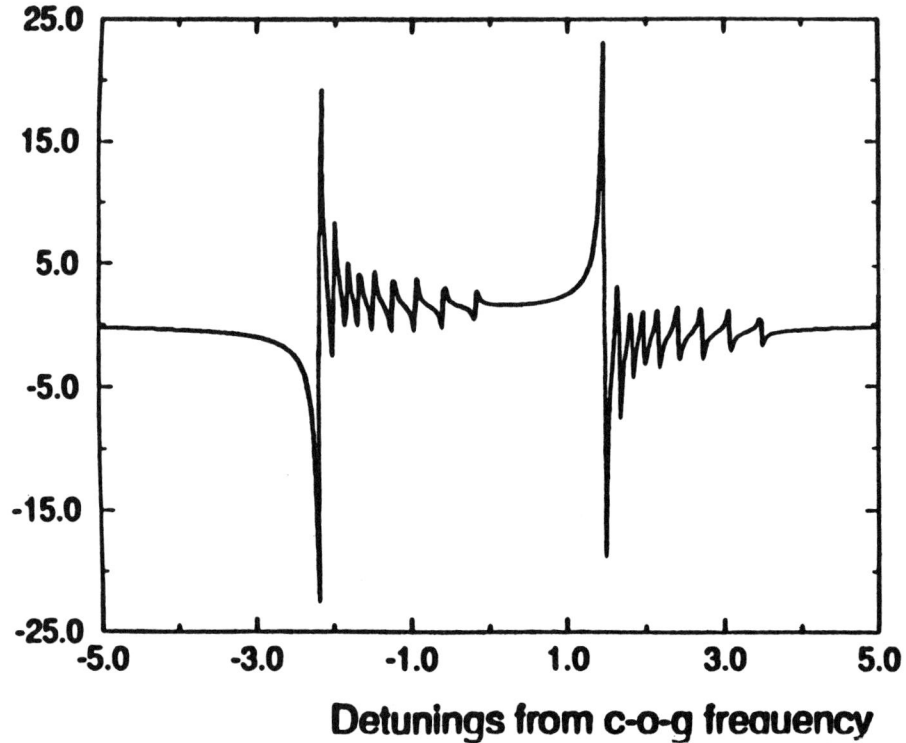

Figure 8: The effect of a high magnetic field, which induces ℓ-mixing, on the refractive index of a high series member, as calculated in reference[18]

is a very powerful experimental method. It is hoped that the present brief review will provide some inspiration to the reader to consider how it may be related to more usual spectroscopic techniques in his field.

REFERENCES

1. Faraday Phil. Trans. Roy. Soc. Series XIX, 1 (1846)
2. Kerr Phil. Mag. **1** 337 (1875)
3. Zeeman P. Verk. der Phys. Ges. zu Berlin 15 Jahrg. No7 128 (1896)
4. lo Surdo A. Acad. Lincei Atti **22** 665 (1913)
5. Stark J. Berlin Akad. Wiss. **40** 932 (1913)
6. Zeeman P. 'Researches in Magneto-optics' Macmillan & Co Ltd (London) 1913
7. Holle A. Main J. Wiebusch G. Rottke H. and Welge K.H.
'*Atomic Spectra and Collisions in External Fields* Edited by K.T. Taylor M.H. Nayfeh and C.W. Clark Plenum Press (New York) (1988) 8. Gay J.C. Comments on At. Mol. Phys. **XXV** Nos 1 – 6 (1990)
9. K. Taylor and D. Delande (Eds) *J. Phys. B: At. Mol. Opt. Phys.* (Special issue) July 1994
10. Garton W.R.S. and Tomkins F.S. *Astrophys. J.* (1969)
11. J.-P. Connerade W.R.S. Garton M.A. Baig J. Hormes T.A. Stavrakas and B. Alexa, J. de Physique **43** C2-317 (1982)

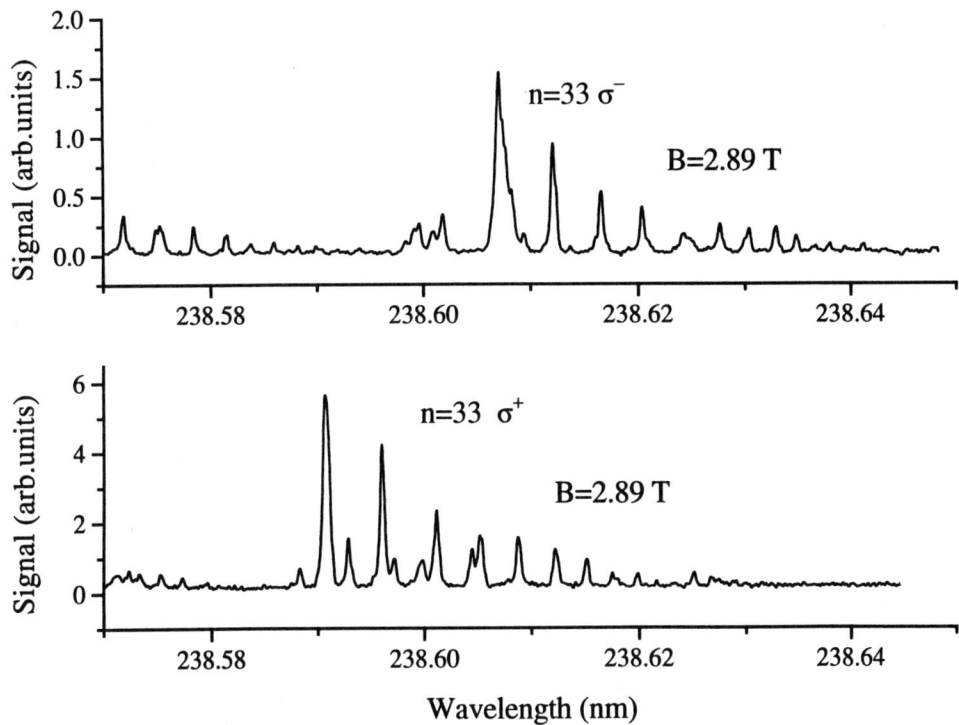

Figure 9: Spectra of barium taken in σ^+ and in σ^- polarisation, showing the differences of structure and intensity which have been observed in an atomic beam experiment, as described in the text.

12. J.-P. Connerade J. Phys. B: At. Mol. Phys. **16** 399 (1983)
13. Hook method
14. J.-P. Connerade W.A. Farooq and M. Nawaz,
J. Phys. B: At. Mol. Opt. Phys: **25** 1405 (1992)
15. W.H. Parkinson E.M. Reeves and F.S. Tomkins
J. Phys. B: At. Mol. Phys. **9** 157 (1976)
16. W.A. Farooq M. Nawaz J.-P. Connerade and J.P. Marangos
J. Phys. B: At. Mol. Opt. Phys. **25** 4141 (1992)
17. J.P. Marangos N. Shen H. Ma M.H.R. Hutchinson and J.-P. Connerade
J. Opt. Soc. Am. **7** 1254 (1990)
18. X.H. He and J.-P. Connerade J. Phys. B: At. Mol. Opt. Phys. **26** L255 (1993)
19. A.J. Warry D.H. Heading and J.-P. Connerade *J. Phys. B: At. Mol. Opt. Phys*
20. J.-P. Connerade G. Droungas R. Elliott X.H. He N. Karapanagioti M.A. Farooq H. Ma J.P. Marangos and M. Nawaz
J. Phys. B: At. Mol. Opt. Phys. **27** 2753 (1994)
21. O'Mahony P.F. and Taylor K.T.
Phys. Rev. Lett. **46** 167 (1983)

STEPWISE MULTIPHOTON STUDY OF YTTERBIUM

Natalia E. Karapanagioti*, George Droungas
and Jean-Patrick Connerade

Blackett Laboratory,
Imperial College of Science, Technology and Medicine,
Prince Consort Road,
London SW7 2BZ
* currently at : Institute of Electronic Structure and Laser,
Foundation for Research and Technology-Hellas,
P.O.Box 1527,
711 10 Heraklion, Greece

INTRODUCTION

Spectroscopy has always been one of the most important experimental tools in the pursuit of an understanding of the atom. The interaction of light with matter is a unique probe of atomic mechanisms because of the added selectivity offered in comparison to other spectroscopic methods. Very complex investigations can often be conducted with relatively simple experimental apparatus, thus adding to the attraction of the field. Through the years, spectroscopic methods progressed enormously by making use of newly available technologies, and in particular tunable lasers. The resulting data created a need for original explanations, guiding the theorists to new directions and leading to the constant refinement of the atomic picture. Simultaneously, advances in computer technology made it possible for theorists to use numerical methods, the application of which would sometimes be very difficult in the past.

In the past, a lot of the spectroscopic research concentrated on the excitation of a single valence electron. Such excitation can often be quite well explained by using the picture of the independent particle model. This is also the case for the excitation of the innermost shells: both cases can often be represented with what is called the 'quasi-particle approximation',[1] i.e. for example Hartree-Fock schemes including corrections for dynamical screening. However, such approximations can fail when one breaks into the outermost inner shells. These spectra can be particularly interesting, since they reveal the non-coulomb nature of the potential inside the atom, and therefore can display considerable deviations from the usual characteristics of Rydberg series with respect to linewidths and oscillator strengths.

Such effects can also demonstrate themselves in the simultaneous excitation of two valence electrons, instead of the usual single excitation spectra. The study of doubly-excited states has been attracting a lot of interest in the past few years, as it is offers the chance to observe many-body effects that arise from the interaction of the two electrons with each other, as well as with the atomic core.

Usually, the above two problems are closely connected with the study of autoionization. Inner shell and doubly excited states almost always occur above the first ionization threshold, which leads to their having characteristically asymmetric lineshapes. The study of this asymmetry is a further way to investigate the effects of series perturbations and interactions, and the autoionizing range of atoms is the subject of much theoretical and experimental work.

Another facet of the non-Rydberg character of atoms are the centrifugal barrier effects, arising from the high orbital momentum of certain states. The centrifugal term in the Schrödinger equation leads to a spatial segregation of high l wavefunctions from the core. These states are then confined to a region where the potential is purely coulombic. Under certain conditions however, the states can be transferred abruptly, or 'collapse', to regions close to the core, and thus display strong effects due to correlations with the other core electrons. Again, such phenomena have been the subject of extensive study[2], which shows no signs of abating in the future.

Multiphoton spectroscopy (MPS) has for a long time been a favorite tool to study the effects mentioned above. Apart from the obvious advantage of being able to add many photons to reach otherwise inaccessible regions of the spectrum, such as the autoionizing range, it offers access to states the transitions to which are forbidden by single photon spectroscopy. Moreover, multiphoton transitions can introduce an extra selectivity of the states attained by the use of incident photon polarization selection rules.

Table 1. Two-photon transition selection rules.

General Rules			
$|\Delta J| \leq 2$			$J_i + J_f$ = integer
Particular Polarization Rules			
Polarization of ω_1	ω_2	Allowed Transitions	Forbidden Transitions
σ^+	σ^-	$\Delta M = 0$	
σ^+	π	$\Delta M = -1$	$\Delta J: 0 \to 0$
σ^-	π	$\Delta M = 1$	$\Delta J: 0 \to 0$
π	σ	$\Delta M = \pm 1$	$\Delta J: 0 \to 0$
π	π	$\Delta M = 0$	$\Delta J: 0 \leftrightarrow 1$
σ^+	σ^+	$\Delta M = -2$	$\Delta J: 0 \leftrightarrow 1, 0 \to 0, 1/2 \to 1/2$
σ^-	σ^-	$\Delta M = 2$	$\Delta J: 0 \leftrightarrow 1, 0 \to 0, 1/2 \to 1/2$
Additional Rules for Equal Frequency Photons			
$\Delta J: 0 \leftrightarrow 1$			Forbidden for all polarizations
If $|\Delta J| = 1$	$\Delta M: 0 \to 0$		Forbidden for all polarizations

Though extremely useful in MPS, it is often hard to find a comprehensive summary of multiphoton transition selection rules. A particularly convenient method has been used by Bonin and McIlrath[3] who obtained these rules for direct two-photon transitions by applying irreducible tensorial set formalism and by expressing the respective transition probability in terms of 3-j symbols. The resulting selection rules are summarized in table 1 and apply for a transition involving two photons of frequencies ω_1 and ω_2.

The selectivity and control added by the two-photon transition polarization selection rules can be used to determine the total angular momentum of the final states. This can be particularly useful for autoionizing states, for which experimental identification methods based on the Zeeman and Stark effects used in the discrete spectrum are not applicable due to the large widths of autoionizing lines. It is possible however to obtain that information by exciting the same part of the spectrum using various combinations of polarization of the two photons. Then, by comparing the spectra and noting which lines are absent for certain combinations, one can ascertain the J value of the final states. Elizarov and Cheperkov[4] determined the total angular momentum of various autoionizing levels of the 6p7p doubly excited configuration of barium in this way. Similar techniques can also be used for a large number of photons and steps, as well as for the discrete spectrum, as long as the magnetic quantum number M of the initial state is known. (see, for example, Jones et al.[5])

Making use of polarization selection rules, multiphoton excitation schemes can also be used to eliminate the Doppler broadening of spectral lines due to the thermal motion of the atoms, in particular using two-photon absorption. For two light waves of equal frequencies $\omega_1, \omega_2 = \omega$ which travel at opposite directions (i.e. $\underline{k}_1 = -\underline{k}_2$), the total Doppler shift of the two-photon absorption becomes zero and all atoms, irrespective of their velocities, absorb at the same sum frequency of $2\omega = \omega_1 + \omega_2$. Hence, the Gaussian Doppler profile of a spectral line reflecting the distribution of atomic velocities is eliminated if the two photons absorbed come from equal frequency, counter-propagating beams. Naturally, the possibility still remains that the atom will absorb two photons from the same beam, also fulfilling the resonance condition. Even though the probability of this occurring is only half as large as that of the previous process, this gives rise to the so-called *Doppler pedestal*. As Cagnac et al.[6] pointed out however, this problem can in some cases be eliminated by the proper choice of laser polarization. If the incident wave has, say, σ^+ polarization, a reflected wave will have σ^- and thus only the absorption of counter-propagating photons will lead to $\Delta M = 0$ transitions. Similar selection of incident light polarization can be implemented to obtain almost any series required, resulting in totally Doppler free even parity spectra.

Additionally however, multiphoton excitation schemes can also be used to study certain states which would not be easily accessible by introducing an additional selectivity through the use of a particular intermediate state. Unusual schemes can be devised to look at the objects of interest, such as isolated core excitation[7] (ICE), which is by now widely used in the study of doubly excited states.

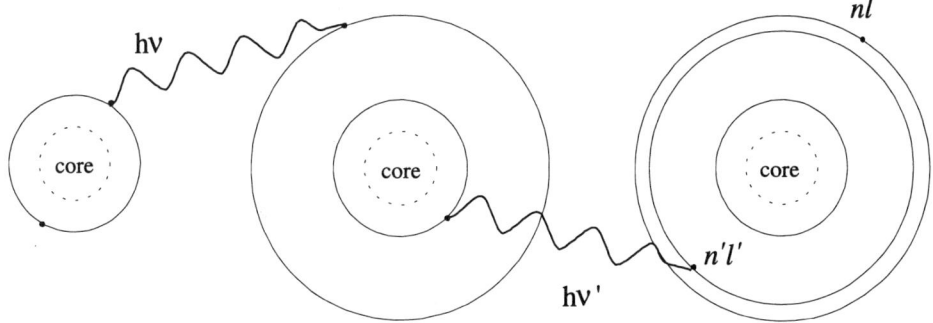

Figure 1. Schematic diagram of an isolated core excitation scheme.

In ICE, one photon is used to excite a valence electron to a high Rydberg state, thus ensuring its minimal interaction with the core, which then resembles the respective ion. Another photon then excites a second valence electron, giving rise to double excitation spectra in which the autoionizing states are particularly symmetric, due to the ionic character of the atom when they were excited. A schematic diagram of ICE is given in figure 1 for an atom with two valence electrons such as an alkaline earth.

The experiment described in this paper made use of a multiphoton excitation scheme very similar to ICE. Unlike ICE however, it depended on the interaction of the excited Rydberg electron with the core to achieve the simultaneous excitation of two electrons in the second step of the excitation. The scheme was similar to the one used for the study of the autoionizing range of mercury by Ding et al.[8] The original scheme also made use of many-body relaxation and was devised to obtain the inner shell spectra of mercury, and in particular to access the high l $5d^9nf$ states

The application of the excitation scheme in mercury was successful and led to the observation of the multiplet splitings for the nf states for the first time. It was also observed that in the spectra obtained, the nf series were dominant, unlike the single photon spectrum which is dominated by the 5d-np excitations. In fact, no such excitations were observable in the Hg multiphoton spectra, a fact which validated the effectiveness of the scheme in the observation of high l states and in probing the spatial extents of atomic orbitals.

DESCRIPTION OF THE EXPERIMENT

In this experiment, the scheme was applied in ytterbium, in which the $6s^2$ subshell gives rise to a complex doubly excited spectrum. The presence of this subshell makes ytterbium resemble the alkaline earths in many of its chemical and spectroscopic properties. In addition, the excitation of the 4f inner subshell produces series which are in themselves of interest for reasons stated above. Finally, the excitation of a 4f electron into the high orbital angular momentum g-states also makes it possible to study potential barrier effects on the atomic wavefunctions. The doubly excited series mentioned have large overlaps with the 4f inner shell spectrum, resulting in series interactions and perturbations which are of particular interest in the understanding of the atomic mechanisms.

The single photon spectra of the Yb autoionizing range had already been obtained both by synchrotron[9] and laser[10,11] (through VUV generation) radiation. The main object of study was the difference between the single-photon and multiphoton spectra. These differences were expected to demonstrate themselves as altered autoionizing lineshapes and relative intensities of the spectral features. One of the particular aims was to enhance the high l g-series and the choice of intermediate state was made with that particular goal in mind.

It may be useful to point out that, since our interest was the inner shell spectrum, direct multiphoton excitation off the ground state did not suit the purposes of this experiment. Whereas the method has the advantage of only requiring only one intense enough laser, it has been pointed out by Lambropoulos[12] that this process is much more likely to couple to the outer electrons of an atomic system and therefore unsuitable for inner-shell spectroscopy. The scheme used can be summed up in the following way:

$$4f^{14}6s^2(^1S_0) \xrightarrow{2\omega_1} 4f^{14}6s8s(^1S_0) \xrightarrow{\omega_2} 4f^{13}(^2F_{7/2})6s^2nd, ng$$
$$4f^{14}5d(^2D_{3/2,5/2})np, nf$$

As can be seen above, the excitation scheme makes use of three linearly polarized photons. Two photons from the fixed frequency dye laser take the atoms from the $4f^{14}6s^2(^1S_0)$ ground state to a state belonging to the $4f^{14}6sns(^1S_0)$ Rydberg manifold. States of this series have the same parity and J-value as the ground state and can be obtained through two-photon excitation off the ground state using linearly polarized light to satisfy selection rules. Employing two photons of σ^+ and σ^- circular polarization respectively, would forbid transitions to $J=2$ states but would introduce complications to the experimental set-up. The data was obtained through the $6s8s(^1S_0)$ state at 41939.9 cm^{-1},[13] as Hartree-Fock calculations indicated that it had the largest radial overlap with the ng series.

The second step of the excitation involved a third photon originating from the other tunable laser which was being scanned. That photon was absorbed by the already excited atoms and brought them to the autoionizing region of interest. The mechanism of the excitation scheme is not straightforward, but enables us to attain one of our objectives, that of exciting the high l states. To reach the states of interest, the second step must involve the participation of two electrons, i.e. a double excitation. For the final state to be of the form of $4f^{14}$ 5d np, nf or $4f^{13}$ 6s nd, ng, both the excited electron and one of the core electrons must take part in the transition.

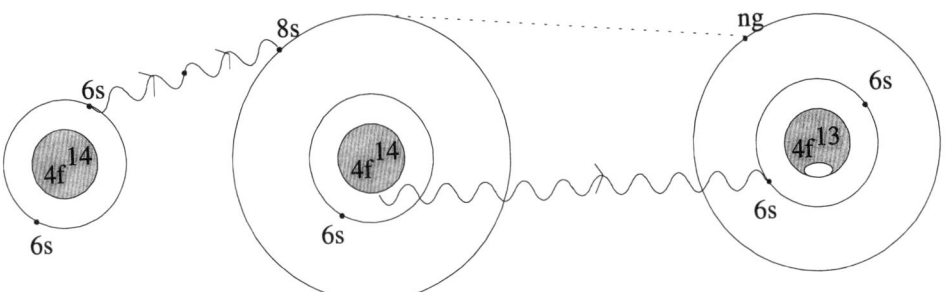

Figure 2. Schematic representation of the excitation scheme.

An independent particle schematic representation of the excitation scheme is given in figure 2. The oncoming photon is expected to couple with an inner-shell 4f electron and excite it to fill the vacancy in the 6s shell, leaving a hole in the 4f shell. At the same time, the Rydberg electron is non-radiatively shaken down to an ng or nd state (monopole shakedown), thus providing the energy necessary for the inner shell to be broken open. This process depends strongly on the radial overlap between the initial and final states' wavefunctions. The limitation of this picture is that an independent particle description is not capable of accounting for electron-electron correlations and configuration mixing which make this excitation possible. Still, this description has been used to explain the process of shakedown and shake-up satellites in X-ray spectra by Åberg[14], who uses the sudden approximation to account for the effect. According to this theory, the creation of the core hole can be regarded as a sudden perturbation to the Hamiltonian, resulting in "anomalous configurations" which give rise to the satellites. In this context, the formation of the satellite depends directly on the radial overlap between the outer electron initial and final state, as is the case for monopole processes.[15] This requirement seems to hold even when the excitation is viewed in terms of more complex pictures, such as excited state configuration interaction. This added constraint was therefore used to enhance the high l series by choosing an intermediate state of similarly large radial extent.

Experimental Description

A diagrammatic representation of the apparatus layout is given in figure 3. The linearly polarized photons required for the excitation were produced by two tunable dye lasers of around 5 GHz bandwidth with pulse energies ranging from ~ 0.1-5 mJ. The resulting signal was detected by a thermionic diode detector, which is effectively a high-gain ion detector.[16,17] Its function is summed-up by the amplification of the diode current between the anode and the cathode when an ion is formed in the active region of the detector. This makes it an excellent device to measure ions produced by autoionization of atomic states above the continuum and also to detect high Rydberg states which get collisionally ionized in the presence of a buffer gas. Since this detector can operate with very small numbers of ions produced, it requires very low vapour pressures, in contrast with other absorption techniques. The spectra obtained were cross-calibrated with the single photon data, resulting in a wavelength accuracy of ± 0.03 Å, and were digitally filtered using a Fast Fourier Transform filter. The relative intensities of the features observed were correct to within ± 10%, a limitation set by the inevitable background ionization signal through competing multiphoton processes.

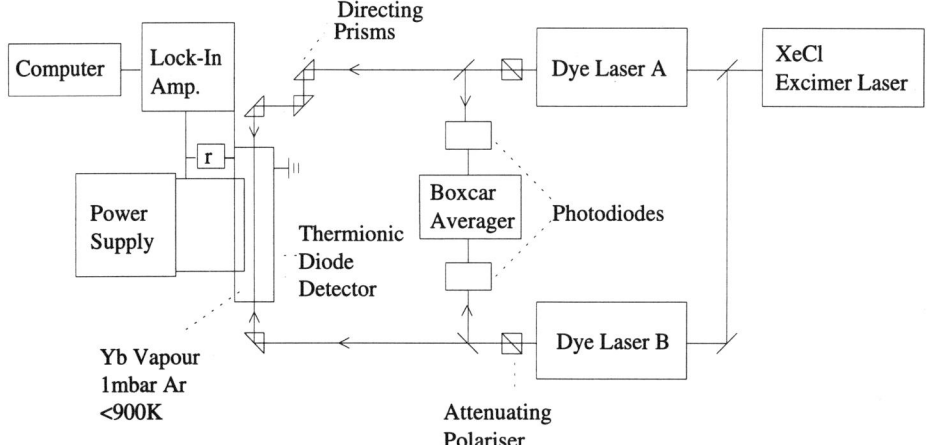

Figure 3. Diagram of the layout of the apparatus.

RESULTS

The autoionizing region studied was the range up to the second and third ionization limits of ytterbium, the $4f^{13}6s(^2F_{7/2})$ and the $4f^{14}5d(^2D_{3/2})$ series limits, located at 71861.75 cm^{-1} and 73403.87 cm^{-1} respectively.

The three inner shell series converging to the second ionization limit are the following:

$4f^{13}(^2F_{7/2})$ nd $[3/2]_1$

and ng $[1/2]_1$

$[3/2]_1$.

However the ng series is K-degenerate, leading to only two observable series of features in the spectra.

In the same range, broader resonances are also found, which arise from double excitation of the $4f^{14}5d(^2D_{3/2,5/2})$ np, nf series, converging to higher limits. Detailed studies of the interactions between the singly and doubly excited series have previously been reported[9], revealing that the double excitations perturb the inner shell spectrum, giving rise to intensity fluctuations, q-reversals and quantum defect variations characteristic of interacting autoionizing series. Lu-Fano plots of the series' quantum defects were calculated using the single photon data and they illustrated very effectively the extent of the perturbations as well as pointing out the identity of the main perturbers.

Table 2. Table of new $4f^{13}(^2F_{7/2})ng[1/2,3/2]$ series members, where n^* is the effective quantum number and μ is the quantum defect.

n	Scanning λ (nm)	Total λ (Å)	Energy (cm^{-1})	n^*	μ
19	337.644	1397.49	71556.86	18.972	0.028
20	337.294	1396.89	71587.60	20.007	-0.007
21	337.003	1396.39	71613.23	21.013	-0.013
22	336.753	1395.96	71635.29	22.013	-0.013
23	336.543	1395.60	71653.77	22.970	0.030
24	336.351	1395.27	71670.72	23.967	0.033
25	336.177	1394.97	71686.13	24.997	0.003

The ng inner shell series is very weak in the single photon spectra, due to the effect of the centrifugal barrier arising from the large orbital angular momentum ($l = 4$) of these states. Moreover, although the series can be followed up to higher n, the series members between $n = 18$ to 25 are missing from the single photon data.[10] The excitation scheme in this experiment was intended to increase the probability of excitation of this series by using an extended Rydberg state as an intermediate.

In fact, in the multiphoton spectra, both the inner shell nd and ng stand out more clearly, and the previously missing members of the ng series are detected. They appear as small side peaks to the nd states in the range between 1394.5-1398 Å, and their identity is confirmed by the very small quantum defects that this series is expected to have, due to the small core penetration of the high l states. A list of the new members' energies, effective quantum numbers and quantum defects is given in table 2. In contrast with the above observation however, the multiphoton data did not display an extension of the ng series to higher n than the single photon data. This must be attributed to the perturbations which affect the higher members of this series, which overlap with the $(^2D_{3/2})12p[3/2]$ and $[1/2]$ doubly excited states.

Turning to the nd series, we observe another difference between the spectra if we examine the line profiles. In the single photon spectra, the nd series members are nearly symmetric or have positive q values, whereas in the multiphoton results, a pronounced asymmetry of negative q is observed (see figure 4). It is interesting to see that the change of symmetry is maintained even in the cases where the series members appear as troughs superimposed on broader doubly excited resonances. Moreover, the q reversal of the nd

series through the single photon data, which occurs at $n = 18$, also appears in the multiphoton work.

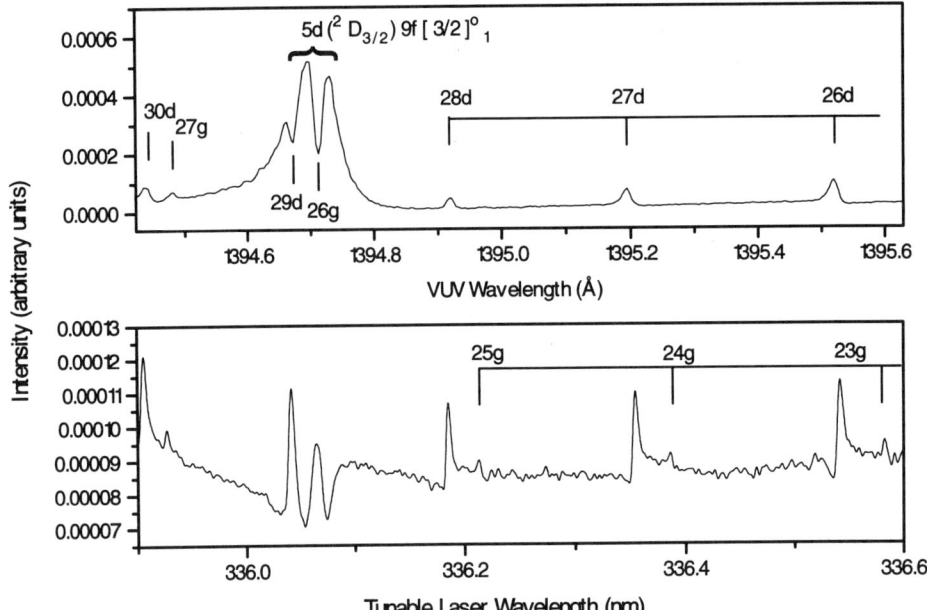

Figure 4. A portion of the spectrum in the region of the 5d ($^2D_{3/2}$) 9f [3/2]$^o_1$ perturber obtained by single photon excitation[10] (top) and multiphoton excitation (bottom). One can observe the symmetry changes, the appearance of the ng members, as well as the depression of the broad doubly excited perturber. This last effect may be due to phase cancellation, i.e. to the possible coincidence of a lobe of the 9f radial wavefunction with a node in its 8s counterpart.

Differences between autoionizing state symmetries according to the mode of excitation are known to occur since Fano's expression for the q parameter includes the excitation probability from the initial state.[18] In the present experiment, we have obtained the q values for various members of the nd series by fitting the single photon and multiphoton experimental profiles to the Fano formula[19] using a least squares method. An example of the fit indicating the reversal is shown in figure 5.

This symmetry change can prove particularly useful in the identification of series in the presence of perturbers, and in our case allows us to reassign the 29d from the peak position of 1394.663 Å in the single photon data to the trough position at 1394.673 Å (see detail of the spectrum in figure 4), as well as confirming other existing assignments.

In fact, the presence and effect of the broad doubly excited states seem to be greatly affected in the multiphoton spectra. For example, the ($^2D_{3/2}$)9f[3/2] state is totally depressed in our spectra, as can be seen in figure 4. On the other hand, broad structures appear which have no counterpart in the single photon spectrum and which have yet to be identified.

The three series converging to the second limit mentioned are the following:

$4f^{14}(^2D_{3/2})$ np $[1/2]_1$
$[3/2]_1$
and nf $[3/2]_1$.

Lu-Fano plots of the quantum defects of these states, obtained by using the single photon data, show all three series to be heavily perturbed by states converging to the higher, $4f^{14}5d(^2D_{5/2})$ series limit, namely the ($^2D_{5/2}$) 6f, 7f, 11p and 12p states.

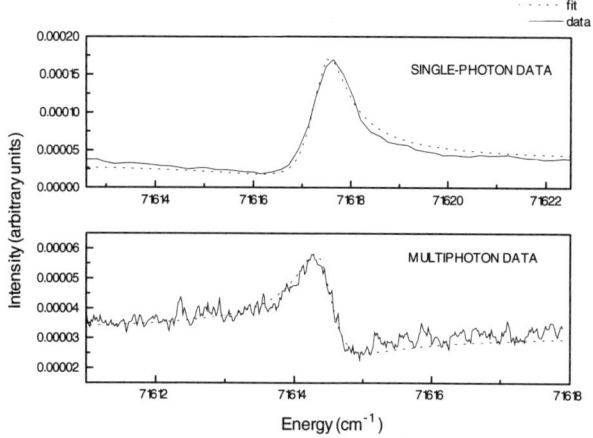

Figure 5. The profile of the $4f^{13}(^2F_{7/2})6s^2 24d$ line obtained through single-photon[10] and multiphoton excitation. Superimposed in dotted lines are fits using the Fano formula for an isolated resonance interacting with one continuum. Note the reversal of the line asymmetry: the fitted q value is 2.9 for the single photon data and -2 for the multiphoton data (unfiltered).

What is immediately evident by comparing the two sets of data is that in the multiphoton spectra, the $nf[3/2]$ series is much more prominent than in the single-photon ones. The structure belonging to this series is narrow and quite symmetric, irrespective of the mode of excitation. In both data sets, the relative intensities and widths of the series members

Table 3. Table of new $5d(^2D_{3/2})nf[3/2]$ series members, where n^* is the effective quantum number and μ is the quantum defect.

n	Scanning λ (nm)	Total λ (Å)	Energy (cm⁻¹)	n*	μ
35	318.815	1364.144	73306.05	33.494	1.506
36	318.760	1364.043	73311.46	34.461	1.539
37	318.709	1363.950	73316.48	35.437	1.563
39	318.619	1363.785	73325.35	37.384	1.616
40	318.579	1363.712	73329.29	38.358	1.642
41	318.543	1363.646	73332.84	39.305	1.695
42	318.509	1363.583	73336.19	40.266	1.734
43	318.478	1363.526	73339.24	41.207	1.793
44	318.447	1363.470	73342.30	42.217	1.783
45	318.419	1363.418	73345.06	43.197	1.803
46	318.394	1363.372	73347.53	44.132	1.868
47	318.370	1363.328	73349.89	45.090	1.910
48	318.346	1363.284	73352.26	46.113	1.887
49	318.325	1363.246	73354.33	47.067	1.933
50	318.306	1363.211	73356.21	47.984	2.016
51	318.286	1363.174	73358.18	49.010	1.990
52	318.271	1363.147	73359.66	49.824	2.176

are very irregular, providing further evidence that the series is heavily perturbed. In the single photon data, the series is followed up to $n = 32$; in the multiphoton data however, members going up to $n = 52$ can be observed, all except the $n = 33, 34$ and 38 which may have lower probability amplitudes due to the perturbations.

A list of the new $nf[3/2]$ series members, their energies, effective quantum numbers and quantum defects is given in table 3. A plot of μ vs. state energies is given in figure 6. For an unperturbed series, μ should be constant as one gets close to the ionization limit. Figure 6 however, indicates a steady increase in μ, consistent with the trend apparent in the Lu-Fano plot for the lower series members. Apart from the new series members, the favoring of the nf series by the excitation scheme leads to the proposed reassignment of the 17f and 20f.

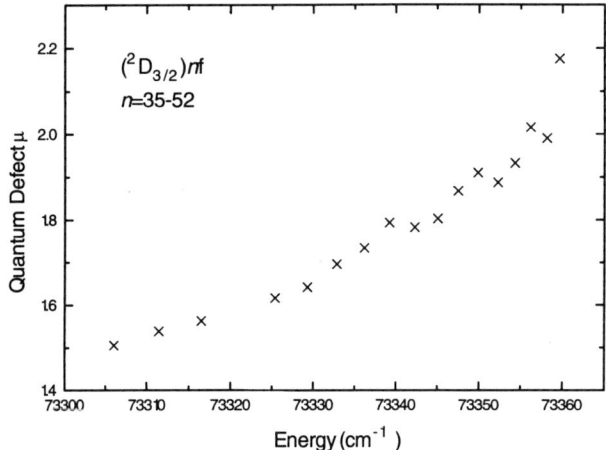

Figure 6. Quantum defects vs. resonance energy for the new $(^2D_{3/2})nf[3/2]$ series members.

Another significant observation when comparing the single photon and multiphoton data, is that the relative intensities of the structure belonging to the $4f^{14}(^2D_{3/2})np\,[1/2]_1$ and $[3/2]_1$ series are reversed in many occasions. In the single photon data, the $np\,[3/2]_1$ series dominates. In the multiphoton spectra however, the $np[1/2]$ series members are more prominent, while the $np[3/2]$ series is often too weak to be observed, as can be seen for the characteristic cases of $n = 18$ and 20 in figure 7.

This observation may be difficult to explain since the coupling used to describe the principal Rydberg series of ytterbium (L-S) is different from that used for the autoionizing series (j_cK or Racah pair coupling[20]). However, it can be understood in terms of the difference in the possible K value designation of the ground state and that of the intermediate state used in the multiphoton work. The $4f^{14}6s^2$ ground state can be well described by L-S coupling, having a core j value of $j_c = 0$. Its K value can then be considered as zero, since $K = j_c + l$, and since there are no excited electrons in the ground state. However, the intermediate $4f^{14}6s8s$ state can be described by j_cK coupling, which in its case may be more appropriate than L-S coupling. The j value of the $4f^{14}6s(^2S_{1/2})$ core is then $j_c = 1/2$, which when coupled to the $l = 0$ of the excited 8s electron gives only one possible K value, i.e. $K = 1/2$. The coupling of K and the spin of the 8s electron gives rise to a $[K = 1/2]\,J = 1$ and $[K = 1/2]\,J = 0$ doublet, consistent with the L-S assignment of $(^3S_1)$ and $(^1S_0)$.[13]

The experimental findings here seem to suggest that transitions to states of the same K designation as the initial state are favored over states with $\Delta K = 1$. This is consistent with atomic theory, which predicts that transitions between j_cK coupled states follow the following rule for line strengths:[21]

strongest: $\Delta K = -1$
↓ $\Delta K = 0$
weakest: $\Delta K = +1$

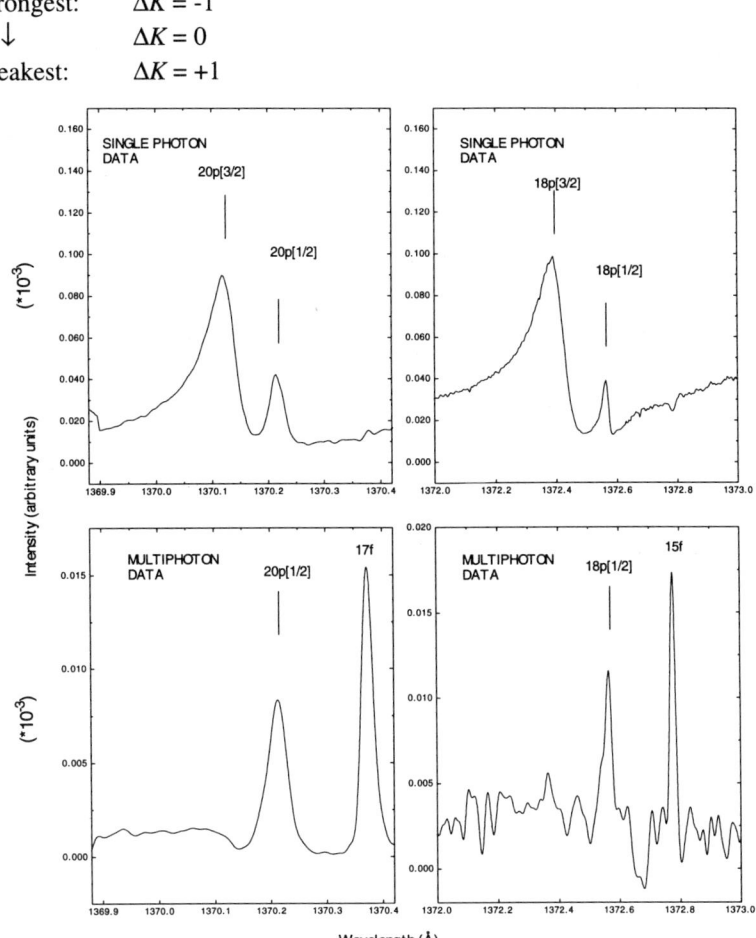

Figure 7. Details of the spectra demonstrating the changes in relative intensities between the nf, np[3/2] and np[1/2] features when accessed by single photon and by multiphoton excitation.

On this point however, theoretical work is required to support the experimental evidence, since the series is perturbed and this generalization cannot be extended to all the series members observed. In the multiphoton spectra, many of the perturbers are suppressed and this may affect differently each of the np series doublets, depending on their proximity to the perturber as well as on the strength of the interaction. It is also worth noting that the features belonging to the np[1/2] series are narrower than the np[3/2] ones, and narrow states seem to be favored by this particular excitation scheme.

CONCLUSIONS

This experiment described was set up in order to study part of the autoionizing range of neutral ytterbium by employing a stepwise multiphoton excitation scheme. The excitation scheme was devised to favor transitions to high orbital angular momentum inner shell states, which have a low probability of excitation from the ground state of the atom.

In fact, we were able to observe and extend two high l series, the inner shell $4f^{13}(^2F_{7/2})6s^2ng[1/2],[3/2]_1$ and the doubly excited $4f^{14}5d(^2D_{3/2})nf[3/2]_1$. The prominence of these series indicates that the excitation scheme used does indeed favor the excitation to extended states, as was originally suggested by Ding et al.[8] Therefore, the scheme can be used for the study of high orbital angular momentum states in other cases of interest, as well as a probe of the spatial distribution of atomic orbitals. However, variations in the relative intensities of these series were observed. These individual effects might be explained by considering the radial overlaps between the states, as the total matrix element could be reduced by phase cancellation between different regions of the radial wavefunctions.

We also observed significant variations in the relative intensities of the doubly excited series, which tend to dominate the single photon spectrum. Although the quenching influenced members from all the detected series, especially below the $(^2F_{7/2})$ ionization threshold, certain series were more affected, particularly the $4f^{14}5d(^2D_{3/2})np[3/2]_1$, many members of which were too weak to be detected. This was attributed to the possible K designation of the intermediate state used, indicating a breakdown of $L-S$ coupling in the high n Rydberg series members. The quenching of certain doubly excited states below the second ionization threshold allowed the observation of members of the inner shells series that were previously obscured by the broad doubly excited profiles. This led to the reassignment of certain inner shell series members and allowed the analysis of their profiles. This effect can be used in other similar cases where the doubly excited spectrum is prominent and thereby obstructs the investigation of overlapping inner shell series.

The comparison between the single photon and the multiphoton spectra indicated considerable differences in the autoionizing profiles of the lines. In particular, the asymmetry of the $4f^{13}(^2F_{7/2})6s^2nd[3/2]_1$ series was reversed in the multiphoton results, with the q parameter being negative instead of positive. This change persisted even when the lines overlapped and interacted with broader double excited resonances. Though not surprising, this difference is also a way to identify members of a perturbed series in complex spectra. In our case, it served to confirm but also to revise existing assignments, as well as to identify some of the multiphoton features.

What needs to be noted however, is the need for a rigorous theoretical treatment of the results obtained. Theoretical calculations of autoionization spectra can be achieved using R-matrix and K-matrix methods, although this is by no means a simple task, especially when a combination of inner shell and doubly excited series is involved. However, this might be useful in order to understand some of the effects observed, such as the symmetry reversal of the inner shell nd series and would also be a good test for theory. Calculations of radial atomic wavefunctions would also be very useful for a more precise understanding of the feature intensities and would serve to fully explain the mode of excitation.

As we can see, multiphoton spectroscopy can still promote the understanding of the atom: by making use of photon polarization and atomic state characteristics it can serve to reveal new atomic features, inaccessible to single photon spectroscopy. More significantly however, in combination with single photon techniques it can utilize its selectivity to allow a more complete grasp of excitation mechanisms, such as double excitation, which are still in need of interpretation.

Acknowledgments

The authors would like to thank W. G. Kaenders and J. P. Marangos for valuable assistance and discussions. Two of the authors (N. E. K. and G. D.) are also grateful to the Greek State Scholarship Foundation for its financial support during the course of this work.

REFERENCES

1. Pines D., *Elementary Excitation in Solids*, 1964, New York: Benjamin.
2. Connerade J. P., *Contemp. Phys.* **19** (1978) 415.
3. Bonin K. D. and McIlrath T. J., *J. Opt. Soc. Am.* **B 1** (1984) 52.
4. Elizarov A. Y. and Cherepkov N. A., *Pis'ma. Za. Eksp. Teor. Fiz.* **44** (1986) 3.
5. Jones R. R., Panming Fu and Gallagher T. F., *Phys. Rev.* **A 44** (1991) 4260.
6. Cagnac B., Grynberg G. and Biraben F., *J. Phys. (Paris)* **34** (1973) 845.
7. Bhatti S. A., Cromer C. L. and Cooke W. E., *Phys. Rev. Lett.* **34** (1981) 161.
8. Ding R., Kaenders W. G., Marangos J. P., Shen N., Connerade J. P. and Hutchinson M. H. R., *J. Phys. B: At. Mol. Opt. Phys.* **22** (1989) L251.
9. Baig M. A. and Connerade J. P., *J. Phys. B: At. Mol. Phys.* **17** (1984) L469.
10. Kaenders W. G., Shen N., Marangos J., Hutchinson M. H. R. and Connerade J. P., *J. Mod. Opt.* **37** (1990) 835.
11. Kaenders W. G., *Imperial College DIC Thesis* (1990) University of London.
12. Lambropoulos P., *Atoms in Strong Fields* (NATO ISI), ed. Nicolaides C., Clark C., Nayfeh M., London Plenum 1989.
13. Martin W. C., Zalubas R. and Hogan L., Atomic Energy Levels of Rare Earth Elements, *Nat. Bur. Stand.* **60** 1978.
14. Åberg T., *Phys. Rev.* **156** (1967) 156.
15. Mansfield M. W. D. and Connerade J. P., *Proc. Roy. Soc. Lond.* **A 359** (1978) 389.
16. Niemax K., *Applied Phys.* **38** (1985) 147.
17. Hermann P. P., Schlumpf N., Telegdi V. L. and Weiss A., *Rev. Sci. Instr.* **62** (1991) 609.
18. Fano U., *Phys. Rev.* **124** (1961) 1866.
19. Shore B. W., *J. Opt. Soc. Am.* **57** (1967) 881.
20. Racah G., *Phys. Rev.* **61** (1942) 537.
21. Cowan R. D., *Theory of atomic structure and spectra*, University of California Press, Berkeley-Los Anegeles-London, 1981.

SPIN POLARISED ELECTRON SCATTERING IN CONDENSED MATTER - AN ATOMIC APPROACH

J A D Matthew

Department of Physics
University of York
Heslington
York YO1 5DD

INTRODUCTION

Electron scattering from atoms is a mature field in which so called 'complete' scattering experiments are now possible, ie spin polarised electron beams may be scattered from spin polarised atomic targets and the spin resolved differential scattering cross sections determined - see other articles in this volume. Spin polarised scattering from ferromagnetic solids has a much shorter history but has parallels with the ideal atomic experiment with spin orientation being provided by the internal magnetic field in the surface region[1,2]. Kirschner[3] and Hopster and Abraham[4] have carried out 'quasi complete' inelastic scattering experiments on Stoner excitations in ferromagnetic metals with electron loss energies of a few eV. However these are excitations from bands of spin anti parallel to the incident spin direction up to parallel bands with the wave vectors of the states influencing the electron energy loss profile. In contrast core-valence excitations in transition metals eg $3p \to 3d$ excitations in Fe, $4d \to 4f$ transitions in rare earths are quasi atomic as opposed to band like in character with the traditional atomic angular momentum labels L, S and J providing an appropriate classification of the excitations. Similarly $4f^n - 4f^n$ transitions in the rare earths and $3d^n - 3d^n$ transitions in transition metal insulating compounds (as opposed to metals) show quasi atomic behaviour in spite of the fact that both the initial and final states lie close to the Fermi level. Under such circumstances it should be possible to adapt electron-atom scattering experience to describing scattering in the solid. It should be noted that:

1. most experiments are at primary energies E_p well above typical excitation energies ΔE and so Born type expansions are appropriate
2. the atomic systems to be considered eg Gd, Mn are much more complicated than those traditionally tackled by atomic physicists
3. solids have much higher density than gaseous scattering and so multiple scattering is endemic. Indeed double scattering is essential in the usual electron energy loss experiment (Fig 1): because normally $E_p \gg \Delta E$ inelastic scattering is strongly forward dominated so that it is usual for a second elastic scattering to accompany the

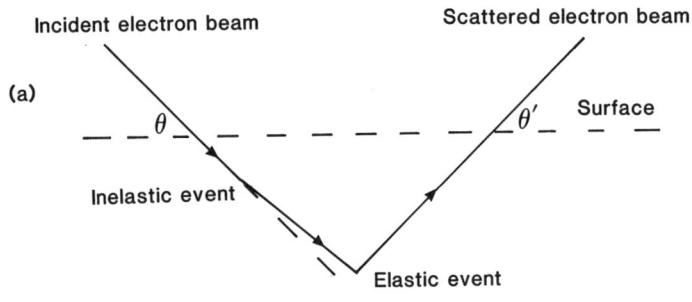

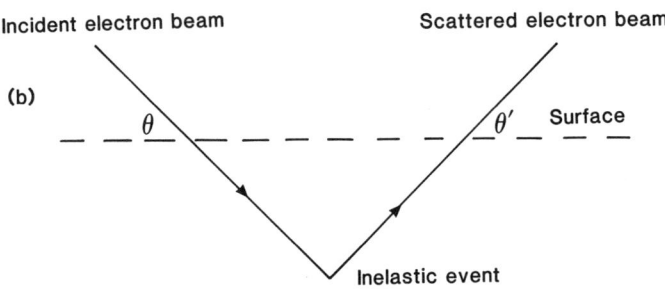

Figure 1. (a) Small angle inelastic scattering combined with large angle elastic scattering. (b) a high angle inelastic scattering event.

inelastic scattering in order to detect the scattered electron in reflection geometry (Fig 1a). Although the *intensity* may well be dominated by such double or multiple scattering, a single stage inelastic event (Fig 1b) is possible, and such high angle processes will be of particular importance if spin change occurs in the target, since spin flip often involves radical redirection of the incident electron.

When incident electrons scatter off atoms with net spin, electrons parallel and anti parallel to the atomic spin direction will have different cross sections for a given excitation resulting in spin polarisation in the scattered beam. Although single scattering events make only modest contributions to the overall scattering intensity, they may be highly influential in determining the spin polarisation of the scattered beam. This balance is not even well understood qualitatively and this paper sets out to review how relatively simple atomic scattering theories give insight into recent observations. Scattering involving spin polarisation may be investigated in different ways:

(a) variation of cross section with primary energy - it is well known that the energy dependence of dipole allowed, non dipole allowed and transitions involving spin change all have different energy variations above threshold

(b) dedicated spin flip channels - certain open shell $4f^n - 4f^n$ or $3d^n - 3d^n$ transitions require spin flip so that induced spin polarisation is implied even in a target which is not spin oriented

(c) measurement of polarisation of loss electrons from an oriented target - unpolarised electron beams incident on oriented eg ferromagnetic targets - both specular and off specular

(d) 'quasi complete' energy loss experiments.

Spin polarisation effects arise both from spin orbit interaction and exchange: here we will consider only exchange effects in order to make an initial interpretation of scattering from systems where exchange is likely to dominate. Born approximation theories may be applied at various different levels. First Born[5] takes no account of exchange, but the $E_p^{-1} \ln(\gamma E_p/\Delta E)$ variation of total cross section it predicts (with γ constant) is a useful yardstick for other scattering processes to be measured against. The lowest order approximation using exchange (Born Oppenheimer (BO)) approximation incorporates exchange and may be simplified through the Born Ochkur approximation (BOCH)[6,7] to give useful guidance, and the distorted wave approximation (DWA)[8] can treat individual multiplets in open shell systems suggesting a complex pattern of cross section variation. Here we review results from BOCH and DWA calculations for a number of quasi atomic systems.

PARITY ALLOWED TRANSITIONS

If the scattering of the incident electron and the promotion of an oriented electron (quantum numbers nl) into an empty state (n'l') are treated as a simple two electron problem without any treatment of multiplet effects, the scattering may be described by a direct amplitude $f(\theta)$ and an exchange amplitude $g(\theta)$ in the spirit of Kessler's original formulation of e-H scattering[9].

In the 1st Born approximation f is given by

$$f_{n'l'nl} = -\frac{2}{q^2} \langle \psi^*_{n'l'} | e^{i\underline{q}\cdot\underline{r}} | \psi_{nl} \rangle ,$$

where ψ_{nl} and $\psi_{n'l'}$ are the one electron wave functions for the initial and final electron states, $\underline{q} = \underline{k}_i - \underline{k}_f$ is the momentum transfer between the initial state (wave vector $\underline{k}_i$) and final state ($\underline{k}_f$) of the scattered electron. In the BOCH approximation

$$g_{nln'l'} = \frac{q^2}{k_i^2} f ,$$

a result obtained by expanding the Born-Oppenheimer amplitude in inverse powers of k_i^2, and retaining the leading term[10]. This rather pragmatically avoids orthogonality problems associated with BO and leads to a very simple formula which gives approximate magnitudes for the effects when $E_p = \frac{1}{2} k_i^2$ greatly exceeds $\Delta E = \frac{1}{2}(k_i^2 - k_f^2)$.

The resulting polarisation P of an unpolarised electron beam incident on a ferromagnetic surface has the following quite general characteristics[11]:

(i) in the forward scattering limit

$$P \approx \frac{g(0)}{f(0)} \approx \frac{1}{4} \left[\frac{\Delta E}{E_p}\right]^2$$

ie in the Born limit P is small, and arises not from *spin flip* processes but mainly from differences in the *non spin flip scattering* amplitudes of electrons parallel and antiparallel to the atomic spin.

(ii) $P \to 1$ at some intermediate scattering angle $\theta \sim 60°$ at all primary energies, and falls off slowly to a substantial value in the backscattering limit. Here spin flip transitions are very significant, but the differential cross sections are then small.

Although neither of these results is quantitatively correct, they establish an important qualitative pattern. In the geometry of typical surface science experiments, strong small polarisation forward inelastic scattering may be accompanied by high angle elastic scattering in competition with weak single scattering inelastic events of high polarisation. It is a delicate balance[12]: in single crystal systems diffraction will concentrate the forward inelastic scattering processes in the vicinity of LEED beams, while in the diffuse region the single scattering events may dominate; for amorphous systems, eg in the pioneering work of Mauri et al[13] on $3p \to 3d$ transitions of Fe in FeB, both kinds of events are certainly intermixed. If only forward scattering (non spin flip) was involved P would vary with primary energy as E_p^{-2}, whilst in practice P falls off more slowly with E_p,[13] indicating that spin flip events are also important.

PARITY FORBIDDEN TRANSITIONS

In some systems with infilled shells, eg rare earths and transition metal compounds, it is possible to observe low energy intra shell excitations, eg $4f^n$-$4f^n$, $3d^n$-$3d^n$ transitions, which are quasi atomic in character, using electron energy loss spectroscopy. In the extreme case of half filled shells, eg $4f^7$-$4f^7$ processes in Gd or Eu, $3d^5$-$3d^5$ in Mn compounds, the initial states (^8S or ^6S respectively) are the only multiplets allowed by the Pauli principle that have all spins parallel, so that all transitions involve spin flip in an LS coupling description. This implies that spin flip transitions can be readily isolated in conventional EELS without recourse to electron spin analysis[14-17] and without the need for ferromagnetic spin orientation. Since now the initial and final state spatial wave functions are identical

$$\frac{d\sigma}{d\Omega} \propto |g|^2 \propto \left| \int |\psi_{nl}|^2 \, e^{i\underline{q}\cdot\underline{r}} d\underline{r} \right|^2$$

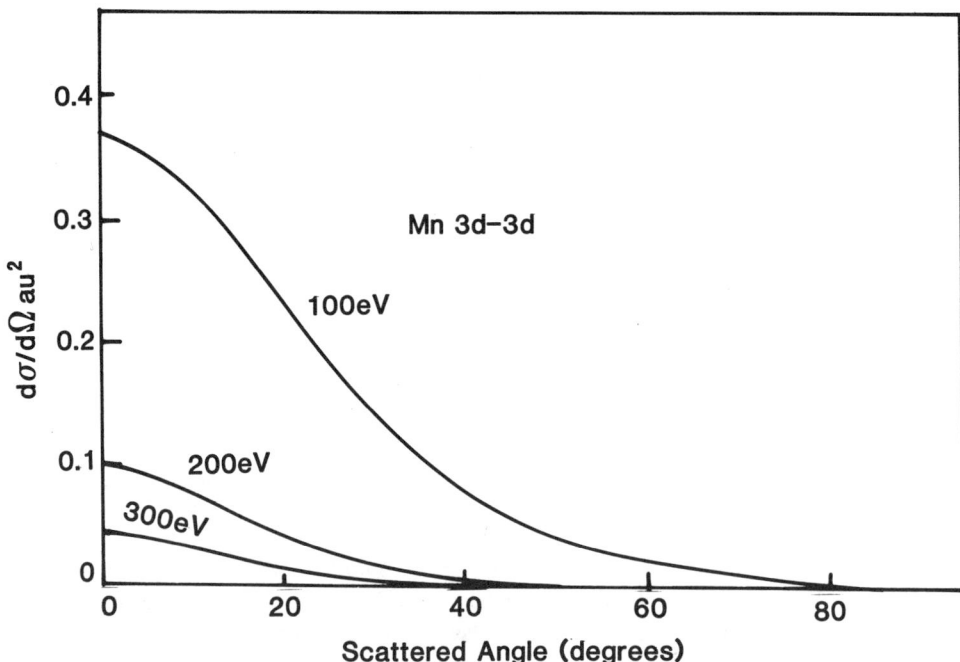

Figure 2. Differential cross section versus scattering angle for 3d - 3d spin flip transitions in Mn^{++} ($\Delta E \approx 3$ eV on average) at E_p = 100, 200 and 300 eV within the Ochkur approximation.

Figure 2 shows the variation of differential cross section with angle for $3d^5$-$3d^5$ transition in Mn: the fall off with increasing momentum transfer follows an "uncertainty principle" style relation

$$\Delta q < r_{nl} > \sim 2$$

where Δq is the momentum transfer (in au) at which the transition amplitude g falls to half its $\theta = 0$ value, and $< r_{nl} >$ is expectation value of r for the orbital concerned. Although high angle processes are not as prominent as for parity allowed transitions, the fall off of cross section with angle is dramatically different from that of dipole allowed transitions of the same exitation energy, where at $E_p = 100$ eV the half width is of order 1°.

Where the shell is not half filled, some transitions require spin flip, others not; Gorschlüter and Merz[18] have made a detailed investigation of localised d - d excitations in NiO(100) and CoO(100), and use the different primary energy dependence and geometry dependence of the structure to discriminate between parity forbidden non- spin flip transitions, eg the $^3A_{2g}$ - $^3T_{1g}$ ($\Delta E = 1.7$ eV) and spin flip $^3A_{2g}$ - ^1E_g ($\Delta E = 1.6$ eV) transitions in NiO vary in prominence with E_p with the non spin flip process dominant for high E_p, and the spin flip more prominent at higher scattering angles. Such interpreations have been confirmed in spin polarised EELS on NiO(100) by Fromme et al[19] and are consistent with a previous quasi complete spin polarised EELS study of Cr_2O_3 by Hopster[20]. These experiments confirm that ferromagnetic ordering is not essential to highlight spin flip transitions, and that spin flip effects are more prominent away from diffraction spots. One of the surprising features of these observations is that spin flip remains competitive with non spin flip processes at primary energies high compared to the excitation energy ($E_p/\Delta E \sim 30$), whereas in gas phase experiments they are only significant near threshold. Although the cross sections for spin flip scattering fall rapidly with primary energy ($\sim E_p^{-3}$) - see Fig 3,

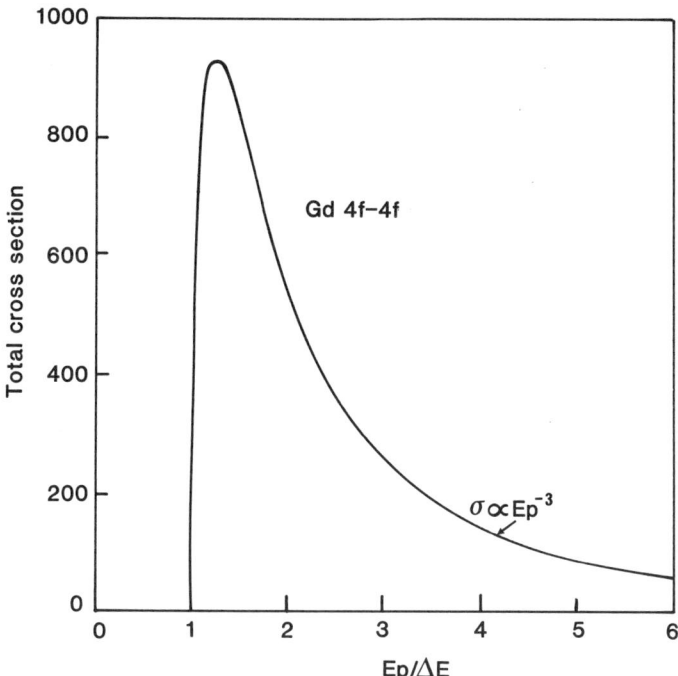

Figure 3. The variation of the total cross section with primary energy in threshold units $E_p/\Delta E$ for 4f - 4f transitions in Gd ($\Delta E \approx 4$ eV on average) within the Born Ochkur approximation.

the double scattering dipole channel also varies rapidly with energy because the elastic reflectivity decreases greatly with energy from low E_p (reflectivity high) to $E_p \sim 100$ eV where it is reduced to a few %, even although the inelastic cross section itself varies slowly with E_p. Eventually at high E_p ($\gtrsim 300$ eV) the rapid fall off of exchange scattering takes its toll, and dipolar scattering wins. In addition there is the possibility of resonant enhancement of spin flip through core excitation, eg 4d - 4f in Gd, and 3s - 3d in Ni.

The Born-Ochkur approximation gives useful semi-quantitative insight into spin flip processes in electron energy loss spectroscopy and to the spin dependence of direct scattering processes. However, there will be detailed differences in the cross sections of different multiplet transitions. To explain this a more sophisticated theory is required. Moser and Wendin[21] have applied the distorted wave approximation in circumstances much more complicated than atomic physicists normally consider - 3d - 4f and 4d - 4f transitions in La and 4d → 5f and 4f - 5f transitions in Th. What is now required are similar LS resolved studies for the intra shell excitation data now emerging.

SUMMARY

Quasi atomic excitations are observed in the electron energy loss spectroscopy of rare earth and transition metal condensed matter systems, and these may be modelled using electron-atom scattering style theories. The very simple Born-Ochkur approximation gives useful insights into the energy dependence and angular dependence of exchange scattering cross sections of both core excitations and intra-shell transitions, but a higher level theory is needed to distinguish the detailed behaviour of transitions to individual crystal field split multiplets.

ACKNOWLEDGEMENTS

The author wishes to thank Stuart Porter and Richard Leggott for their help in developing this perspective of spin polarised EELS.

REFERENCES

1. J. Kirschner. "Polarised Electrons at Surfaces", Springer Verlag (1985)
2. H Hopster, Spin-polarised electron energy loss spectroscopy, *Surf. Rev. Letts.* 1:89 (1994)
3. J. Kirschner, Direct and exchange contributions in inelastic scattering of spin-polarised electrons from iron, *Phys. Rev. Lett.* 55:973 (1985)
4. H. Hopster and D.L. Abraham, Spin-dependent inelastic electron scattering on Ni(110), *Phys. Rev. B* 40:7054 (1989)
5. B.L. Moiseiwitsch and S.J. Smith, Electron impact excitation of atoms, *Rev. Mod. Phys.* 40:238 (1968)
6. V.I. Ochkur, The Born Oppenheimer method in the theory of atomic collisions, *Soviet Phys. JETP*, 18:503 (1964)
7. V.I. Ochkur, Ionisation of the hydrogen atom by electron impact with allowance for exchange, *Soviet Phys. JEPT*, 20:1175 (1965)
8. N.F. Mott and H.S.W. Massey, 1965, "The Theory of Atomic Collisions" (3r ed) Clarenden Press, Oxford
9. J. Kessler, 1976, "Polarised Electrons" Springer-Verlag, Berlin
10. C.J. Joachain, 1975, "Quantum Collision Theory", North Holland
11. S.J. Porter, J.A.D. Matthew and R.J. Leggott, Inelastic exchange scattering in electron-energy-loss spectroscopy: localised excitations in transition-metal and rare-earth systems, *Phys. Rev. B* 50:2638 (1994)
12. S.J. Porter, 1991, "The Scattering of Polarised and Unpolarised Electrons from atoms and Ferromagnetic Surface, DPhil Thesis, University of York, UK
13. D. Mauri, R. Allenspach and M. Landolt, Electron spin polarisation in inner-shell excitations in a solid, *Phys. Rev. Lett.* 52:152 (1984)

14. J.A.D. Matthew, W.A Henle, M.G. Ramsey and F.P. Netzer, $4f^7$ - $4f^7$ transitions in Gd, oxidised Gd and epitaxial Gd silicide. *Phys. Rev. B* 43:4897 (1991)
15. F. Della Valle and S. Modesti, Exchange-excited f-f transitions in the electron-energy-loss spectra of rare earth metals, *Phys. Rev. B* 40:993 (1989)
16. J. Kolaczkiewicz and E. Bauer, Low-energy electron loss spectroscopy of Eu, Gd and Tb. The valence region, *Surf. Sci.* 265:39 (1992)
17. A. Gorschlüter, R. Stiller and H. Merz, Dipole forbidden f-f excitation in ytterbuim oxide, *Surf. Sci.* 251/252:272 (1991)
18. A. Gorschlüter and H. Merz, Localised d-d excitations in NiO(100) and CoO (100), *Phys. Rev. B* 49:17293 (1994)
19. B. Fromme, M. Schmitt, E. Kisker, A. Gorschlüter and H. Merz, Spin-flip low-energy electron-exchange scattering in NiO(100), *Phys. Rev. B* 50:1874 (1994)
20. H. Hopster, electron depolarisation by inelastic exchange scattering from Cr^{3+} magnetic moments, *Phys. Rev.* 42:2540 (1990)
21. H.R. Moser and G. Wendin, Theoretical models for intensities of d → f transitions in electron-energy-loss spectra of rare earth and actinide metals, *Phys. Rev. B* 44:6044 (1991)

MAGNETO-OPTIC KERR EFFECT STUDIES OF ULTRATHIN MAGNETIC STRUCTURES

J A C Bland

University of Cambridge
Cavendish Laboratory
Madingley Road
CB3 OHE
UK

SUMMARY

In this article the application of the magneto-optic Kerr effect to the study of the magnetic properties of ultrathin metallic structures prepared by molecular beam epitaxy is described. The differing magneto-optical effects which arise according to the geometrical orientation of the magnetisation with respect to the scattering plane are first considered and experimental arrangements for their measurement described. In-situ measurements of the evolution of magnetic order (the two dimensional magnetic phase transition) and of the evolution of magnetic anisotropies are discussed, illustrated by results for Co/Cu(001) and Fe/GaAs(001) epitaxial overlayer systems during the film growth. Finally, it is demonstrated that it is possible to make vectorial magnetometry measurements using the longitudinal Kerr effect in different applied field configurations and to determine the magnetic anisotropy and interlayer exchange coupling strengths using polar MOKE.

INTRODUCTION

The magneto-optic Kerr effect (MOKE) was first reported by Kerr in 1887, following a suggestion from Faraday to search for an optical polarisation change induced by a magnetic field.[1] However it was not until the late 20th century that the extraordinary sensitivity of the effect in probing the magnetic properties of magnetic films was appreciated. In 1985 Moog and Bader[2] demonstrated that the magnetic hysteresis loop of sub-monolayer Fe/Au films could be detected in UHV. A particular advantage is that a polarised light beam can be readily passed through the windows of a UHV chamber and the reflected beam can also be easily analysed using an external polariser. Thus, in contrast with the relative complexity of polarised electron beam techniques described elsewhere in this volume, the magneto-optic Kerr effect can be readily adapted to the sensitive probing of the magnetic properties of ultrathin films (typically 1-30 monolayer thickness for the case of Fe). Since the middle 1980's, the number of groups using the MOKE to probe the properties of ultrathin magnetic films has greatly increased,[3] although most of these use the effect ex-situ in studying the magnetic properties of ultrathin films, particularly the magnetisation loop. In such studies, the saturation and coercive fields can be accurately obtained as function of the applied field orientation but the loop amplitude is frequently not analysed. Many of these studies are made at fixed wave-length using a laser source, although a number of groups have carried out spectroscopic measurements in which the Kerr rotation and ellipticity are studied as a function of wavelength (described below). Many exciting new advances in the field of ultrathin magnetic structures have been provided by Kerr effect studies, of which the observation of a 2D Ising behaviour in the magnetic phase transition of Fe/Pd films[4] and the reorientation transition in Fe/Ag films[5] are just two important examples. The development of metallic multilayer materials with perpendicular anisotropy for magneto-optic recording is an important area of applied research. For example rare-earth (RE) – transition metal (TM) alloy films[6] are used commercially for erasable magnetic recording and recently, Co/Pt multilayers[7] have become available for this application. Co/Pt multilayers offer the advantage of a large Kerr rotation at the short wavelengths needed for high bit-packing densities. These important aspects lie outside the scope of the current article although they have stimulated a vast research effort addressing the magneto-optic properties of magnetic multilayers.

In this article no attempt is made to provide a comprehensive survey of the application of the MOKE to the study of ultrathin magnetic structures, but rather it is intended to illustrate its usefulness in studying their magnetic properties, drawing extensively from the results of the Cambridge research group.

1. THE MAGNETO-OPTIC KERR EFFECT

The magneto-optic Kerr effect can be described as the magnetisation-induced change in polarisation state and/or intensity of light upon reflection from a magnetised medium. The Kerr effects are proportional to the magnetisation, thus distinguishing them from other magneto-optical effects. At normal incidence the plane of polarisation is rotated (polar Kerr effect) by a sizeable fraction of a degree[8] for saturated[9] ferromagnetic transition metals and is therefore easily measurable. Indeed, using a relatively simple arrangement, the effect was first demonstrated by reflecting light from the polished poles of an electromagnet. Since light penetrates metals only the short distance determined by the optical skin depth (typically 15–20nm in metals[10]) the Kerr rotation is determined by the surface vicinity of the medium on this lengthscale. For this reason ultrathin films can yield a measurable Kerr rotation and ellipticity.

The origin of magneto-optical effects in ferromagnetic metals lies in the spin-orbit interaction between the electron spin and the orbital angular momentum.[11] The electric field of the light couples to the electron dipoles via the orbital wavefunctions which are in turn influenced by the electron spin via the spin-orbit interaction.[12] In ferromagnets a net electron spin polarisation leads to an overall rotation of the polarisation of the light. Calculation of the magneto-optic response of metals requires that the spin-resolved bandstructure is known. Spin dependent optical transitions are calculated for the appropriate photon energy for right and left circularly polarised light. For the ferromagnetic transition metals, for example, the magneto-optic contribution to the conductivity (off-diagonal components of the conductivity tensor) has been computed in the visible range.[13,14] Since the electronic structure of films beneath typically 5ML thickness departs from the bulk structure, the magneto-optic response is modified accordingly.[15] Surprisingly, bulk constants do appear to describe well experimental measurements of Kerr spectra obtained for Fe/Ag[16] films of few monolayers thickness, but recent studies for Co/Au films[17] show that an interface induced Kerr rotation exists. Recent experiments on Fe/Au/Fe and Fe/Ag/Fe trilayer structures[18] have confirmed the presence of new optical transitions associated with spin polarised quantum well states confined to the non-magnetic spacer layers. Following recent advances in computational and measurement techniques, obtaining accurate agreement between the measured and calculated magneto-optic constants is now in prospect and thus magneto-optical studies of ultrathin films may soon be placed on a more quantitative footing. Interest is now rapidly growing in magnetic circular dichroism (MCD) in the X-ray region applied to the study of ultrathin magnetic layers and interfaces.[3] MCD can be used to probe spin split atomic levels thus providing element specificity together with magnetic information. It is also possible to separate the spin and orbital contributions to the magnetic moment while high sensitivity can be achieved by exploiting an absorption edge and using a synchrotron source.

The following treatment is not concerned with calculation of the appropriate magneto-optic constants but rather describes a phenomenological description based on the dielectric tensor, appropriate for the visible range.

1.1 A macroscopic description of the Kerr effect

The macroscopic description of magneto-optical effects is based on the dielectric tensor in which off-diagonal terms arise due to the magneto-optical transitions.[19,20] Such off-diagonal terms can easily be understood to arise from the Lorentz force on the electrons giving rise to a component of the dielectric tensor proportional to $B \times E$. For a crystal with cubic symmetry the dielectric tensor is given by:

$$[\varepsilon] = \varepsilon_0 \varepsilon_r \begin{pmatrix} 1 & iQ_z & -iQ_y \\ -iQ_z & 1 & iQ_x \\ iQ_y & -iQ_x & 1 \end{pmatrix} \qquad (1)$$

where $Q = \gamma B/\varepsilon_r$ (B is the magnetic induction and γ is a constant), the y axis is the intersection of the scattering plane with the sample surface and z is the normal to the interface. Since the magnetic induction inside a ferromagnetic medium is approximately equal to the magnetisation M, to a very good approximation the off-diagonal terms of the matrix are proportional to the appropriate components of M. The wave equation can be written down using (1) and the resulting eigenvalue equation used to yield the eigenmodes of the medium. Wave solutions can then be written in terms of the forward and backward propagating eigenmodes. This requires a 4 component vector since each mode in general has to be defined by 2 components of electric field. By satisfying the boundary conditions on the electric field components it is possible to obtain the reflectivity matrix. This approach can be extended for an arbitrary number of layers: Yeh[21] has described a matrix approach to the problem and a similar approach appropriate for ultrathin film multilayers has been described by Zak et al.[19] The reflectivity matrix can be written as:

$$[R] = \begin{pmatrix} r_{pp} & r_{sp} \\ r_{ps} & r_{ss} \end{pmatrix} \qquad (2)$$

where p,s refer as usual to the components of the light polarisation in the scattering plane and perpendicular to it. The components of the matrix depend upon the appropriate component of Q according to the orientation of M with respect to the scattering plane.

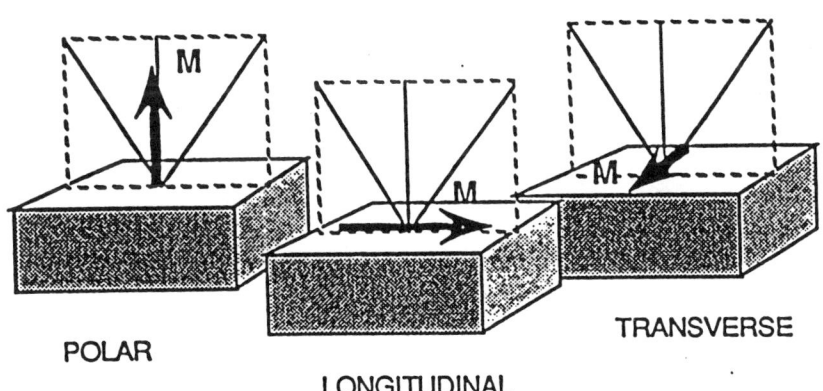

Figure 1. The orientation of the magnetisation with respect to the scattering plane for each of the three principal geometries for the Kerr effect. The polar and longitudinal geometries describe the magnetisation M in the scattering plane, normal and parallel to the sample surface respectively. The transverse (equatorial) geometry describes the case where the magnetisation is perpendicular to the scattering plane but in the plane of the sample surface.

In Fig. 1 the orientation of the magnetisation with respect to the scattering plane is shown for each of the three principal geometries for the Kerr effect and in Table 1 the corresponding components of the reflectivity matrix for a single interface are given. For the polar and longitudinal geometries (magnetisation M in the scattering plane, normal and parallel to the sample surface respectively) the off-diagonal terms are proportional to Q and the diagonal terms are unchanged. For the transverse (or equatorial) geometry (M perpendicular to the scattering plane but in the plane of the sample surface) no off-diagonal term arises and the diagonal reflectivities contain a Q dependent term. It is important to recognise that, as with the usual Fresnel coefficients, the reflectivity matrix depends on the incidence angle. Thus for normal incidence, the longitudinal effect vanishes while the transverse effect becomes purely second order, as can be seen from table 1. The longitudinal Kerr effect is maximised at around 60 degrees from the surface normal. The polar effect on the other hand is maximised at normal incidence. For general orientations of the magnetisation vector, the geometries considered above are not appropriate. In this case a mixture of Kerr effects arises, considerably complicating the interpretation of the resulting magnetisation loops. The dependence of the reflectivity on the orientation of the magnetisation vector with respect to the scattering plane allows vectorial information to be extracted, as discussed below.

	r_{pp}	r_{ps}
Polar	$\dfrac{\sqrt{\varepsilon_2}\alpha_1 - \sqrt{\varepsilon_1}\alpha_2}{\sqrt{\varepsilon_2}\alpha_1 + \sqrt{\varepsilon_1}\alpha_2}$	$\dfrac{Q_P \sqrt{\varepsilon_1 \varepsilon_2}\alpha_1}{i(\sqrt{\varepsilon_2}\alpha_1 + \sqrt{\varepsilon_1}\alpha_2)(\sqrt{\varepsilon_1}\alpha_1 + \sqrt{\varepsilon_2}\alpha_2)}$
Long.	$\dfrac{\sqrt{\varepsilon_2}\alpha_1 - \sqrt{\varepsilon_1}\alpha_2}{\sqrt{\varepsilon_2}\alpha_1 + \sqrt{\varepsilon_1}\alpha_2}$	$\dfrac{Q_L \sqrt{\varepsilon_1 \varepsilon_2}\alpha_1 \tan\theta_2}{i(\sqrt{\varepsilon_2}\alpha_1 + \sqrt{\varepsilon_1}\alpha_2)(\sqrt{\varepsilon_1}\alpha_1 + \sqrt{\varepsilon_2}\alpha_2)}$
Trans.	$\dfrac{\sqrt{\varepsilon_2}\alpha_1\sqrt{1-Q_T^2/\alpha_2^2} - \sqrt{\varepsilon_1}\alpha_2 - i\sqrt{\varepsilon_2}Q_T\alpha_1\tan\theta_2}{\sqrt{\varepsilon_2}\alpha_1\sqrt{1-Q_T^2/\alpha_2^2} + \sqrt{\varepsilon_1}\alpha_2 - i\sqrt{\varepsilon_2}Q_T\alpha_1\tan\theta_2}$	0

Table 1. The Fresnel coefficients for p-polarised light incident on a single interface non-magnetic/magnetic layer system. The complex dielectric constants of each layer are ε_1 and ε_2, and the angle to the normal of the interface that the light makes in each layer is given by θ_1 and θ_2. The substitutions $\alpha_1 = \cos\theta_1$ and $\alpha_2 = \cos\theta_2$ have been used.

The magnetisation-induced change in the light polarisation for the longitudinal Kerr effect can now be described. For p polarised incident light the reflected light polarisation state corresponds to an ellipse with its major axis rotated by an angle θ_k (the Kerr angle) with respect to the incident polarisation direction and an ellipticity ε_k where:

$$\frac{r_{sp}}{r_{pp}} = -\theta_k + i\varepsilon_k \tag{3}$$

Both the Kerr rotation and ellipticity are therefore proportional to the magnetisation to first order. For the polar effect, at normal incidence only rotation occurs, i.e. r_{sp} is real. For an ultrathin magnetic overlayer structure with a non-magnetic part of the refractive index N supported by a non-magnetic substrate with refractive index N_{sub} the Kerr angle and ellipticity are approximated by:

$$\theta_P = \frac{4\pi}{\lambda}\left(\frac{N^2}{1-N^2_{sub}}\right)Q_P t \text{ and } \theta_L = \frac{4\pi}{\lambda}\left(\frac{N_{sub}}{1-N^2_{sub}}\right)\beta Q_L t \tag{4}$$

where t is the film thickness and β the incidence angle. An important feature of these expressions is that they are proportional to t and thus vanish at zero thickness. For ultrathin multilayers satisfying $2\pi/\lambda \, \Sigma N_i t_i \ll 2\pi$ an additivity law approximately holds in which the Kerr rotations of successive layers add.[19]

1.2 Experimental arrangements

A large number of arrangements have been used for measuring the Kerr effect in ultrathin films, involving both modulation techniques[22] and dc methods.[3,23] Polarisation modulation techniques offer the advantage of very high sensitivity and the possibility of making direct measurements of the Kerr polarisation (ellipticity and rotation). However, for in-situ measurements, the windows of the vacuum system depolarise the beam and thus additional procedures are required to separate the sample and apparatus-induced polarisation.

A simple, but widely used arrangement shown in Fig. 2 consists of a plane polarised laser beam (usually a HeNe laser, $\lambda = 632.8$nm) reflected at an incidence angle β from the sample placed between the poles of an electromagnet before passing through an analyser to enter an optically screened Si photodiode. A Hall sensor can be attached to one of the pole pieces of the electromagnet and calibrated so that the measured Hall voltage can be converted into the value of the field at the centre of the pole pieces where the sample was located. The crystal axis G can be rotated in-plane with respect to the applied field H in order to study in-plane magnetic anisotropies (see Fig.2). In Fig. 2 the geometry for p-polarised incident light is shown schematically. The analyser is set to transmit light with the electric vector inclined at an angle α with respect to the normal to the scattering plane. For in-situ measurements the optical components are placed outside the vacuum system and the beam enters the vacuum chamber via windows. In this case the strain induced birefringence has to be minimised,

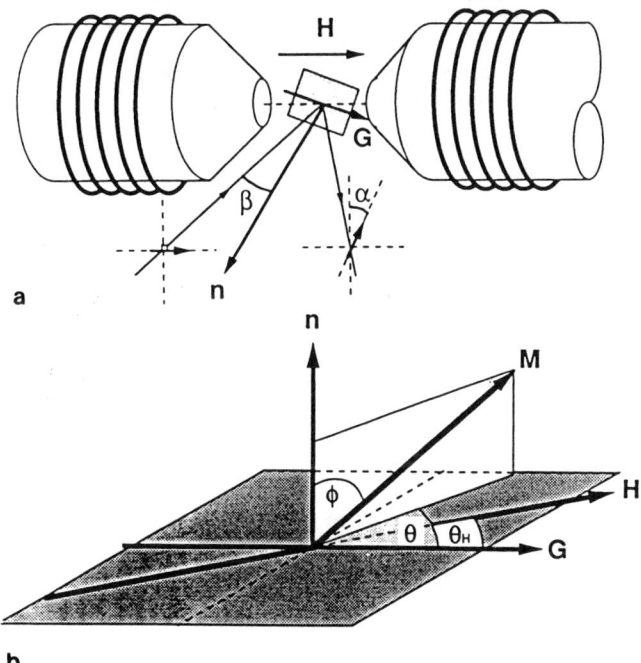

Figure 2. A schematic of a dc arrangement for observing the Kerr effect. The incident (laser) beam is p-polarised and incident at an angle β (typically 20°) with respect to the sample normal n. The sample can be rotated about the normal so that the angle between the applied field H and a specific crystal axis G can be varied. The polarisation is analysed by a polariser with transmission direction aligned at an angle α with respect to the scattering plane normal. The detector is not shown.

either by ensuring that the stress introduced by the bolts on the flange is as uniform as possible or by using a special strain free window. The angle of incidence is usually constrained by the window available to the range 20-30°. The detector responds to an intensity which is increased or reduced according to the magnetisation induced ellipticity and rotation. Since both quantities are proportional to the magnetisation, by setting the analyser angle α to an appropriate value (close to extinction), a component of the detected signal which is proportional to the magnetisation can be observed. The experiment is usually controlled by a computer used to vary the field produced by the electromagnet, while recording the Hall voltage and photodiode signal.

An explicit expression[23] for the dependence of the transmitted intensity on analyser setting and the Kerr angle and ellipticity is readily obtained by evaluating the quantity $|u.E_r|^2$ where $u = (sin\alpha, cos\alpha)$ and E_r is the reflected polarisation obtained from (2). An additional intensity must be added corresponding to the depolarised beam intensity and the background signal. The analyser setting is important: too close to the extinction position the intensity change which is first order in M vanishes and thus this setting cannot be used for probing the hysteresis loop. Moreover, unless a photomultiplier is used, the background intensity due to the combination of the detector dark noise and electronic noise becomes a limiting factor. At too large an analyser setting, the non-magnetic signal increases and its associated optical noise (related to the intensity fluctuations of the incident laser beam) dominate. Commercially available HeNe lasers of typically 10mW output power offer sufficient intensity stability to allow monolayer thickness films to be probed without additional measures being taken. However, the vibrational stability of all components, particularly that of the sample is of the greatest importance and all electrical and stray field induced noise must be suppressed as far as possible. The above arrangement also requires the polarising elements to have a high extinction ratio since this quantity usually limits the sensitivity which can be achieved.[3] For highly sensitive measurements internally intensity stabilised lasers are available (typically 0.1% intensity stability) or an external intensity stabiliser can be used. An appropriate intensity stabiliser has been described by Bland et al.[23] consisting of an electro-optic cell, quarter-wave plate, polariser, beamsplitter and reference photodiode. The signal from the photodiode is fed into a differential amplifier which has an adjustable reference voltage on its other input. The resulting amplified error signal is applied to an electro-optic cell. Careful adjustment of the parameters of the feedback loop leads to an intensity stability of better than 0.1%.

For ex-situ measurements, the arrangement can be combined with a superconducting magnet or electromagnet depending on the applied field range and orientation required. Polar MOKE measurements can be conveniently carried out at 300K using a split coil superconducting magnet without windows and a concave mirror for focusing the laser beam,[24] thus eliminating problems due to Faraday rotation or birefringence associated with windows or lenses. By scanning the spot position it is possible to investigate wedged samples in which the layer thickness varies along the sample. For the vector magnetometry measurements described below, the sample is mounted on a rotary stage adjusted so that the reflected beam does not move as the sample stage is rotated through a full 360°. The orientation of the magnet can also be changed without disturbing the sample stage. For temperature dependent measurements, a cryostat is required, but windows with a low Verdet constant are essential.

For in-situ measurements where vectorial information is not required, a highly sensitive arrangement has been described by Kerkmann et al.[25] Here the incident light is polarised at 45 degrees with respect to the normal to the scattering plane and the light is analysed very close to extinction. The field is applied to the sample using a small Helmholtz coil along a direction perpendicular to the scattering plane (TF geometry). Much of the non-magnetic background signal which arises in the pure transverse Kerr effect is optically attenuated. However the arrangement is sensitive to both components of the magnetisation vector and thus is appropriate for measurements of the switching field and saturated loop amplitude close to the easy axis.

2. MOKE MAGNETOMETRY

Measurement of the magnetisation curve is important in gaining an understanding of the magnetic behaviour of ultrathin film structures, providing valuable information on magnetic anisotropies and magnetisation reversal processes. The magnetic anisotropy behaviour of ultrathin epitaxial films[26] is strikingly different from that of the bulk material due to the break of symmetry by the interface, with which additional anistropies are associated. Such interface anisotropies can be very large and in some cases sufficient to overwhelm the dipolar (shape) anisotropy at small thickness as occurs for example in Fe/Ag,[5,26] Fe/Cu,[3,26] Co/Pt[3] structures: this is an important phenomenon for magneto-optic recording applications which require a remnant magnetisation vector perpendicular to the film. Moreover the possibility of carefully controlling the crystal structure and lattice strains via MBE allows artificial magnetocrystalline anisotropies to be studied. Recent advances in computational techniques and preparation techniques have culminated in close agreement between the predicted and measured interface anisotropies in some cases.[26]

In ultrathin films with planar anisotropy the magnetisation vector is confined to the film plane by the dipolar forces and most frequently a single component of the magnetisation is measured, usually parallel to the applied field. For in-situ studies such information is readily obtained by MOKE and is difficult to get by other means: for example polarised electron techniques are not compatible with measurements at finite applied fields. Measurements of the magnetisation loop by MOKE can be used to investigate the evolution of the magnetic order and of the anisotropy strength. An understanding of the in-plane magnetisation reversal process can be gained by measuring the magnitude and direction of the total magnetisation in-plane as a function of the applied field. Such vector magnetometric information is particularly valuable in studies of exchange coupling in trilayers. In vector magnetometric measurements using MOKE, the quantity M/M_s is determined quantitatively for a wide range of field orientations. A single additional measurement of the saturation magnetisation M_s using SQUID magnetometry, for example, permits such measurements to be placed on an absolute scale. While the in plane magnetisation reversal process is frequently complicated by the presence of magnetic domains at low fields, the reversal process for fields applied along the sample normal proceeds by rotation only for planar anisotropy films. Thus the polar MOKE magnetisation curve can often be used to directly obtain information concerning the strength of interface anisotropies and exchange coupling as discussed below.

2.1 Comparison of MOKE with conventional magnetometry techniques

It is instructive to compare the loops obtained for MOKE with conventional techniques. In Fig. 3 shows the magnetisation loop obtained for a Si(100)/60ÅFeNi/60ÅCu/40ÅCo 'spin valve' structure[27] using longitudinal MOKE, SQUID and VSM mag-netometry for the applied field close to the easy axis. For this orientation an antiparallel alignment of the magnetic layers is achieved at low field due to the uniaxial anisotropies of the Co and FeNi layers which arises in these specific samples. These structures are of interest for the giant magnetoresistance behaviour which arises from the antiparallel orientation.[27] All the techniques give the same values of the switching field, as expected, but the loop amplitude is distorted for the MOKE curve due to the different magneto-optic constants for the different magnetic layers. Thus in MOKE the vertical scale does not give a true representation of the magnetisation loop. The SQUID and VSM loops agree, although the VSM is much noisier. It is interesting to note that the noise level of the SQUID and MOKE loops are comparable.

It should be mentioned that for single films the Kerr loop does not provide a truly accurate measure of the amplitude of the magnetisation loop even in the ultrathin limit. For finite values of the analyser setting α an asymmetry is introduced by the finite value of the polariser angle β: accordingly the positive and negative loop amplitudes are inequivalent. This distortion is usually small but can be significant in certain circumstances.

2.2 The two dimensional magnetic phase transition

The role of magnetic fluctuations in 2D is much greater than in 3D. Accordingly, in comparison with 3D magnets, 2D magnetic systems display a critical region which extends over a much larger temperature range, expressed as a fraction of T_c. For this reason, the observation of the critical behaviour associated with the phase transition is more easily studied in 2D. Many studies of the 2D phase transition, for example of the thickness dependent Curie temperature and the critical exponent, have been carried out by MOKE.[3,4]

Recent experiments by Schumann et al.[28-30] on fcc Co/Cu(100) films during growth at 300K in the vicinity of the magnetic phase transition illustrate the sensitivity of in-situ MOKE measurements. The MOKE measurements are made using an arrangement described by Kerkmann.[25] For ultrathin paramagnetic Co/Cu(001) films the paramagnetic susceptibility χ increases sharply upon Co deposition within a narrow thickness range close to the critical thickness d_c at which the onset of ferromagnetism occurs.[28] Beneath d_c no ferromagnetic signal is observed and above d_c (corresponding to approximately 1.6ML) the onset of ferromagnetism is evidenced by a non-vanishing remanence in the Kerr loop. In Fig. 4a, the first detectable M-H loop is displayed. A straight line can be seen to provide a good fit to the data over the entire field range. With increasing thickness the signal-to-noise ratio is increased and the slope of the curve increases. In Fig. 4b, the last paramagnetic loop is given and the region for which the linear fit is employed is reduced to a few gauss. The distinction between paramagnetic and ferromagnetic loops close to the transition point can be made by plotting the thickness dependence of χ and comparing the observed behaviour with a power law. The measured thickness dependence of χ is shown in Fig. 5 for two separate experiments (as indicated by circles and triangles). The main feature revealed by this plot is that within a narrow thickness regime (~6% of d_c), χ increases very sharply, indicating a well-defined magnetic transition. However the determination of the critical exponent does not require knowledge of the absolute value of the critical point. The measured thickness-dependent χ data for the two experiments are fitted to a power law of the form $(1 - d_c/d)^{-\gamma}$ yielding $\gamma = 2.41 \pm 0.07$ (circles) and $\gamma = 2.38 \pm 0.07$ (triangles). The observed behaviour is attributed to a geometrical, or percolation phase transition, since the corresponding 2D critical exponent $\gamma = 43/18 = 2.39$ agrees well with the experimentally determined value of 2.40 ± 0.07. In the percolation transition, at a critical 2D concentration (corresponding to the critical thickness in our experiment), exchange interactions between the atoms can extend across the whole sample. At low coverage, superparamagnetic islands are present and no long range magnetic order occurs. The possibility of measuring the paramagnetic response of an ultrathin film is in itself remarkable and can be explained in terms of giant spin blocks aligned by the applied field.[31]

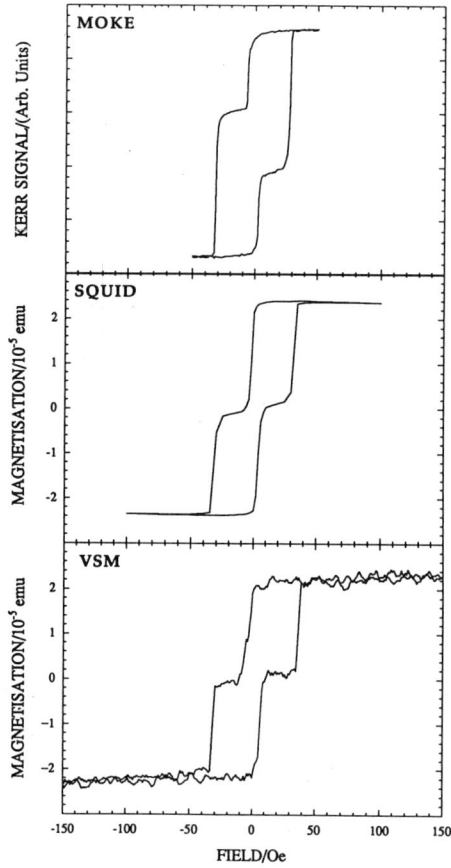

Figure 3 The magnetisation loop obtained for a Si(100)/60ÅFeNi/60ÅCu/40ÅCo sample using MOKE, SQUID and VSM, for the applied field close to the easy axis.

The coercivity is found to increase sharply with thickness beyond the critical value at which ferromagnetism develops.[29] In Fig. 6 the hysteresis loop obtained for a fcc Co/

Cu(001) film with (a) the Co thickness close to the critical thickness d_c and (b) after a further Co deposition corresponding to $0.09d_c$ shown. A dramatic increase in coercivity is seen to occur in this small thickness range. In Fig. 7 the measured coercive field is shown as a function of reduced Co thickness for Co/Cu(001) films deposited at 300K. The coercive field is found to vary as $(d/d_c - 1)^\alpha$ where $\alpha = 0.58 \pm 0.07$ within the thickness range $1 < d/d_c < 1.09$. Such a power law behaviour suggests that the behaviour is thermodynamic in origin and cannot be described as the simple admixture of surface and bulk anisotropies in this thickness range. In such a description the anisotropy strength has two parts, one thickness independent describing the volume dependent term, and the other which scales as the inverse thickness d describing the interface or surface term.[26] Recent theoretical calculations suggest that the anisotropy behaviour is likely to be more complicated with even oscillatory behaviour occurring in certain cases for ultrathin films.[32]

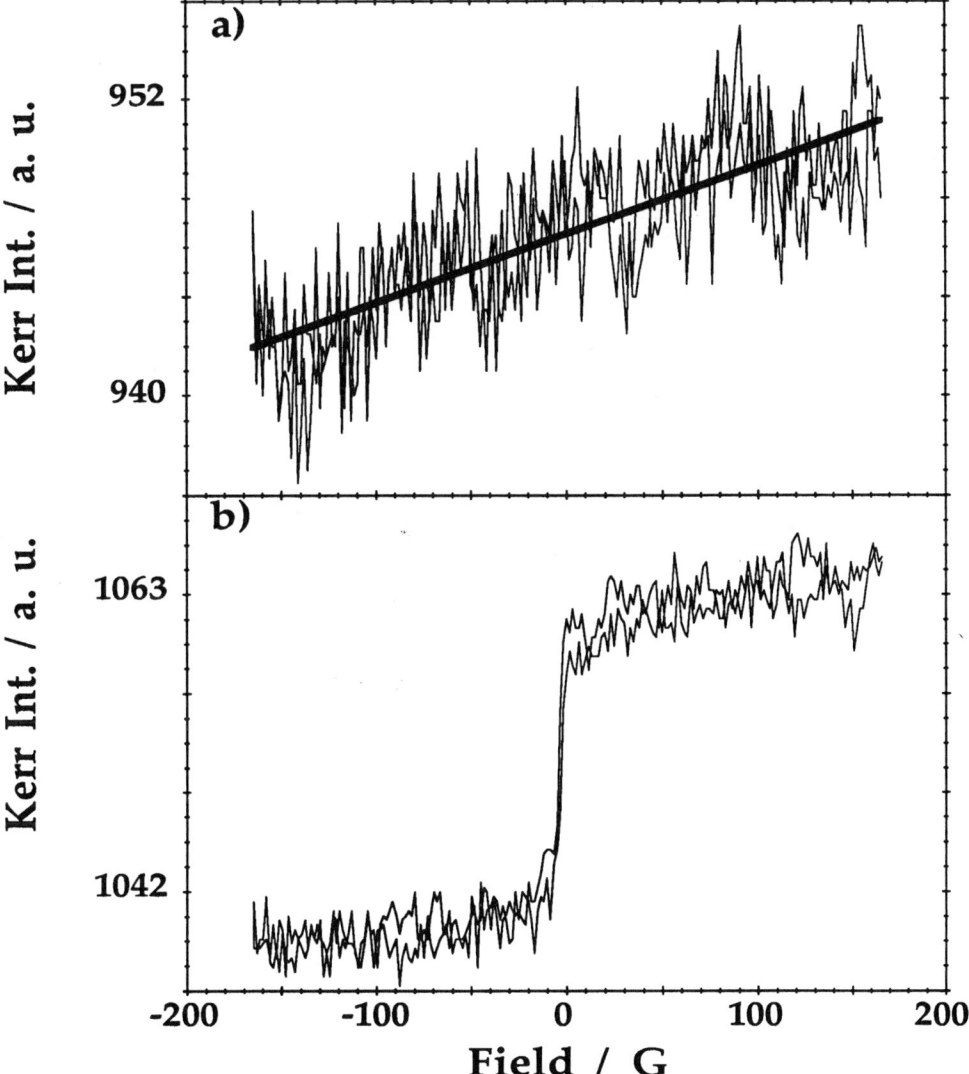

Figure 4(a) The upper panel shows the first paramagnetic M-H loop with a linear fit, the slope of which is proportional to the susceptibility χ. **(b)** The last loop of the growth sequence is displayed in the lower panel.

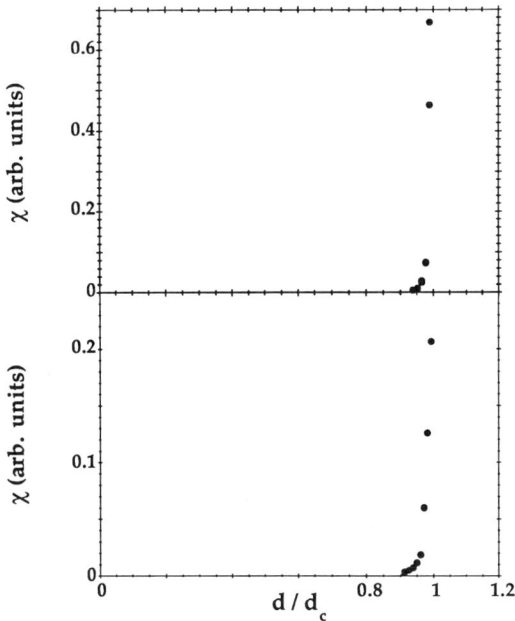

Figure 5. The evolution of χ is shown as a function of the thickness in reduced units: for the first experiment (upper panel) and for a second experiment (lower panel). The critical thickness was determined (in arb. units) as described in the text.

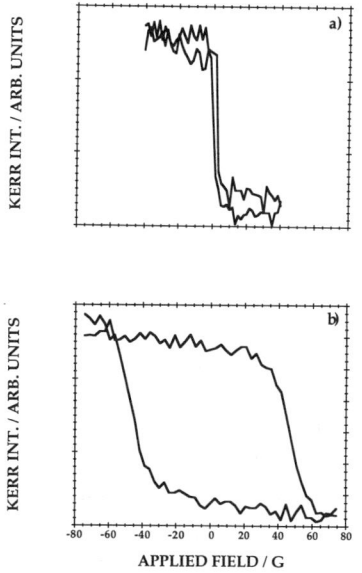

Figure 6. The hysteresis loop obtained for a fcc Co/Cu(001) film with **(a)** the Co thickness close to the critical thickness d_c and **(b)** after a further Co deposition corresponding to $0.09 d_c$.

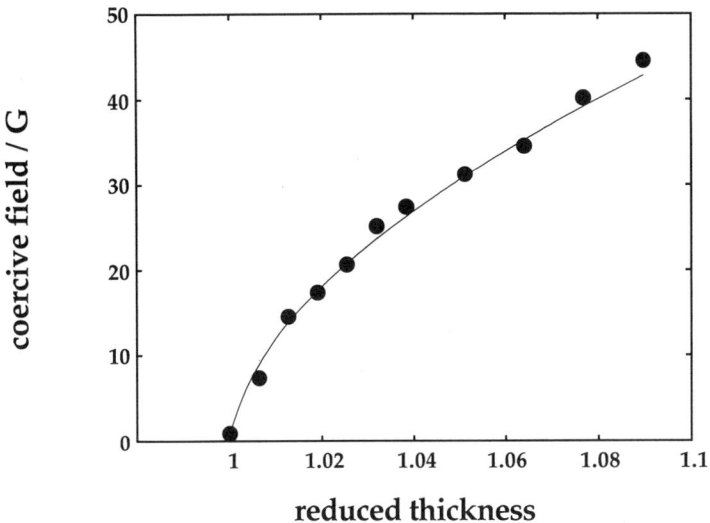

Figure 7. The coercive field as a function of reduced Co thickness for Co/Cu(001) films deposited at 300K.

Schneider et al.[33] studied the thickness dependence of the Curie temperature of fcc Co/Cu(001) films and found that overcoating the sample with Cu resulted in a marked decrease in Curie temperature and that moreover the coercivity was strongly temperature dependent. Careful structural studies by Egelhoff et al.[34] showed that surface segregation of Cu occurs at elevated temperatures. In Fig. 8 M-H loops are shown for a $1.6d_c$ Co/Cu(001) film obtained during growth of the Cu overlayer at 300K in an attempt to simulate the effect of surface segregation.[30] Clearly minute coverages reduce H_c drastically and change the loop squareness S. The squarest loop is obtained for a small Cu overlayer thickness of approximately 0.2 ML. A careful analysis of all M-H loops during this growth sequence revealed a non-monotonic dependence of S, M_{sat} (the loop amplitude) and H_c on Cu thickness d_{Cu} as shown in Fig. 9. In particular the coercivity exhibits a sharp minimum at the thickness at which S and M_{sat} peak; with increasing Cu layer thickness, S displays a weak peak before reaching a constant value in the range in which H_c and M_s continue to vary. For all Co thicknesses studied in the range 1–2 ML, qualitatively similar variations in each of these quantities are observed where the features are virtually the same except that the non-monotonic behaviour of S is less pronounced for thinner films for which S is closer to unity than for the thicker Co films. For all the structures studied H_c always rapidly drops by roughly a factor of ~3, and the enhancement of the peak saturated magnetic signal is always of the order of 20% upon the deposition of only 0.2 ML of Cu. While this behaviour is as yet not fully explained it is likely that the evolving interface electronic structure induces a change in the magnetic properties causing a change in both the magnetic anisotropy and magneto-optic constants. In the case of Co/Pt for example[3] the interface induced spin polarisation of the Pt leads to an increased Kerr rotation. So-called 'anomalous' anisotropy behaviour has been observed by Engel et al. for perpendicular Co/Pd films.[35] In this case covering the Co films with a non-magnetic overlayer results in a non-monotonic variation in the coercivity and the evolving electronic structure is also believed to be the most likely origin of this behaviour. For the Co/Pd structures the maximum effect is obtained for a non-magnetic overlayer thickness of around 1.8ML. Studies of different combinations of overlayer/magnetic film on Cu(001) will also provide more insight into the physical mechanism for in-plane anisotropy and are currently in progress. Nonetheless the studies of in-plane Co films illustrate both the sensitivity of in-situ magneto-optic measurements (comparable with surface electron scattering techniques) and the delicate dependence of magnetic effects on the presence of minute coverages.

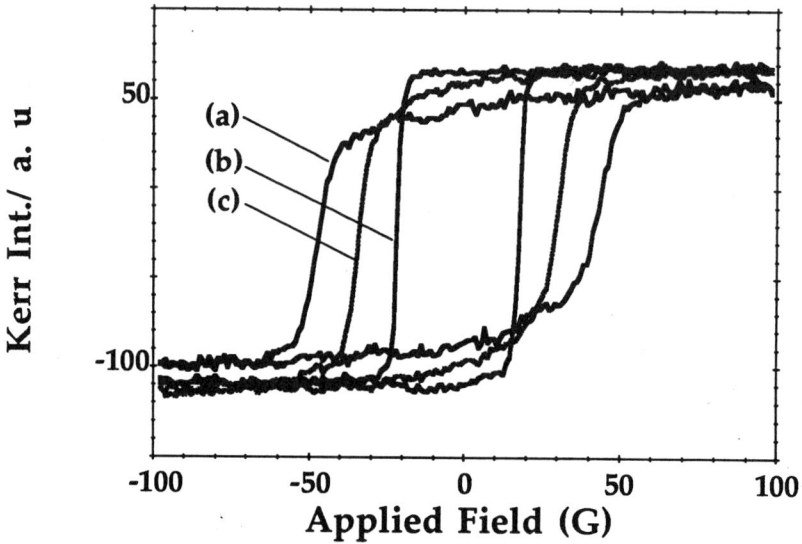

Figure 8. M-H loops for a $1.6d_c$ thick Co/Cu(001) film during the growth of a Cu overlayer, with Cu overlayer thickness d_{Cu} of 0.04 ML, 0.16 ML and 0.36 ML for a)-c).

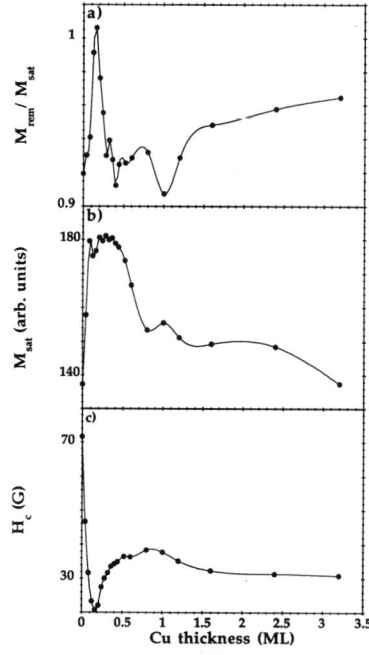

Figure 9. The results of a careful analysis of the Cu overlayer growth sequence of a $1.6d_c$ thick Co/Cu(001) film are shown, revealing the nonmonotonic behaviour of the squareness S in (a), saturation magnetisation M_{sat} in (b) and the coercive field H_c in (c). The noise of the M-H loop results in a peak value for S larger than 1.

2.3 Evolution of anisotropies

Studies of the thickness dependence of the anisotropy behaviour have been carried out for a wide range of epitaxial systems using in-situ MOKE.[3] Such investigations reveal a highly sensitive dependence of the magnetic anisotropy behaviour on the growth conditions and film morphology.

Epitaxial bcc Fe/GaAs substrates illustrate the complex thickness dependence which can arise. Considerable interest in the magnetic properties of this epitaxial system derives from its well defined, controllable magnetic properties and its technological potential in device applications for example.[36] A strongly thickness dependent uniaxial anisotropy arises which has been attributed to the 2 fold symmetry of the unsatisfied Ga or As at the Ga or As rich surface terminations resulting in equivalent strains along the two in-plane <110> directions. Chemical interaction and interdiffusion between Fe and As also occurs which results in a magnetisation and cubic anisotropy strength which increases with thickness.[36] The magnetic anisotropy energy for the (001) surface can be written in the form:

$$E_{(001)} = \frac{1}{4}K_1 \sin^2(2\phi) + K_u \sin^2(\phi + \pi/4) \tag{5}$$

where the magnetisation vector is oriented in-plane at an angle ϕ with respect to the in-plane [110] crystal axis and where the 4-fold and 2-fold anisotropy strengths K_1, K_u are both thickness dependent. Thus for $K_u = 0$ and $K_1 > 0$ the easy axes are aligned along the <100> directions. The corresponding expression for the (110) surface has an effective 2 fold anisotropy contribution from K_1 due to the lower surface symmetry and has easy and hard axes along [100] and [111] respectively.[36]

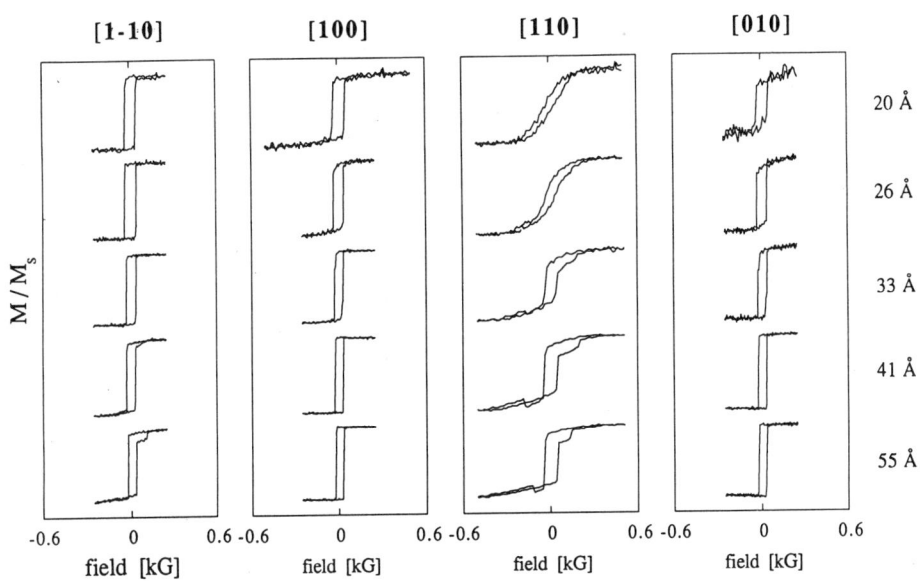

Figure 10. Magnetisation loops for Fe/GaAs films during growth for the (001) surface. The loops are shown for the field applied along principal in-plane crystal axes.

In-situ magneto-optic Kerr effect (MOKE) measurements[37] allow an evaluation of the complete thickness dependence of the magnetic anisotropy of a film on a single surface, avoiding any uncertainties due to varying substrate morphologies which arises in studies of different samples. On both the (001) and (110) substrates Fe initially grows in a paramagnetic phase up to a thickness of ~15 Å. On the (001) substrate, beyond the paramagnetic region a uniaxial anisotropy with the principal axes along the <110> directions dominates the magnetic response up to ~60 Å of Fe. Above this Fe thickness a cubic anisotropy develops, eventually dominating the magnetic response after ~100 Å of Fe, with a strength close to the bulk value, and with easy axes along <100> directions. Fig. 10 shows M-H loops obtained by in-situ MOKE for Fe/GaAs (001) as a function of azimuth angle ϕ with respect to the [1$\bar{1}$0] direction. At large thickness, 'overshoot' features can be seen for the hard <110> directions which is discussed further below. Contrasting results are obtained for the (110) surface: a dominant in-plane anisotropy first develops with easy axis along [1$\bar{1}$0] and the hard axis along [001]. Beyond ~100ÅFe, the easy axis reorients to lie along the [001] direction.

A detailed analysis of the M-H curves requires knowledge of the magnetisation reversal mechanism. However the form of the curves is sensitive to the ratio of K_u to K_1 and the hard axis saturation fields can be used to estimate the anisotropy strengths. Detailed analysis of the MOKE loops and of Brillouin light scattering (BLS) measurements of the spin wave frequencies[38] yield values for K_1 which increase with thickness for both surfaces. For the (100) surface, K_u decreases monotonically with thickness whereas for the (110) surface, both the MOKE and BLS results are consistent with a change of sign in K_u with increasing thickness, in agreement with previous ex-situ measurements.[39] Such a behaviour cannot be readily explained as arising from a single mechanism only and suggests that competing mechanisms arise, e.g. strain-induced, step-induced or surface anisotropy.

Another important example of the use of MOKE in-situ is provided by studies of the perpendicular anisotropies.[5,26] At this point it is first useful to consider the various contributions to the effective demagnetising field perpendicular to the surface which determines the remanent orientation of the magnetisation. Neglecting cubic anisotropies, in the presence of a surface uniaxial anisotropy K_s the effective demagnetising field for a single magnetic film is given by two terms due to shape and interface anisotropy respectively:

$$\mu M_{eff} = \mu D M_s - 2K_s/M_s d \tag{6}$$

where M_s, d and D are the saturation magnetisation, layer thickness and perpendicular demagnetising factor respectively. For thick films $D=1$ but for monolayer thickness films D is slightly reduced from unity according to the crystal structure of the film.[26] For positive M_{eff} the magnetisation lies in plane and for negative M_{eff} the magnetisation is perpendicular. At the so-called reorientation transition, the two terms on the right of eqn. 6 cancel, whereupon, in the absence of higher order anisotropy terms, spin wave theory predicts that the magnetisation should vanish.[40] Qiu et al.[5] have studied both the perpendicular and in-plane components of the magnetisation in the Fe/Ag(001) system as a function of temperature and Fe layer thickness and show that the transition is asymmetric with respect to temperature and, moreover, that the magnetisation does not completely vanish at the transition point. This result is consistent with the effect of the residual cubic anisotropy which remains when the shape and perpendicular anisotropies cancel.

The complexity of the surface anisotropy behaviour for ultrathin films is well illustrated by the data of Liu and Bader for p(1 x 1) Fe/Cu(100) films.[41] In this work the stability of the perpendicular phase is mapped out as a function of temperature and thickness – see Fig. 11. At very small thicknesses the perpendicular phase collapses - in contradiction with theoretical predictions. Such behaviour demonstrates the need for careful preparation and control of structure at the atomic level, such as can be provided by a combination of LEED, RHEED, angle resolved electron spectroscopies (Auger, photoelectron) and scanning tunneling microscopy (STM). Interlayer roughness, interdiffusion and agglomeration all greatly influence the behaviour at this level.

2.4 Vector magnetometry

In MOKE, since the longitudinal magnetisation component produces a change in r_{ps} only, while the transverse magnetisation component produces a change in r_{pp} only (see Table 1), it is possible to measure these components independently of each other, allowing a

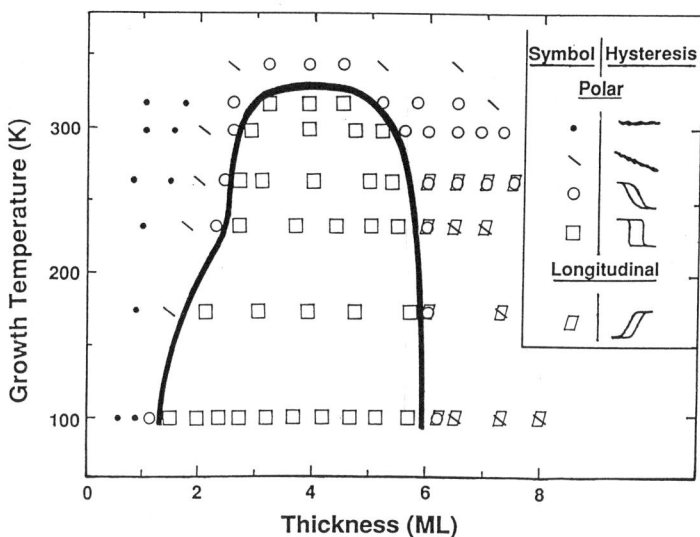

Figure 11. Hysteresis loop phase diagram for p(1 x 1) Fe/Cu(001) showing how the magnetic behaviour depends on growth conditions. From ref. 41.

determination of components parallel and perpendicular to the applied field. There are therefore two possible ways to measure both $M_\parallel$ and $M_\perp$: one method relies on the direction of the applied magnetic field remaining fixed while the analyser is rotated;[42] the other uses a fixed analyser but the applied field and sample are rotated together (as illustrated in Fig. 12). In contrast to the first method the second method, described here, allows a direct comparison between the $M_\parallel$-H and $M_\perp$-H loops since they are both measured using the same longitudinal MOKE effect. In this case the relative amplitude and direction of M can immediately be obtained. With the second method it is necessary to distinguish between the applied field and magnetisation orientations, remembering that it is the orientation of the magnetisation with respect to the scattering plane which determines the Kerr effect. For the applied field the terms PF, LF and TF are used for the polar, longitudinal and transverse geometries respectively. For the magnetisation components the terms PM, LM and TM are used to describe the orientation of the component being detected. Thus it is possible to have the magnet in the transverse geometry while detecting the longitudinal component of the magnetisation (TF/LM), for example.

For a system with a single magnetic layer with a magnetisation of magnitude M (which may not be constant), the two MOKE loops are approximately proportional to $M\cos\theta$ and $M\sin\theta$ where θ is the angle between the magnetisation and the applied field direction. Using the two loops it is therefore possible to determine the relative magnitude and direction (with respect to the applied field) of the magnetisation. For a system with two or more layer dependent magnetisations M_i, the MOKE loops are now proportional to the sum of each component, i.e. $\Sigma M_i \cos\theta_i$ and $\Sigma M_i \sin\theta_i$. In this case it is not possible to determine the magnitude and orientations of each layer dependent magnetisation explicitly. However, the coherent rotation process can be observed by comparing the magnitude of the signal for the saturated state, i.e. ΣM_i, with the total magnitude along some direction for different field strengths. When the total magnitude is equal to the total saturation magnetisation then the magnetisations of each layer are pointing in the same direction and have their full saturation magnetisation value.

MOKE vector magnetometric measurements[42] on a Cu/30Å fcc Co(111)/27Å fcc Cu(111) structure grown by MBE on a Ge(110)/GaAs(110) substrate are shown in Fig. 13. The sample is uncoupled and displays a significant uniaxial anisotropy. The absence of coupling is attributed to large interface roughness in this structure associated with the Ge layer. The data are compared with the results of calculations of the magnetisation orientation assuming coherent rotation and appropriate anisotropy constants, magnetisation and layer thicknesses. The TF/LM and LF/LM results indicate that the full magnetisation lies in a single domain state along the easy anisotropy axis for zero applied field, since the magnitude of the $M_\perp$–H loop in the TF/LM geometry at zero field is equal to that of the $M_\parallel$–H loop in the LF/LM geometry when the sample is saturated. The agreement between the theoretical and the measured results is good, confirming that coherent magnetisation rotation actually occurs and that there is no significant coupling between the two magnetic layers.

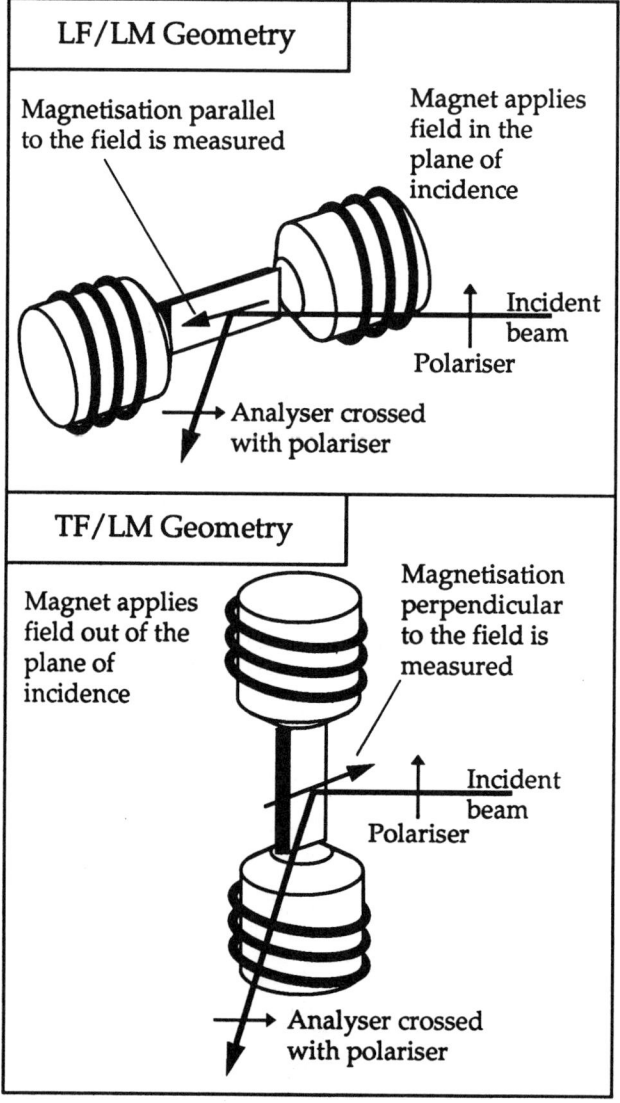

Figure 12. An illustration of the two experimental geometries used to measure the components of magnetisation parallel and perpendicular to the applied field.

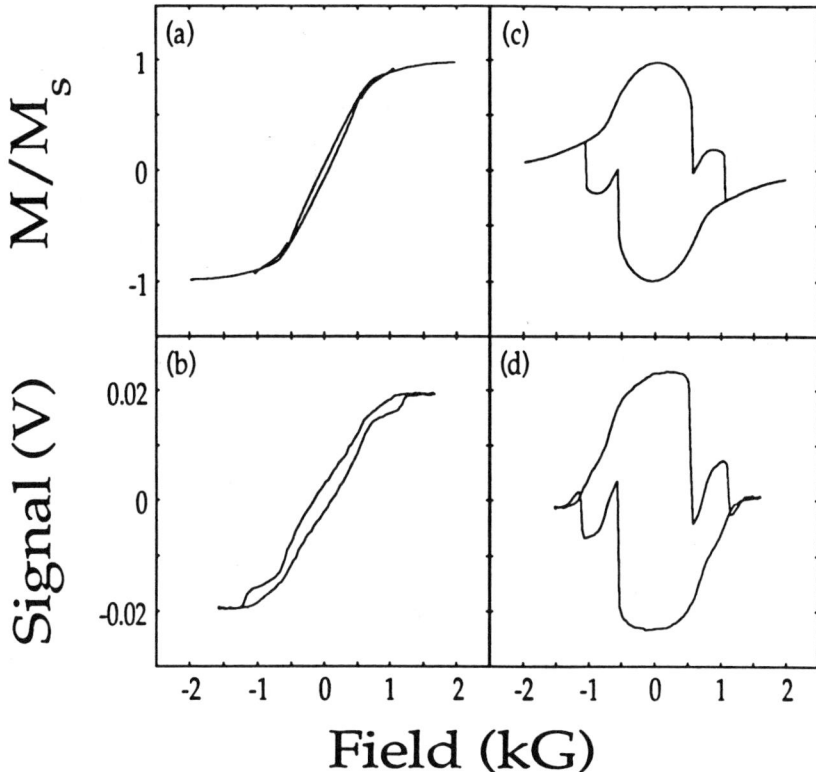

Figure 13. A comparison of the calculated $M_\parallel$-H (a) and $M_\perp$-H (c) loops, and the measured $M_\parallel$-H (b) and $M_\perp$-H (d) loops for the Co/Cu(27Å)/Co sample described in the text.

It is possible to plot out the orientations of the magnetisations from the calculated results and to examine how they contribute to each loop. This is shown in Fig. 14 which shows the calculated $M_\parallel$-H and $M_\perp$-H loops with insets indicating the orientations of the two magnetisation components for various points on the curves. The sharp features occurring at fields of ±0.5kG and ±1kG are associated with an abrupt transition in which the magnetisations in the top and bottom Co layers respectively 'flip' over their own hard axis. From this it can be seen that measurements of $M_\perp$–H loops for switching in the vicinity of the hard axis provide a sensitive means of determining the layer dependent anisotropy fields K_i/M.

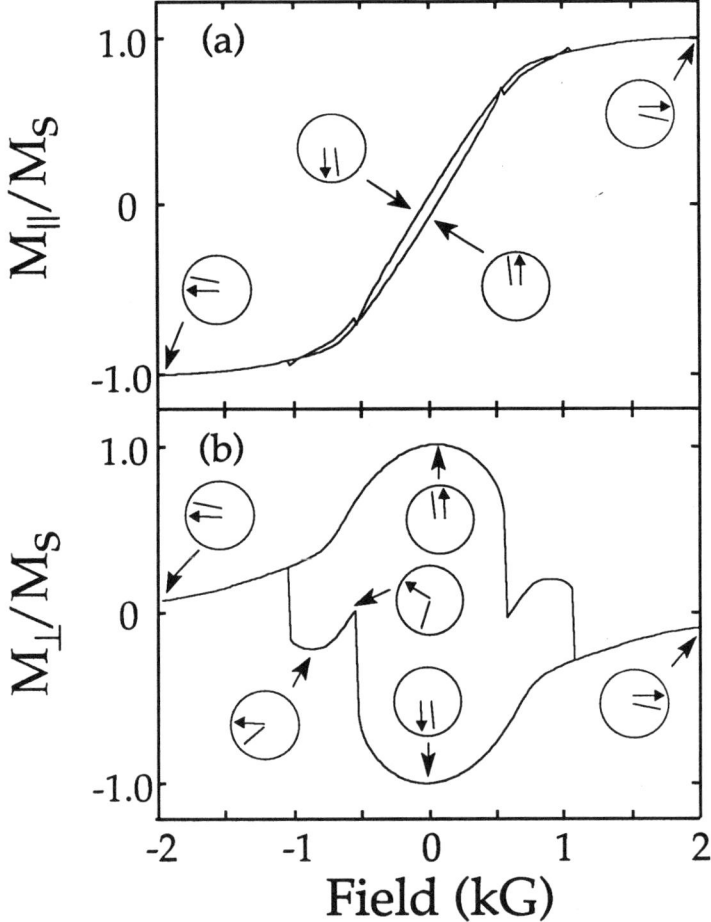

Figure 14. The calculated $M_\|$-H (a) and $M_\perp$-H (b) loops for the Co/Cu(27Å)/Co system described in the text, showing the directions of the magnetisations at various points on the loops as indicated by the insets.

Recently the vector magnetometric technique has been applied to the study of switching processes in epitaxial Fe/GaAs(001) films.[43] The combination of the uniaxial and cubic anisotropies results in the two <110> directions being inequivalent and it is found that the interplay of these two terms strongly affects the magnetisation reversal process. Fig. 15 shows vector M–H loops obtained by MOKE for an Fe film with a significant anisotropy ratio $r = K_u/K_1$ as a function of the angle ϕ between the applied field and the hard uniaxial anisotropy axis. Figs 15(a) and (b) show the components of M parallel and perpendicular to the applied field for $\phi = 75^\circ$. Both loops show an irreversible jump at the same field. For $\phi = 20^\circ$ two jumps can be seen as shown in Figs 16(c) and (d). These two jumps occur when the magnetisation traverses each of the two hard axis directions that exist in a sample with $|r| < 1$.

The double 'jump' switching gives rise to 'overshoots' in the MOKE loops along only one of the two <110> axes as seen for example in Fig. 10 at large thickness. The 'overshoot'

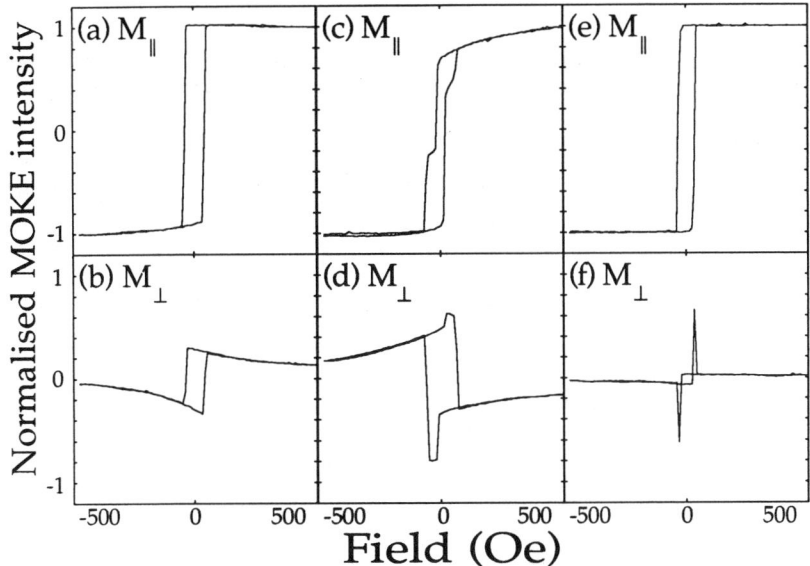

Figure 15. *M–H* loops obtained by MOKE for an Fe/GaAs (001) film with a ratio of uniaxial to cubic anisotropy strengths given by $r = 0.4$ as a function of azimuth angle ϕ with respect to the hard axis in-plane. (a) and (b) show the parallel and transverse components for $\phi = 75°$; (c) and (d) show the parallel and transverse components for $\phi = 20°$; (e) and (f) show the parallel and transverse components for $\phi = 60°$.

features appear in this case to arise from the component of the magnetisation transverse to the applied field which is also detected in the MOKE signal shown in Fig. 10 because of the analyser setting used in this case. For appropriate analyser settings, the relative contributions of the transverse and parallel magnetisation components to the MOKE signal can be varied and hence separated.

The exact number of jumps and the detailed form of the transverse magnetisation loop is found to depend sensitively upon the orientation of the applied field and the ratio r of the uniaxial anisotropy K_u to the cubic anisotropy K_1. Minimum energy calculations, which assume the values of K_1 and K_u determined by BLS, provide a good qualitative description of this reversal process, although the magnetic microstructure appears to influence the exact values of the switching fields.[43] Moreover, the transverse component of the magnetisation provides a much more accurate means of determining the saturation field than the parallel component for which a gradual approach to saturation occurs. Saturation fields determined in this way closely agree with those determined from the BLS measurements. It should be noted that the form of the transverse loops are extremely sensitive to the orientation of the applied field and that the theoretical curves closely reproduces features seen in the experimental curves for the appropriate orientation of the applied field.

It is clear from the above studies that vector MOKE reveals considerable detail in the magnetisation reversal processes. Such information is important in interpreting *M-H* loops from exchange coupled films discussed in the next section.

2.5 Oscillatory exchange coupling in trilayer structures

MOKE has been used extensively in studying the exchange coupling which arises between ultrathin ferromagnetic transition metal layers separated by a non-magnetic material of appropriate thickness – e.g. Fe/Cr/Fe,[44] Co/Cu/Co,[45] Fe/Mo/Fe.[46] The origin of the thickness-dependent coupling is associated with spin density oscillations in the spacer layer, related to the RKKY interaction which arises in dilute alloys.[3] In this case short period oscillations in the bilinear coupling with thickness are predicted. Long period oscillations associated with the detailed Fermi surface features have also been predicted and observed in many systems.[44,45] The coupling energy for a trilayer can in general be written as:

$$E = -A_{12} m_1.m_2 + B_{12}(m_1.m_2)^2 \tag{7}$$

where A_{12} and B_{12} are the bilinear and biquadratic coupling strengths and m_1, m_2 are unit vectors parallel to the magnetisation of each layer. Both A_{12} and B_{12} are thickness dependent and for $A_{12} < 0$ an antiferromagnetic alignment of the layer magnetisations is favoured in zero field. B_{12} is always positive and favours a 90º alignment of the spins in zero field. Thus vector magnetometric measurements are particularly useful in determining the presence of each type of coupling. A particular advantage of spatially resolved MOKE (100µm is sufficient) in this context is that it can be used to probe wedge structures to yield thickness dependent data from one sample.[3] A difficulty with MOKE is that ferromagnetic coupling strengths cannot be determined from the saturation field (see section 2.5 on polar MOKE) since in this case the coupling does not contribute to the torque on the magnetic layers in an applied magnetic field.

In the case where both crystal anisotropies and exchange coupling appears a variety of spin orientation behaviours result. Fig. 16 shows *M-H* curves obtained by MOKE for epitaxial Fe/Ag/Fe(001) structures by Celinski et al.[47] In these structures A_{12} corresponds to anti-ferromagnetic coupling for the 10ML spacer and to ferromagnetic coupling for the 6ML spacer. The biquadratic coupling strength exceeds the bilinear coupling strength in the second sample and is roughly half the bilinear coupling strength in the first sample. Two critical fields are observed for the 10ML Ag spacer structure. The upper critical field corresponds to the point that both layers depart from ferromagnetic alignment along the applied field. At the second field the onset of the antiferromagnetic configuration occurs. At zero field a remanence corresponding to a 90 degree alignment of the spins stabilised by the cubic anisotropy of the sample is observed. For the sample with strong biquadratic coupling only one transition field is seen. The observed loops can be compared with the results of simulations which assume that the reversal process corresponds to minimum energy or coherent rotation. Thus the form of the loops and the critical fields can be used to deduce values for A_{12} and B_{12} or to place limits on their values when only one critical field is observed. Further information can be obtained from vector magnetometric loops.[42] The strength of the biquadratic coupling is found to be strongly dependent on the terrace size which varies strongly according to the growth temperature used.

Exchange coupling has been extensively studied in the Fe/Cr/Fe system.[44] These studies illustrate the extreme sensitivity of the coupling behaviour to the interface morphology and the difficulties encountered in comparing theory with experiment. Initially only long period oscillations were observed[48] but by growing the Cr layers at elevated temperatures optimised to produce the sharpest interfaces, short period oscillations were found.[49] By using films prepared on Fe whiskers it was demonstrated that the phase of the short period oscillation is opposite to both that theoretically predicted and the results obtained on structures grown on single crystal substrates.[44] This is surprising given that STM and RHEED studies confirm that the terrace sizes achieved for growth on Fe whiskers greatly exceeds that which can be produced on other surfaces.

Epitaxial Fe/Cr/Fe(001) systems can be grown on Ag(001)/GaAs substrates[50,51] and the interplay of coupling and magnetic anisotropy leads to a complex phase-diagram for the spin orientation behaviour. The range of Cr thickness in which the biquadratic coupling dominates has been reported by Ruhrig et al.[51] for Cr wedge structures to be extremely narrow in the first region in which it appears (approximately 0.25Å) and to correspond to points at which the bilinear term is small or vanishes: however studies by Daboo et al.[50] find that the step-like loop corresponding to dominant biquadratic coupling is obtained over a significantly larger thickness range in which the bilinear term remains significant. This finding contrasts markedly with the data reported by Ruhrig et al. for the Cr thickness range of only 0.55Å in the second region in which biquadratic coupling occurs. According to a model recently proposed by Slonczewski[52] the biquadratic coupling arises from a small variation in Cr thickness on a lateral length scale smaller than the exchange length. As a result, the system 'averages' over the parallel and antiparallel configurations that would be favoured by the two extremes of thickness of the Cr film and adopts the 'compromise' configuration of a 90º relative orientation. Monolayer thickness variations are sufficient to introduce such behaviour since it is known from recent studies by Unguris et al.[53] that the coupling oscillates on a length-scale close to that of the vertical separation of adjacent Cr layers. In contrast to the structures prepared on Fe whisker samples, only long period oscillations are clearly seen in the structures prepared by Daboo et al.[50] because of the low Cr deposition temperature used during the growth. Heinrich et al.[26] have presented the results of detailed analyses of the stepped loops obtained by Kerr effect for Fe/Cr/Fe samples prepared on Fe whiskers. The strength of the biquadratic coupling is sensitively dependent on the ratio of the film thickness to the terrace length of the Fe film and therefore is only visible in high quality films for which this ratio is small, but non-vanishing. The larger thickness range over which biquadratic coupling dominates in the films prepared by Daboo et al.[50] in comparison with those reported by Ruhrig et. al.[51] may be indicative of a large terrace length in the samples prepared by Daboo et al.

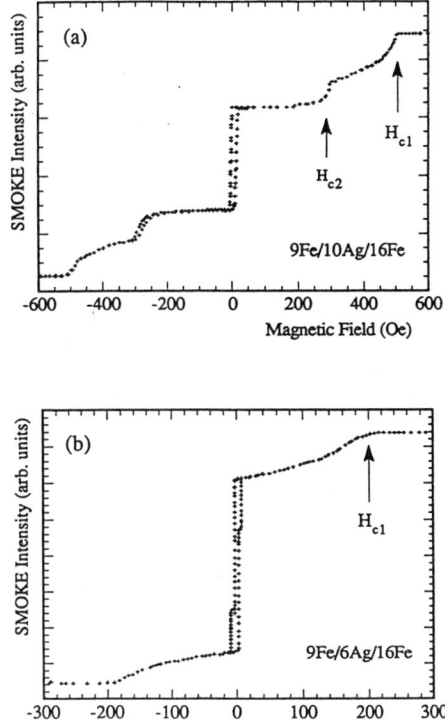

Figure 16. M–H loops obtained by MOKE for 9MLFe/Ag/16Fe/Ag(001) structures with Ag spacer layer thicknesses of (**a**) 10ML and (**b**) 6ML. The field was applied along an easy [100] axis in-plane. From ref. 47.

2.6 Polar Kerr magnetometry at high fields

For planar films a great deal of information can be obtained from polar M-H loops[54] since the magnetisation reversal process can be readily calculated for a given set of magnetic parameters assuming that coherent rotation occurs. For demagnetising and surface anisotropy fields only, the polar M-H curve is linear. Polar MOKE measurements can be used to determine the effective demagnetising field, which in the absence of cubic anisotropies and coupling is given by the perpendicular saturation field. Measurements of the perpendicular saturation field H_s can yield useful information for trilayer samples since the presence of higher order anisotropies and exchange coupling modify the value of H_s. By measuring the fully saturated curve the perpendicular susceptibility can be measured normalised to M_s. From the polar MOKE magnetisation curves it is possible to measure the perpendicular saturation field $H_s^\perp$ and a saturation field extrapolated from the initial magnetisation gradient, χ_0, defined as $H_s^{\chi,\perp}=1/\chi_0$, as shown in Fig. 17. In the absence of coupling and for fixed layer thickness, these fields are constant.

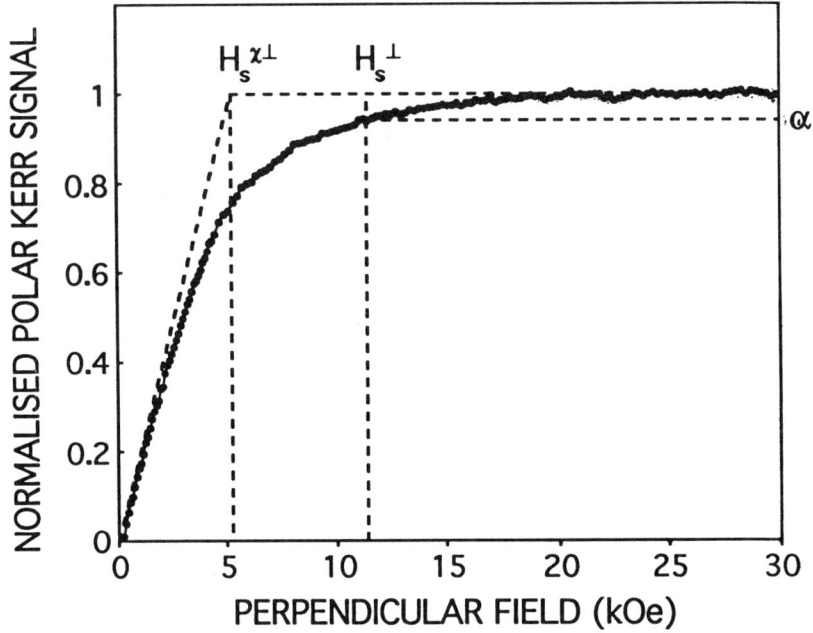

Figure 17. A polar MOKE magnetisation curve from the Co/Cu/Co(111) wedge corresponding to a thickness of 7.4Å. See text for an explanation of the saturation fields.

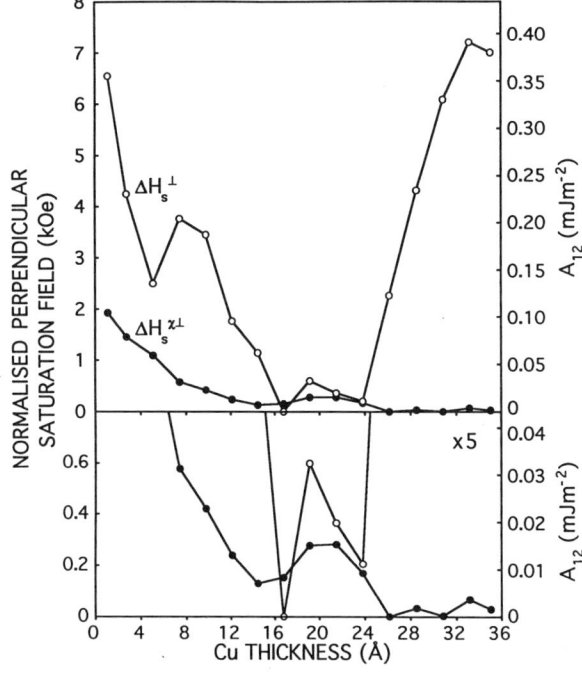

Figure 18. $\Delta H_s^\perp$ (open circles) and $\Delta H_s^{\chi\perp}$ (closed circles) are plotted versus Cu thickness for the Co/Cu/Co(111) wedge sample. The right hand scale gives corresponding estimates of the coupling constant A_{12}.

For a Co(15 Å)/Cu(0–35 Å)/Co(15 Å) wedged trilayer, measurements of the perpen-dicular saturation field as a function of Cu spacer thickness reveal an oscillatory component superimposed on large background variations.[54] The results from the polar MOKE measurements on the wedged trilayer are shown in Fig. 18. The quantities $\Delta H_s^\perp$ and $\Delta H_s^{\chi\perp}$ representing the variation in these quantities from the isolated limit (large spacer thickness) on the left hand scale versus Cu layer thickness, with corresponding estimated values of the bilinear coupling strength shown on the right hand scale. Both the first and second AFM coupling peaks, at about 9 Å and 20 Å Cu respectively can be seen. As stated above, in addition to being sensitive to the coupling, the perpendicular saturation field is sensitive to variations in the interface anisotropy field and demagnetising field which can result in large background variations.[54] From the oscillatory part a coupling period of 11±2 Å is estimated in good agreement with previous results and it is deduced from the background variations that, in addition to pinhole defects, there are large variations in interface quality at the magnetic-nonmagnetic interfaces as the interlayer thickness is varied along this wedge sample. Thus polar MOKE is particularly useful in assessing the effect of interface quality and the sample homogeneity on the magnetic properties in such samples. The in-plane saturation field, in contrast shows, no clear oscillatory component can be observed as a result of the dominant ferromagnetic character of the loops.

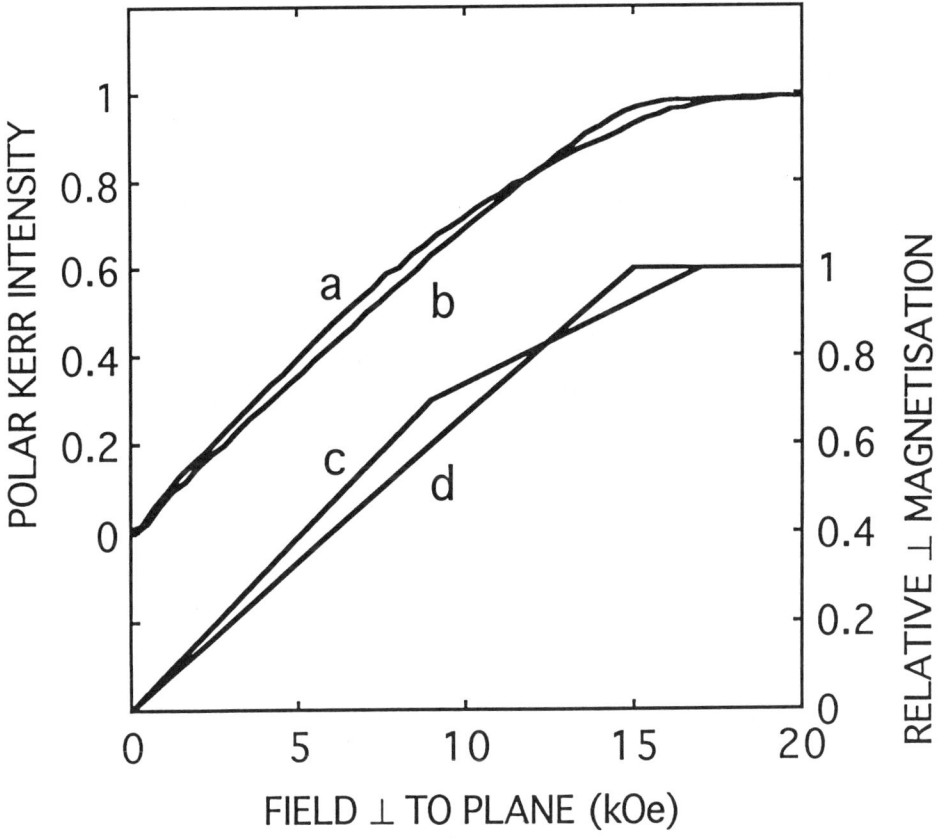

Figure 19. The normalized Kerr intensity for Fe/Pd/Fe samples described in the text plotted as a function of the magnetic field strength H for polar MOKE measurements in which the field was applied parallel to the film normal.

Fig. 19 shows the polar MOKE curve obtained for (a) an uncoupled trilayer of the form 10Å Fe/120Å Pd/20Å Fe and for (b) a wedged structure of the form 10Å Fe/d_{Pd} Å Pd/20Å Fe (14Å < d_{Pd} < 30 Å). For (b) the loop corresponds to a Pd thickness yielding a significant ferromagnetic coupling strength.[55] In these samples a perpendicular anisotropy field arises which lowers the effective demagnetising field in the layers. The uncoupled structure exhibits a clear kink at a field of 6.1 kOe at which the thinner Fe layer saturates. The fact that the loop is unrounded and exhibits this sharp kink is clear evidence that the two Fe layers are decoupled in this sample. If the bulk value for the magnetisation is assumed then the values of the kink and saturation fields imply positive values of the average surface anisotropy constant K_S for both the thick and thin Fe layers, as calculated in curve (c). In principle the height of the kink could also be used to deduce the ratio of moments in the two Fe layers but since there will be a substantial optical skin depth effect for a Pd spacer layer of thickness 150 Å, and since the optical constants of the thin film are not well known, this ratio has not been estimated. Curve (d) corresponds to the coupled film case as a fit to the data of curve (b). The effect of the coupling is to round the M-H curve and to introduce a saturation field intermediate between that of the separate layers. MOKE can therefore be used to probe ferromagnetic coupling in the case where a different perpendicular anisotropy field arises for the two layers and introduces a torque on the layers during magnetisation reversal.

Thus polar MOKE provides a powerful probe of the coupling in trilayer structures, with much important information being provided directly by the loop shape.

CONCLUSION

MOKE provides an extremely powerful probe of the magnetic properties of supported ultrathin films in UHV with sub-monolayer sensitivity (comparable with SQUID magnetometry). Such sensitivity is comparable with that of SQUID or in-situ polarised electron techniques. Quantitative magnetometric information can be extracted as a function of applied field: for example the orientation of the magnetisation vector and the value of saturation fields can be extracted in studies of magnetisation reversal, magnetic anisotropies and exchange coupling for example. MOKE offers the further advantage of spatial resolution (submicron scale) and so can be used to probe inhomogeneous samples and magnetic domain structure.[56] Dynamic measurements on picosecond timescales can also be performed with high sensitivity in order to probe the magnetisation dynamics.[57] The loop amplitude is determined by the magneto-optic interaction and therefore cannot be used for absolute magnetic moment measurements nor can the usual Kerr effect distinguish between bulk and surface effects, in contrast to second harmonic generation (SHG) studies of Kerr rotation. For this reason it is particularly valuable in studying ultrathin films supported by a non-magnetic substrate. For the study of the surface of bulk magnetic materials, polarised electron techniques are required, as described elsewhere in this volume. The richness of the behaviour so far observed suggests that many new phenomena are still to be discovered and it is certain that MOKE will continue to yield important new discoveries in the field of ultrathin magnetic structures, paralleled by exciting new developments in X-ray optical techniques applied to the study of magnetism in ultrathin magnetic structures.

ACKNOWLEDGEMENTS

I would like to thank several members of my research group, many of whose results I have included in this article: Dr. C. Daboo, Dr. J. Hicken, Dr. F. O. Schumann, Dr. M. Gester, A. J. R. Ives, M. Buckley, S. J. Gray, T. Fujimoto, Dr. M. Patel and U. Ebels. The support of the SERC, Toshiba Corporation and the Royal Society is gratefully acknowledged.

REFERENCES

1. J. Kerr, *Phil. Mag.* 3:321 (1877).
2. E R Moog and S D Bader, *J. Superlatt. and Microstruc.*, 16 (1985).
3. S D Bader and J L Erskine, Magneto-Optical Effects *in:* "Ultrathin Magnetic Structures", Vol.2, B Heinrich and J A C Bland, eds, Springer Verlag, Berlin (1994).

4. C Liu and S D Bader, *J. Appl. Phys.* 67:5758 (1990).
5. Z Q Qui, J Pearson and S D Bader, *Phys. Rev. Lett.* 70:1006 (1993).
6. B S Krusor and G A N Connel, Thin Film Rare Earth-Transition Metal Alloys for Magneto-optical Recording *in:* "Physics of Thin Films", Vol. 15, M H Franscombe and J. L. Vossen, eds, Academic, Boston (1991), pp 143–217.
7. P F Carcia, *J. Appl. Phys.* 63:5066 (1988).
8. K H J Buschow, P G van Engen and R Jongebreur, *J. Magn. and Magn. Mater.* 38:1 (1983).
9. In this case the material must be magnetised perpendicularly to the surface.
10. E. D. Palik, ed., *Handbook of Optical Constants of Solids*, Academic Press, Orlando (1985).
11. P N Argyres, *Phys. Rev. A* 97:334 (1955).
12. H S Bennett and E A Stern, *Phys. Rev. A* 137:448 (1965).
13. J M Sexton, D W Lynch, R L Benhow and N V Smith, *Phys. Rev. B* 37:2879 (1988).
14. N V Smith and S Chiang, *Phys. Rev. B* 19:5013 (1979).
15. G Y Guo and H Ebert, *Phys. Rev. B*, submitted.
16. Z Q Qui, J Pearson and S D Bader, *Phys. Rev. B* 45:7211 (1992).
17. S Visnovsky, M Nyvlt, V Prosser, J Ferré, G Pénnisard, D Renard and G Sczigel, *J. Magn. and Magn. Mater.* 128:179 (1993).
18. T Katayama, Y Suzuki, H Hayshi and A Thiaville, *J. Magn. and Magn. Mater.* 126:527 (1993).
19. J Zak, E R Moog, C Liu and S D Bader, *J. Magn. and Magn. Mater.* 89:107 (1990).
20. S Visnovsky, Czech. *J. Phys. B* 37:218 (1987), *ibid B* 36:1049 (1986).
21. P Yeh, *Surf. Sci.* 96:41 (1980); "Optical Waves in Crystals", Wiley, New York (1988).
22. R M A Azzam and N M Bashara, "Ellipsometry and Polarized Light", North Holland, Amsterdam (1989).
23. J A C Bland, M J Padgett, R J Butcher and N Bett, *J. Phys. E: Sci. Instrum.* 22:308 (1989).
24. A J R Ives, R. J. Hicken, J. A. C. Bland, C. Daboo, M. Gester and S. J. Gray, *J. Appl. Phys.* (1994).
25. D. Kerkmann, *J. Appl. Phys. A* 49:523 (1989).
26. B Heinrich and J F Cochran, *Advances in Physics* 42:523 (1993); J G Gay, R Richter, G H O Daalderop, P J Kelly, M F H Schuurmans, W J M de Jonge, P J H Bloemen and F J A de Broeder, Magnetic Anisotropy, Magnetisation and Band Structure *in:* "Ultrathin Magnetic Structures", Vol. 1, J A C Bland and B Heinrich, eds, Springer Verlag, Berlin (1994).
27. M Patel, T Fujimoto, E Gu, C Daboo and J A C Bland, *J. Appl. Phys.* 75:6528 (1994).
28. F O Schumann, M E Buckley and J A C Bland, *Phys. Rev. B* (in press).
29. F O Schumann and J A C Bland, *J. Appl. Phys.* 73:5945 (1993).
30. F O Schumann, M E Buckley and J A C Bland, *J. Appl. Phys.* (in press).
31. H. C. Siegmann and E Kay, Spin-polarised Spectroscopies *in:* "Ultrathin Magnetic Structures", Vol. 1 J A C Bland and B Heinrich, eds, Springer Verlag, Berlin (1994).
32. M Cinal, D M Edwards and J Mathon, *J. Magn. and Magn. Mater.* (in press).
33. C M Schneider, P Bressler, P Schuster, K Kirschener, J J de Miguel and R Miranda, *Phys. Rev. Lett.* 64:1059 (1990).
34. M T Kief and W F Egelhoff Jr., *Phys. Rev. B* 47:10785 (1993).
35. B N Engel, M H Wiedmann, R A Van Leeuwen and C M Falco, *J. Appl. Phys.* 73:6192 (1993).
36. G A Prinz and J J Krebs, *Appl. Phys. Lett.* 39:397 (1981); G. A. Prinz, *Phys. Rev. Lett.* 54:1051–1054 (1985); G A Prinz, C Vittoria, J J Krebs, K B Hathaway, *J. Appl. Phys.* 57:3672 (1985).
37. M Gester et al., to be published.
38. R J Hicken, D E P Eley, M Gester, S J Gray, C Daboo, A J R Ives and J A C Bland, *J. Magn. and Magn. Mater.* submitted.
39. K T Riggs, E D Dahlberg and G A Prinz, *Phys. Rev. B* 41:7088 (1990).
40. M D Mermin and H Wagner, *Phys. Rev. Lett* 17:1133 (1966).
41. C Liu and S D Bader, *Phys. Rev. Lett.* 60:2422 (1988).
42. C Daboo, J A C Bland, R J Hicken, A J R Ives, M J Baird and M J Walker, *Phys. Rev. B*. 47:11852 (1993)
43. C Daboo, R J Hicken, D E P Eley, M Gester, S J Gray, A J R Ives and J A C Bland, *J. Appl. Phys.* 75:5586 (1994).
44. B Heinrich and J F Cochran, *Advances in Physics* 42:523 (1993).
45. M T Johnson, R Coehorn, J J de Vries, N W E McGee, J van de Steggeand and P H Bloemen, *Phys. Rev. Lett.* 69:969 (1992).
46. Z Q Qiu, J Pearson, A Berger and S D Bader, *Phys. Rev. Lett.* 68:1398 (1992).
47. Z Celinski, B Heinrich and J F Cochran, *J. Appl. Phys.* 73:5966 (1993).
48. S S P Parkin, N More and K Roche, *Phys. Rev. Lett.* 64:2304 (1990).
49. J Unguris, R J Celotta and D T Pierce, *Phys. Rev. Lett.* 67:140 (1991).
50. C. Daboo et al., to be published
51. M Rührig, R Schäfer, A Hubert, R Mosler, J A Wolf, S Demokritov and P Grünberg, *Phys. Stat. Solidi A* 125:635–656 (1991).
52. J C Slonczewski, *Phys. Rev. Lett.* 67:3172–3175 (1991).
53. J Unguris, R J Celotta and D T Pierce *Phys. Rev. Lett.* 67:140-143 (1991) and *Phys. Rev. Lett.* 69:1125 (1992).
54. A J R Ives, R J Hicken, J A C Bland, C Daboo, M Gester and S J Gray, *J. Appl. Phys.* 75:6458 (1994).

55. R J Hicken, A J R Ives, D E P Eley, C Daboo, J A C Bland, J Childress and A Schuhl, *Phys. Rev.* B (in press).
56. J Pommier, P Meyerm, G. Pennisard, J. Ferre, P. Bruno and D. Renard, *Phys. Rev. Lett.* 65:2054 (1990).
57. N Bontemps, J C Rivoal, M Billardon, J Rajchenbach and J Ferre, *J. Appl. Phys.* 52:176 (1981).

MAGNETIC X-RAY DICHROISM.
AN EFFECTIVE WAY TO STUDY THE SPIN AND ORBITAL
MAGNETIZATION IN MAGNETIC MATERIALS

Gerrit van der Laan

Daresbury Laboratory
Warrington WA4 4AD
United Kingdom

1. INTRODUCTION

Novel spectroscopic tools for magnetic materials have recently emerged from the use of circularly and linearly polarized x-rays. These studies would have been impossible without the use of synchrotron radiation. This radiation emitted by the relativistic electrons confined in an electron storage ring is naturally polarized, i.e. linear in the plane of the electron orbit, and left- and right circularly polarized above and below the plane.

Circular dichroism, which is the intensity difference between left and right circularly polarized radiation, can be observed in many types of high-energy spectroscopies where electrons are excited by photons, such as in x-ray scattering, (inverse) photoemission and x-ray fluorescence,[1-11] but the most straightforward example is found in x-ray absorption of magnetic materials, where the effect is called magnetic x-ray dichroism (MXD). The effect originates from electric dipole transitions, similar as in the Kerr effect in the visible region of the optical spectrum, where it gives materials such as tourmaline a colour change when the direction of the linear polarization of the light is altered. MXD has several interesting advantages compared to its counterparts in the visible region. Since the wavefunction of the involved core state is strongly localized and therefore well-defined, MXD is more straightforward to analyse but most importantly it is element, site and symmetry selective.

Consider a ferromagnetic $3d$ transition metal with an unbalanced weight over the two spin states in the valence band, such as nickel, cobalt or iron. In x-ray absorption a core p electron is excited by an electric dipole transition into the magnetically polarized $3d$ states. Electric dipole transitions, irrespective whether this is in the optical or x-ray region, change the parity and the orbital quantum number of the excited electron ($\Delta l = \pm 1$). Circularly polarized light has an electric vector which rotates around the direction of propagation, and changes the orbital magnetic moment, which is determined by the azimuthal charge distribution, by an amount $\Delta m = \pm 1$. Because the electric vector of the light interacts only with the orbital momentum, there is no preference for excitation of electrons with up or down spin.

However, in the presence of spin-orbit interaction, the orbital momentum is coupled to the spin momentum, resulting in a difference in transition probability for up and down spin electrons depending on the polarization of the light. A p electron with magnetic quantum number $m_j = -3/2$ can be excited into a d state with $m_j = -5/2$ using right-circularly polarized light ($\Delta m = +1$) but with left-circularly polarized light ($\Delta m = -1$) no excitation into the $m_j = -5/2$ is possible. Thus the x-ray absorption in a magnetic atom is polarization dependent if there is spin-orbit interaction.

Erskine and Sterne [12] indicated already in 1974 the possibility of a magneto-optical effect in the $3p$ absorption edge of ferromagnetic nickel from augmented-plane-wave (APW) calculations. They predicted that the spin-orbit splitting of the $3p$ core level in combination with the $3d$ final state spin polarization results in a magneto-optical Kerr effect (MOKE) of ~10%. However, early attempts to observe MOKE in the Gd $2p \rightarrow 5d$ transition of amorphous Gd-Fe only gave an upper limit of 0.02%. [13] The field was laying dormant till in 1985 Thole, van der Laan, and Sawatzky [14] predicted a strong MXD in the $3d$ absorption edges of rare-earth materials on the basis of atomic multiplet calculations. It was suggested that complicated magnetic structures could be studied using circular and linear polarization, which are related to the average value $\langle M \rangle$ and $\langle M^2 \rangle$ of the local magnetization, respectively. The first experimental evidence of this effect was on terbium iron garnet [15] in excellent agreement with the theoretical predictions. As an interesting application Goedkoop et al. [16] proposed to exploit this effect to change linearly polarized radiation into circularly polarized radiation by means of a magnetic transmission foil. MXD has also been reported in rare earth L edges [17] and transition metal K edges, [18,19] where the results were confirmed by first principles calculations. [20,21] For these deeper core levels Carra and Altarelli [22] pointed the existence of electric quadrupole transitions out. However, experimental evidence from the angular dependence, which should be different for dipole and quadrupole transitions, is still not conclusive, and the latter are up to the present day disputed.

The Ni $2p$ MXD was measured by Chen et al. [23] who found that the dichroic behaviour of satellite structure could not be explained by tight-binding band structure calculations. They suggested that this structure is due to many-body effects. [24] This has been confirmed by Jo and Sawatzky [25] and van der Laan and Thole [26] on the basis of Anderson impurity model calculations for ferromagnetic nickel where the ground state is a mixture of d^8, d^9 and d^{10} configurations. Thus electron correlation effects are crucial to understand the dichroic behaviour of the satellite structure. For the same reason the experimental data of the Ni $3p$ MXD by Koide et al. [27] was different from the aforementioned predictions by Erskine and Stern, [12] who had not taken electron correlation and $3d$ spin-orbit interaction into account.

It became clear from the calculated $2p$ MXD spectra of the $3d$ transition metal ions in crystal field symmetry that there was a relation between the $3d$ spin-orbit interaction and the integrated signal. [28] The breakthrough came when Thole et al. [29] derived a sum rule which relates the orbital magnetic moment to the integrated MXD intensity. This allows the separation of the orbital and spin part of the magnetic moment, which has always been a major experimental challenge using techniques such as neutron diffraction and resonant magnetic scattering. [9] The orbital magnetization sum rule can provide a useful insight into the microscopic origin of anisotropic magnetic properties, such as the magneto-crystalline effect, easy-direction of magnetization, magnetostriction and coercivity of magnetic materials. Especially, it gives a unique probe to study magnetic materials containing $3d$, $4f$ and $5f$ elements. These materials show a wide variety of magnetic properties, e.g. Fe, Co, and Ni are ferromagnetic metals, Fe_3O_4 is an ferrimagnetic insulator-metal, MnO is an antiferro-magnetic insulator, Cr shows spin density waves, rare earth elements display complicated spiral spin structures, $SmCo_5$ is a hard magnet, $CeCu_2Si_2$ is a heavy fermion system and garnets, such as $Fe_5Gd_3O_5$, are ferrimagnets. Experimental studies [30-34] on these materials which require soft x-rays to assess the relatively shallow core levels are facilitated by the excellent agreement with calculated results for $3d$ transition metal, [28] rare earth, [35,36] and actinide ions. [37]

The outline of this Chapter is as follows. First we will discuss the practical aspects of MXD measurements in the soft x-ray region using the available facilities at Daresbury as example. Next we will give a simple proof of the orbital magnetization sum rules using the expression for the angular part of the dipole transition probability for excitation of a core electron into the unoccupied states which are magnetically polarized. As an application we discuss the experimental results for magnetic multilayers with perpendicular magnetization, such as Co/Pd and Co/Ni multilayers where we demonstrate the use of the sum rules for spin and orbital magnetization. Finally, some further prospects and conclusions are mentioned.

2. PRACTICAL ASPECTS

Spin resolved photoemission has always been restricted to model compounds due to the low count rate and the cumbersomeness of spin detectors. The situation for circular dichroism in x-ray absorption and photoemission is entirely different. A high degree of circular polarization can easily be obtained by using the off-plane synchrotron radiation from a bending magnet, which is linearly polarized in the plane of the electron storage ring and elliptically polarized above and below the plane. Since the intensity decreases rapidly away from the horizontal plane, special insertion devices, such as helical and crossed undulators, have been constructed which provide an intense on-axis photon flux of almost complete polarization.[38]

At several synchrotron radiation facilities beamlines have been either modified or designed for experiments with circularly polarized soft x-rays. At the synchrotron radiation source (SRS) at Daresbury (UK) there are currently two soft x-ray beam lines routinely in operation for circular dichroism. This enables the study of all major types of magnetic materials. Beamline 1.1 covers the $L_{2,3}$ edges of the $3d$ metals and beamline 3.4 covers the $L_{2,3}$ edges of $4d$ metals, the $M_{2,3}$ edges of the $5d$ metals, and the $M_{4,5}$ edges of the rare earths. Polarization choppers have been installed on both beamlines to allow fast data acquisition. The high flux and stable polarization have made it possible to do several novel experiments at Daresbury, such as the circular dichroism in the Ni $3p$ photoemission, which indicates an enhanced orbital magnetic moment at the surface of ferromagnetic nickel[39] in agreement with theoretical predictions.[40]

Beamline 1.1

This beamline contains a high-energy spherical grating monochromator (HESGM) covering the energy region from 250 to 1000 eV.[41] The lay-out of the optical elements is given in Fig. 1. The absence of an entrance slit maximizes the light throughput. Two mrad of horizontal radiation from a bending magnet is incident at a grazing angle of 2° on a Pt coated CVD-SiC premirror, which has a meridian cylinder of radius 299 m and which is water cooled via a gallium-indium eutectic interface. The premirror focuses the light horizontally (the non-dispersive direction) to the exit slit position and reduces the power loading on the gratings by absorbing all radiation above 2 keV. The x-rays are vertically diffracted by one of the three interchangeable lamellar ion etched spherical gratings of nominal radius 148 m operating in negative order (deviation angle 172°) which focus onto the moveable exit slit. The gratings have densities of 1050 (V coating), 1500 (Au coating) and 1800 lines/mm (Ni and Au coated stripes). Finally, the light diverging from the exit slit is refocused to the sample position by an ellipsoidal Pt-coated mirror. Selection of the photon energy is by a simple rotation of the grating. An external sin bar of 412 mm length is

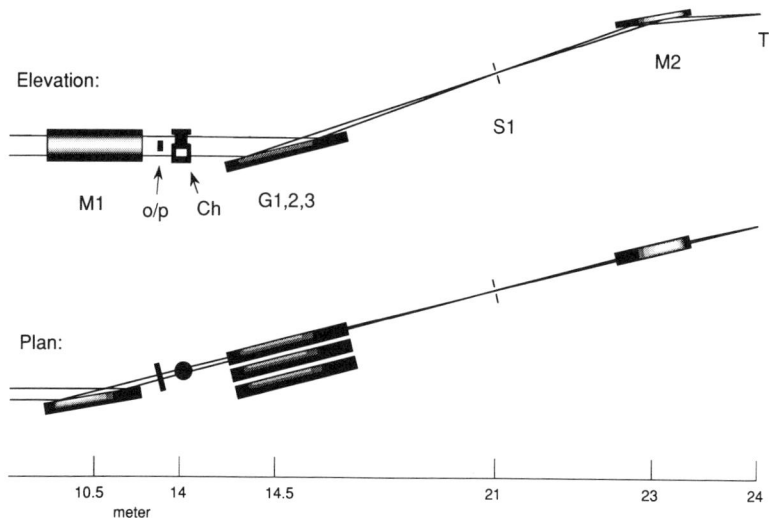

Fig. 1: Geometrical arrangement of beamline 1.1 at the SRS. Premirror (M1), obscuration plate (o/p), chopper (Ch), interchangeable gratings (G1,2,3), exit slit (S1), postfocusing mirror (M2), and target (T). The distance to the source point is indicated underneath.

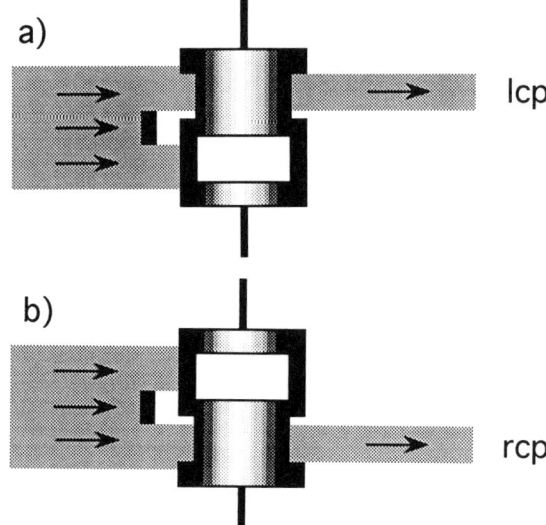

Fig. 2: Polarization chopper of beamline 1.1 consisting of an obscuration plate which removes the central part of the beam and a rotating drum. Position (a) provides left-circularly polarized x-rays and position (b), which is reached after a rotation of 90°, provides right-circularly polarized x-rays.

coupled to the rotation axis of the gratings and the free end can be displaced by an encoder to a precision of 1 µm. The focal point of the spherical grating is wavelength dependent, which is accommodated by a translation of the exit slit. The spot size at the sample position is ~1 mm^2. The flux is in the order of 10^{11} photons/100 mA. The resolving power is ~2000.

The degree of circular polarization is conserved over the optical elements because of the grazing incidence of the x-rays. As seen in Fig. 1 a polarization chopper is located between the premirror and the monochromator. A schematic illustration of this device is given in Fig. 2. The assembly consists of an obscuration plate which removes the middle part of the x-ray beam, which is linearly polarized, and a vertically rotating drum which alternately obscures the upper and lower part of the beam. In this way left and right circularly polarized radiation is produced with a switching frequency of several Hz. The spectra for the two helicities are separately accumulated using the signals of a synchronization plate which is mounted on the axis of the chopper. In choosing the optimal width of the obscuration plate there is a trade-off between a high degree of circular polarization and a high photon flux. Fortunately, it is not necessary to have 100 % circular polarization since the contribution of the linear polarization should cancel in the difference spectrum. It can be shown that the figure of merit of the chopper is proportional to $P\sqrt{I}$, where P is the weighted average of the circular polarization and I is the throughput of the device.[42] A larger figure of merit will result a better signal to noise ratio. The degree of circular polarization near 800 eV photon energy is 70 and 80 % for a centered obscuration of 0.1 and 0.2 mrad, respectively.

Beamline 3.4

This beamline contains a double crystal monochromator which covers the energy region from 800 to 3500 eV.[43] The range of 800 to 1550 eV, containing the $M_{4,5}$ edges of the rare earths and the $L_{2,3}$ edges of Ni and Cu, is accessible with the beryl (10$\bar{1}$0) crystals. With the Ge(111) or InSb(111) crystals covering the 1800 to 3500 eV range, the $L_{2,3}$ edges of the 4d transition metals and the $M_{2,3}$ edges of the 5d transition metals can be investigated.[44] The Bragg diffraction on the crystals has a pronounced effect upon the degree of circular polarization throughout the available energy range due to the attenuation of the p component at non-grazing angles of incidence on the monochromator crystals. This reduces the achievable degree of circular polarization to less than 0.5. The choice of the crystals used above 2000 eV depends upon the edge to be studied. From 2000 to ~2500 eV it is necessary to use the Ge(111) crystals, whereas InSb is the better choice for higher photon energies. E.g. the Pt $M_{2,3}$ (3027, 2645 eV) and Pd $L_{2,3}$ (3330, 3173 eV) are measured with InSb(111). Below 2000 eV there is no usable overlap between the crystal sets and so therefore no choice possible. Unfortunately there are certain regions in the photon spectrum which are not accessible for MXD work, for example from 1000 to 1200 eV as this coincides with the Brewster angle for the beryl crystals. This prevents the Gd $M_{4,5}$ edge from investigation.

The horizontal acceptance angle of the beam is 5 mrad. This is much larger than on beamline 1.1, which makes a vertically rotating drum as chopper impractical. Instead the polarization chopper in front of the monochromator of 3.4 consists of an off-center paddle mounted on a horizontal rotation axis as illustrated in Fig. 3. Counter weights are attached to provide a smooth rotation.[45]

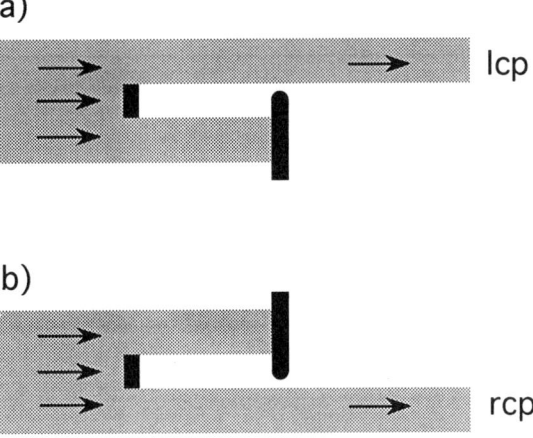

Fig. 3: Polarization chopper of beamline 3.4 consisting of an obscuration plate which removes the central part of the beam and a rotating plate. Position (a) provides left-circularly polarized x-rays and position (b), which is reached after a rotation of 180°, provides right-circularly polarized x-rays.

Modes of Measurement

There are two distinctly different ways to measure the MXD spectrum, *viz.* one can swop either the helicity of the light or the magnetization direction of the sample. Both methods should give the same results in the case of x-ray absorption (but not in the case of photoemission). Changing the helicity of the light requires that both the upper and lower beam must be comparable in every aspect except for their helicity. This is in practise difficult to achieve since the upper and lower beam are travelling over other paths through the optical system. This inevitably causes small differences in degree of polarization, intensity, resolution, and energy between the two beams. These effects can be corrected for a great deal with proper I_0 monitoring together with an energy and polarization calibration using known absorption edges and MXD signals. The second method of measurement, where we change the magnetization direction, is often easier. This can be done with a short current pulse of a few hundred Amps through a magnetization coil around the sample or by mounting identical specimen with opposite magnetization directions onto the sample holder.

The absorption spectra are usually obtained in total electron yield mode using drain current or channeltron detection. The soft x-ray absorption cross-sections are large, which makes transmission measurement, which are common in the hard x-ray region, impractical. Also fluorescence detection is possible, although the yield is low (< 1 %). An advantage is that the applied magnetic field does not interact with the detected particles. A disadvantage is that the signal is not linear with the effect, barring a proper use of the sum rules.

The magnetic field and temperature dependence can be measured by applying an external magnetic field. This gives the possibility to measure the element specific magnetic hysteresis, which is not possible by other methods, such as VSM. Paramagnetic ions in e.g. metallo-proteins can be oriented in a strong magnetic field of several Tesla. At the SRS this is done using a split coil superconducting magnet which allows the incident beam to be either parallel or perpendicular to the magnetic field. Since the soft x-ray region requires high vacuum, the magnet system is outside a vacuum chamber that has the shape of a six-way cross.

3. A SPOT OF THEORY

Fundamental Spectra

A useful way to describe x-ray absorption transitions, such as $2p^63d^n \rightarrow 2p^53d^{n+1}$ is by the hole representation: $l^n(g) \rightarrow cl^{n-1}$, where $|g\rangle$ is the ground state of the configuration l^n with n holes in the l shell. Dipole transitions are allowed to a final state which contains any state f of the configuration l^{n-1} and a core hole with orbital momentum $c = l \pm 1$ and orbital components γ. Omitting the reduced matrix elements, the transition probability for q polarized light is obtained by summing over the m levels of the ground state as

$$I_q = \sum_m \langle g|l_m|f\rangle \langle f|l_m^\dagger|g\rangle \sum_\gamma \begin{pmatrix} l & 1 & c \\ -m & q & \gamma \end{pmatrix}^2 , \qquad (1)$$

where $l_m^\dagger$ and l_m create and destroy a hole from the shell l with azimuthal quantum number m. We assume cylindrical symmetry so that there are no cross terms mm'. The components of the polarization vector of the light are $q = -1, 0, +1$ which denote right-circularly, z linearly, and left-circularly polarized radiation, respectively. The coefficient in the round brackets is a 3-j symbol that gives the coupling of the three angular momentum vectors lm, $1q$ and $c\gamma$.[46-48] The rotation invariance of the cylinder symmetry requires that $\gamma = m-q$.

In Eq.(1) we can consider

$$I(m) \equiv \langle g | l_m | f \rangle \langle f | l_m^\dagger | g \rangle , \qquad (2)$$

as the fundamental properties of the system which contain the information about the one-particle properties connected with the shell l of the atom in the state $|g\rangle$ and also the one-particle excitations to the final states $|f\rangle$. Thus we can write Eq. (1) as

$$I_q = \sum_m I(m) \, t_{mq} , \qquad (3)$$

where t_{mq} is the transition probability for a magnetic sublevel m with light of polarization q:

$$t_{mq} \equiv \sum_\gamma \begin{pmatrix} l & 1 & c \\ -m & q & \gamma \end{pmatrix}^2 = \begin{pmatrix} l & 1 & c \\ -m & q & m-q \end{pmatrix}^2 . \qquad (4)$$

The squared 3-j symbol can be written in powers of m as [49]

$$t_{mq} = \tfrac{1}{3} A_{0lc} + A_{1lc} \frac{qm}{2l} + A_{2lc} (\tfrac{3}{2} q^2 - 1) \frac{m^2 - \tfrac{1}{3} l(l+1)}{l(2l-1)} . \qquad (5)$$

where the coefficients are given as

$$A_{x,l,l-1} = \frac{1}{2l+1} \qquad \forall\, x , \qquad (6)$$

$$A_{0,l,l+1} = \frac{1}{2l+1} , \qquad (7)$$

$$A_{1,l,l+1} = -\frac{l+1}{l(2l+1)} , \qquad (8)$$

$$A_{2,l,l+1} = \frac{(l+1)(2l+3)}{l(2l-1)(2l+1)} . \qquad (9)$$

Thus e.g. A_{xlc} is 1/5 for the $3d^n \to 2p^5 3d^{n+1}$ and 1/7 for the $4f^n \to 3d^9 4f^{n+1}$ transition. Using Eq. (3) and (5) we can make new linear combinations J^x for the transition probabilities

$$J^0 \equiv I_1 + I_0 + I_{-1} = A_{0lc} \sum_m I(m) \equiv A_{0lc} I^0 , \qquad (10)$$

$$J^1 \equiv I_1 - I_{-1} = \frac{A_{1lc}}{l} \sum_m m\, I(m) \equiv A_{1lc} I^1 , \qquad (11)$$

$$J^2 \equiv I_1 - 2 I_0 + I_{-1} = \frac{A_{2lc}}{\tfrac{1}{3} l(2l-1)} \sum_m [m^2 - \tfrac{1}{3} l(l+1)] I(m) \equiv A_{2lc} I^2 . \qquad (12)$$

The fundamental spectra I^0, I^1 and I^2 are the isotropic spectrum, the circular dichroism and the linear dichroism, respectively.

Integrated Signals

We can now derive the sum rules for the I^x spectra. The one-electron properties of the ground state are obtained by summing over all $|f\rangle$ in Eq. (2). Using the closure relation this yields the occupation numbers n_m of the m levels

$$\rho(m) \equiv \sum_f I(m) = \sum_f \langle g|l_m|f\rangle\langle f|l_m^\dagger|g\rangle = \langle g|l_m l_m^\dagger|g\rangle$$

$$= \langle g|n_m|g\rangle = \langle n_m\rangle \ . \tag{13}$$

Thus, the $\rho(m)$'s tell us which electron states in $|g\rangle$ are occupied or unoccupied. Using Eq. (3) and (13) the integrated photoemission for q polarized light is given as

$$\rho_q = \sum_{fm} \rho(m)\, t_{mq} = \sum_m \langle n_m\rangle\, t_{mq} \ . \tag{14}$$

We can make the linear combinations $\rho^x = \sum_f I^x$ as presented below

$$\rho^0 \equiv \frac{\rho_1 + \rho_0 + \rho_{-1}}{A_{0lc}} = \sum_m \langle n_m\rangle = \langle n_l\rangle \ , \tag{15}$$

which means that the integrated isotropic signal is equal to the number of holes in the l shell. The importance of the isotropic signal is that it can be used for normalization [c.f. Eq. (21) and (22)]. Because normally we cannot perform absolute intensity measurements we can only obtain properties per hole.

$$\rho^1 \equiv \frac{\rho_1 - \rho_{-1}}{A_{1lc}} = l^{-1} \sum_m \langle n_m\rangle\, m = l^{-1} \langle L_z\rangle \ , \tag{16}$$

thus the integrated circular dichroism is equal to the orbital magnetic moment,[29] and

$$\rho^2 \equiv \frac{\rho_1 - 2\rho_0 + \rho_{-1}}{A_{2lc}} = \frac{1}{\frac{1}{3}l(2l-1)} \sum_m \langle n_m\rangle [m^2 - \tfrac{1}{3}l(l+1)] = \frac{1}{\frac{1}{3}l(2l-1)} \langle Q_{zz}\rangle \ , \tag{17}$$

thus the integrated linear dichroism is proportional to the quadrupole moment

$$Q_{zz} = \sum_i l_z^2(g) - \tfrac{1}{3}l(l+1) \ . \tag{18}$$

where l_z is the one-electron operator acting only on l-shell functions.

There are also sum rules which relate the isotropic branching ratio to the expectation value of the spin-orbit operator [50,51] and the branching ratio of the MXD to the spin magnetic moment.[52] The derivation of the latter sum rule is too complicated to present here. The orbital and spin magnetic moments obtained from MXD and the use of the sum rules are in good accord [53] with polarized neutron scattering results [54] and Einstein-de Haas gyromagnetic ratio measurements.[55]

Illustration of the sum rules

Here we will illustrate the orbital magnetization sum rules for the transition $3d^n \to 2p^5 3d^{n+1}$.[56,57] Fig. 4 shows the allowed transitions for a hole from the magnetic sublevels m_d (=m) of the valence shell to the sublevels m_p of the core level. The transition probabilities can be obtained from the squared 3-j symbol given in Eq. (4) and the results are collected in Fig. 5a. The total intensities are obtained by multiplying these transition probabilities with the occupation numbers $\langle n_m \rangle$ of the ground state.

It is clear from Fig. 5a that a ground state with a positive value of $M = \Sigma m$ has a preference for left circularly polarized light because the selection rule $q = -\Delta m = +1$ gives a higher intensity for levels of high m value. A ground state with a negative value M has a preference for right circularly polarized light because the selection rule $q = -\Delta m = -1$ gives a higher intensity for levels of low m value. A ground state with a small M value has a preference for Z-linearly polarized light since $q = -\Delta m = 0$ gives more intensity for the levels where |m| is small.

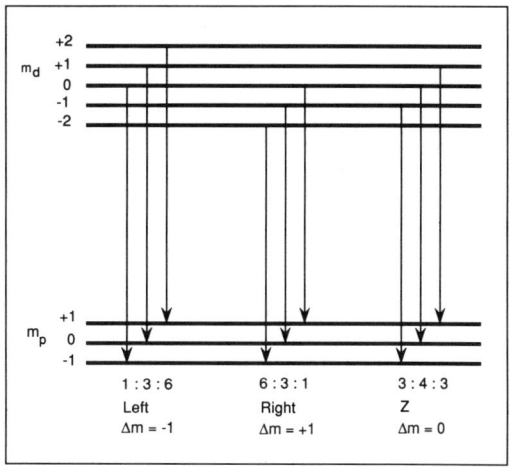

Fig. 4: Transition probabilities for the magnetic sublevels of the d valence hole to the p core sublevels with left, right and Z-linear perpendicular polarized x-rays.

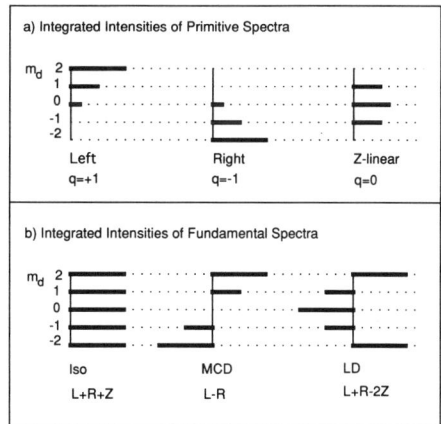

Fig. 5: a) The integrated intensities of the magnetic ground state levels m for $p \to d$ absorption with left-circular (L), linear perpendicular (Z), and right-circular (R) polarized x-rays. b) The resulting integrated intensities of the isotropic spectrum (Iso), magnetic circular dichroism (MCD) and linear dichroism (LD).

For the fundamental spectra the transition probabilities can be obtained by taking the linear combinations, the sum over all polarization directions for the isotropic spectrum, the difference between left and right circularly polarized light for the MCD, and the sum of left and right minus twice the z polarization for the linear dichroism. The results are shown in Fig. 5b. The isotropic spectrum has a constant intensity. It measures the occupation number $\langle n_l \rangle$ of an arbitrary initial state, and is independent of the magnetic moment [c.f. Eq. (15)]. The signal of the MCD is linear in m, hence for the initial state it is proportional to the orbital magnetic moment [c.f. Eq. (16)]. The intensity distribution of the linear dichroism has the shape of a parabola, and is thus quadratic in m [c.f. Eq. (17)], measuring the quadrupole moment. The integrated intensities of the fundamental spectra give the expectation values of the orbital magnetic moment and the quadrupole moment. E.g. for a single hole in the $m = -2, -1, 0, +1, +2$ level Eq. (16) and (17) give $L_z = 2, 1, 0, -1, -2$ and $Q_{zz} = 2, -1, -2, -1, 2$, respectively. When all m levels are occupied the dichroism is zero. As for the primitive spectra, the total intensities are obtained when we multiply the transition probabilities by the occupation numbers $\langle n_m \rangle$ of the ground state. E.g. values for a ground state d^8 $^3P_4$ ($M_J = -4$) can be obtained by substituting $\langle n_{-2} \rangle = \langle n_{-1} \rangle = \langle n_0 \rangle = 0$ and $\langle n_1 \rangle = \langle n_2 \rangle = 1$ into Eqs. (15-17) which yields $\langle n_l \rangle = \Sigma \langle n_m \rangle = 2$, $\langle L_z \rangle = \Sigma \langle n_m \rangle m = -3$, and $\langle Q_{zz} \rangle = \Sigma \langle n_m \rangle [m^2 - \frac{1}{3}l(l+1)] = 1$. The negative value of L_z corresponds to an alignment parallel with S_z, which is negative for more than half filled shells. The positive value of Q_{zz} means that the high values of $|m|$ are occupied which makes the hole distribution flat in the xy plane. Note that Q_{zz} can be non-zero in ferro- and antiferromagnetics as well as in an anisotropic crystal field.

4. APPLICATION OF MXD

Co/Pd Multilayers

The understanding of the magnetic anisotropy in thin films and multilayers is an interesting scientific problem which has great technological importance in areas such as magnetic and magneto-optical recording. A spectacular development has been the finding of perpendicular magnetic anisotropy in magnetic multilayers. This was observed for the first time in 1985, when magnetic multilayers prepared by alternately depositing a few atomic layers of Co and Pd were found to exhibit a perpendicular orientation of the magnetization.[58] Since then this phenomenon has been observed in many multilayered systems. Co/Pd multilayers have already been shown to be promising candidates for magneto-optic recording media. When the thickness of the individual Co layers is increased, the magnetization eventually rotates to an in-plane direction. The origin of the perpendicular magnetocrystalline anisotropy (PMA) of metallic multilayers is not well understood. The influence of reduced symmetry at the interface as suggested by Neel, the interface structure and strain have been considered as possible origins.[59] Since the microscopic origin of magnetocrystalline anisotropy is the spin-orbit interaction, which depends on both the spin and orbital angular momenta, an independent measurement of these quantities is highly desirable.

MXD can be used to understand the origin of the perpendicular magnetization and to find a possible relation with the orbital magnetic moment. Fig. 6 shows the strong MXD in a [Co(2Å)Pd(10Å)]$_{20}$ multilayer at the Co $L_{2,3}$ edges measured on beamline 1.1. Since the method is element selective we also have a unique probe to measure the local hysteresis at each site.[60] This is a large advantage over a vibrating sample magnetometer (VSM) which can not distinguish between different elements.

The values of the orbital and spin magnetic moment and the spin-orbit interaction of magnetic multilayers and the corresponding alloys with perpendicular magnetization have been determined with the sum rules,[50-51] and compare well with the values obtained by Wu *et al.*[61] The value of the orbital magnetic moment of Co in the Co/Pd multilayers is much larger than in Co metal and depends on the Pd layer thickness. The corresponding alloy with the same composition as the Co/Pd multilayers with perpendicularly orientated magnetization shows a magnetic orbital moment similar in value as for Co metal. It was also found that L_z had a relation with the magnetocrystalline anisotropy energy.[62]

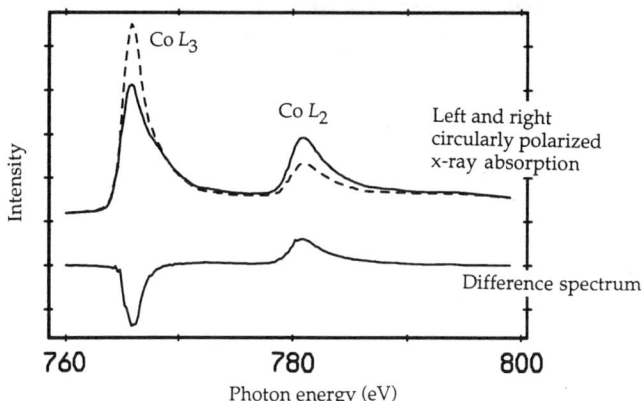

Fig. 6: X-ray absorption for left and right circularly polarized x-rays at the Co $L_{2,3}$ edge of a (2ÅCo + 10ÅPd)x20 multilayer with perpendicular magnetization and the resulting MXD.

Co/Ni multilayers

PMA has been observed in particular for several Co/X multilayers, where X is a nonmagnetic metal such as Pd, Pt, Au, Ir. A slight Pd thickness dependence was found which can be interpreted as being due to the magnetic polarization of the Pd layers close to the interface. This led to the question as to whether a multilayer with Pd replaced by ferromagnetic Ni, which has about the same number of valence electrons, will also have a perpendicular magnetization. The magneto anisotropic energy (MAE) has to be sufficiently large to overcome the demagnetization energy favouring an in-plane magnetization, which is increased compared to the case that only one type of layer is magnetic. For this reason primarily multilayers consisting of one of the magnetic elements Fe or Co together with a nonmagnetic late $3d$, $4d$ or $5d$ transition metal element have been studied extensively, and the possibility of finding a perpendicular orientation of the multilayer magnetization in which both elements are magnetic has been neglected or dismissed. However recently PMA has been predicted and confirmed in Co/Ni multilayers.[63-65]

Fig. 7 shows the Ni $L_{2,3}$ absorption spectra of a perpendicularly magnetized multilayer consisting of 20 layers of 1.8 Å Co/6 Å Ni and capped with 20 Å Au. These layers were deposited on a Si(111) substrates covered with a 200 Å Au buffer layer. The spectra were taken on beamline 3.4 at the SRS using the beryl monochromator providing a degree of circular polarization of $P = 0.5$ and an energy resolution of 0.4 eV at the Ni edge. The sum and difference signals for right and left circularly polarized light integrated over the L_2 or L_3 edge,

$$T_{2,3} = \int_{L_{2,3}} I_1 + I_{-1}, \tag{19}$$

$$D_{2,3} = \int_{L_{2,3}} I_1 - I_{-1}, \qquad (20)$$

are given in Table 1. For the sum signals we used a step function for the background with the step located at the maximum of the peak (see Fig. 7).

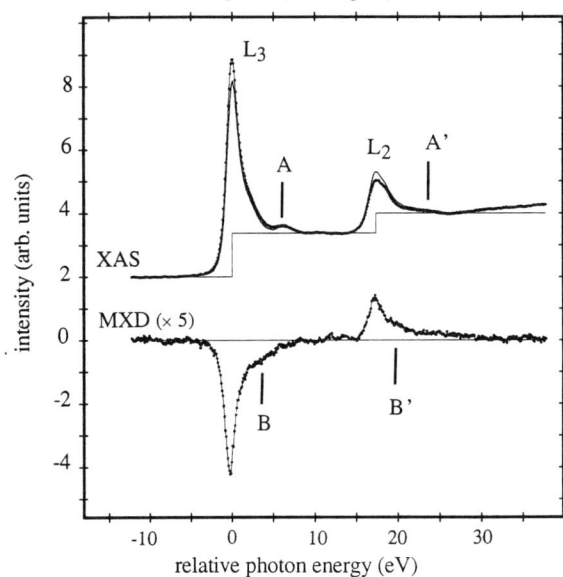

Fig. 7: X-ray absorption for left and right circularly polarized x-rays at the Ni $L_{2,3}$ edge of a (1.6ÅCo + 6ÅNi)x20 multilayer with perpendicular magnetization and the resulting MXD.

The isotropic spectrum has a branching ratio $T_3/(T_2+T_3)$ of 0.74 and a satellite $A(A')$ at an energy of 6.4 eV. The L_3 and L_2 structures have opposite signs in the MXD spectrum where the intensity ratio is -2.12 : 1. From comparison with calculated results it can be shown that the L_3 and L_2 structures consist of a main peak with predominantly $\underline{p}d^{10}$ character and a satellite peak of $\underline{p}d^9$ character. The multiplet structure of the satellite shows a doublet with maxima at the energy positions of peak B and A. There may be extra intensity due to p to s transitions, especially at peak $A(A')$. A small background contribution may be due to the Co $L_{2,3}$ MXD at 60 and 80 eV lower photon energy.

Assuming that the linear dichroism is small, i.e. $I_0 = (I_1+I_{-1})/2$, the orbital magnetic moment per hole for the $p \to d$ absorption is obtained using Eq. (15-16) as

$$\left\langle \frac{L_z}{n_l} \right\rangle = \frac{4}{3} \frac{D_2+D_3}{T_2+T_3} \frac{1}{P}. \qquad (21)$$

For the spin magnetic moment S_z together with the magnetic dipole moment T_z per hole it can be shown that [52]

$$\left\langle \frac{S_z}{n_l} \right\rangle + \frac{7}{2}\left\langle \frac{T_z}{n_l} \right\rangle = \frac{2D_2-D_3}{T_2+T_3} \frac{1}{P}. \qquad (22)$$

The quantities derived from the integrated intensities are given in Table 1. The magnetic dipole operator T provides a measure of the anisotropy of the field of the spins when the atomic cloud is distorted, either by the spin-orbit interaction or the crystal field. In spherical symmetry the component along the field axis T_z is small but it becomes large when there is a crystal field. The value of the orbital magnetic moment for the multilayer is somewhat larger than the bulk value of metallic nickel (Table 1). We can assess the influence of the crystal field from the value of the magnetic dipole operator by assuming that $S_z/n_l = 0.5$, which is the value when the holes are only in the minority band. Then we find that $T_z/n_l = -0.075$, indicating a strong anisotropy.

The quantities derived from Eq. (21) and (22) depend on the degree of circular polarization and the degree of the perpendicular magnetization. Furthermore, we have neglected the influence of linear dichroism and the uncertainty in the background subtraction of the sum spectrum. For the ratio of the orbital to spin magnetic moment we can obtain a more accurate value, which is independent of the number of holes:

$$R = \left\langle \frac{L_z}{S_z + \frac{7}{2}T_z} \right\rangle = \frac{4}{3}\frac{D_2+D_3}{2D_2-D_3} \,. \tag{23}$$

Table 1: Measured integrated intensities for the Ni $L_{2,3}$ edges together with the physical quantities derived using the sum rules for Co/Ni multilayer compared to Ni metal where $n_l = 0.6$.

	integrated intensities				quantities derived		
	T_3	T_2	D_3	D_2	$\left\langle \frac{L_z}{n_l} \right\rangle$	$\left\langle \frac{S_z}{n_l} \right\rangle + \frac{7}{2}\left\langle \frac{T_z}{n_l} \right\rangle$	R
1.6ÅCo/6ÅNi	13.54	4.78	-1.132	0.535	0.087	0.241	0.361
Ni metal					0.083	0.430	0.193

5. FURTHER PROSPECTS

Another important application is magnetic dichroism in photoemission which gives information about the correlation between the valence spin, the core hole orbital moment and the core hole spin.[66-68] For photoemission from an open shell the sum rules give the expectation values of the orbital magnetic moment and quadrupole moment. Analysis of the angular distribution gives the higher magnetic moments. Therefore, polarized photoemission measurements are important for the understanding of the magnetic properties of transition metal and rare earth materials, such as for ferromagnetic nickel where measurements indicate an enhanced orbital magnetic moment at the surface. The circular dichroism gives also additional information about the spectra which allow us to distinguish between features which originate from exchange interaction and hybridization, such as in the Fe 3s photoemission. A further development is magnetic circular polarization of resonant photoemission, where the photoemission decay of each structure in the x-ray absorption spectrum is measured with circular polarized x-rays.[69]

Existing techniques, such as EXAFS, reflexafs, photoelectron diffraction, x-ray standing wave and photoelectron holography can be easily made sensitive for specific magnetic atoms by using circularly polarized x-rays. These techniques are not limited to (anti)-ferro(i)magnetic ions. E.g. applying a 5 Tesla magnetic field enables us to distinguish between paramagnetic and diamagnetic transition metal ions. Clearly, the possibilities are almost unlimited.

6. CONCLUSIONS

The unique attraction of polarized photon probes is the ability to separate spin and magnetic moments, which has long been recognized as the 'Holy Grail' of magnetism. This division is essential for a complete understanding of many classes of materials, including many actinide mixed-valence, and heavy fermion systems, where the spin and orbital part of the magnetic moment play an important role in the anisotropic magnetic properties, such as the magneto-crystalline effect, easy-direction of magnetization, magnetostriction and coercivity. MXD can be applied generally to transition metal, rare earth and actinide metals, alloys and compounds with their large variety of magnetic structures. The surface sensitivity combined with the element and site specificity of MXD make it immensely valuable in the study of magnetic thin films and multilayers. The special magnetic properties characteristic of thin films derive from modifications in symmetry and coordination of the atoms primarily at the surface and interfaces. Materials with enhanced spin and orbital moments, perpendicular anisotropy and antiferromagnetic coupling between ferromagnetic layers across a non-magnetic spacer are of great interest to the magnetic recording industry. The determination of the polarization of a non-magnetic metal at an interface with a ferromagnetic layer will provide a useful insight into the coupling mechanism in such materials. Further research applications are found in systems where the localized character of the valence electrons changes due to dimensionality effects associated with reduced coordination of the magnetic atoms, e.g. surface magnetic moments are enhanced over the bulk values due to narrowing of the $3d$ band width, or by substrate effects due to interaction between differing materials.

REFERENCES

1. P.M. Platzman and N. Tzoar, Phys. Rev. B 2, 3356 (1970).
2. F. deBergevin and M. Brunel, Phys. Lett. 39A, 141 (1972).
3. M. Blume and D. Gibbs, Phys. Rev. B 37, 1779 (1988).
4. G. van der Laan, J. Phys. Condens. Matter 3, 1051 (1991); Phys. Rev. Lett. 66, 2527 (1991).
5. J. Bonarski and J. Karp, J. Phys. Condens. Matter 1, 9261 (1989).
6. D.P. Siddons, M. Hart, Y. Amemiya and J.B. Hastings, Phys. Rev. Lett. 64, 1967 (1990).
7. J.P. Hannon, G.T. Trammell, M. Blume and D. Gibbs, Phys. Rev. Lett. 61, 1245 (1988).
8. B.T. Thole and G. van der Laan, Phys. Rev. Lett. 67, 3306 (1991).
9. D. Gibbs, D.R. Harshman, E.D. Isaacs, D.B. McWhan, D. Mills, and C. Vettier, Phys. Rev. Lett. 61, 1241 (1988).
10. Jin Luo, G. T. Trammell, and J.P. Hannon, Phys. Rev. Lett. 71, 287 (1993).
11. P. Strange, P.J. Durham, and B.L. Gyorffy, Phys. Rev. Lett 67, 3590 (1991).
12. J.L. Erskine and E.A. Stern, Phys. Rev. B 12, 5016 (1975).
13. E. Keller and E.A. Stern, *EXAFS and Near Edge Structure III* (Springer, Berlin, 1984), p. 507.
14. B.T. Thole, G. van der Laan and G.A. Sawatzky, Phys. Rev. Lett. 55, 2086 (1985).
15. G. van der Laan, B.T. Thole, G.A. Sawatzky, J.B. Goedkoop, J.C. Fuggle, J.M. Esteva, R.C. Karnatak, J.P. Remeika, and H.A. Dabkowska, Phys. Rev. B 34, 6529 (1986).
16. J.B. Goedkoop, J.C. Fuggle, B.T. Thole, G. van der Laan, and G.A. Sawatzky, J. Appl. Phys. 64, 5595 (1988).
17. F. Baudelet, E. Dartyge, A. Fontaine, C. Brouder, G. Krill, J.P. Kappler and M. Piecuch, Phys. Rev. B 43, 5857 (1991).
18. G. Schütz, W. Wagner, W. Wilhelm, P. Kienle, R. Zeller, R. Frahm, and R. Materlik, Phys. Rev. Lett. 58, 737 (1987).
19. S.P. Collins, M.J. Cooper, A. Brahmia, D. Laundy, and T. Pitkanen, J. Phys. Condens. Matter 1, 323 (1989).
20. H. Ebert, P. Strange, and B.L. Gyorffy, Z. Phys. B 73, 77 (1988).
21. H. Ebert and R. Zeller, Phys. Rev. B 42, 2744 (1990).
22. P. Carra and M. Altarelli, Phys. Rev. Lett. 64, 1286 (1990).
23. C.T. Chen, F. Sette, Y. Ma, and S. Modesti, Phys. Rev. B 42, 7262 (1990).
24. C.T. Chen, N.V. Smith, and F. Sette, Phys. Rev. B 43, 6785 (1991).
25. T. Jo and G.A. Sawatzky, Phys. Rev. B 43, 8771 (1991).
26. G. van der Laan and B.T. Thole, Phys. Rev. Lett. 60, 1977 (1988).

27. T. Koide, T. Shidara, H. Fukutani, K. Yamaguchi, A. Fujimori, and S. Kimura, Phys. Rev. B 44, 4697 (1991).
28. G. van der Laan and B.T. Thole, Phys. Rev. B 43, 13401 (1991).
29. B.T. Thole, P. Carra, F. Sette, and G. van der Laan, Phys. Rev. Lett. 68, 1943 (1992).
30. P. Kuiper, B.G. Searle, P. Rudolf, L.H. Tjeng, and C.T. Chen, Phys. Rev. Lett. 70, 1549 (1993).
31. M. Sacchi, O. Sakho, and G. Rossi, Phys. Rev. B 43, 1276 (1991).
32. F. Sette, C.T. Chen, Y. Ma, S. Modesti, and N.V. Smith, AIP Conf. Proc. No. 215, 787 (1990).
33. P. Rudolf, F. Sette, L.H. Tjeng, G. Meigs, and C.T. Chen, J. Magn. Magn. Mat. 109, 109 (1992).
34. Ph. Sainctavit, D. Lefebvre, Ch. Cartier dit Moulin, C. Laffon, Ch. Brouder, G. Krill, J.Ph. Schillé, J.P. Kappler, and J. Goulon, J. Appl. Phys. 72, 1985 (1992).
35. J.B. Goedkoop, B.T. Thole, G. van der Laan, G.A. Sawatzky, F.M.F. de Groot, and J.C. Fuggle, Phys. Rev. B 37, 2086 (1988).
36. T. Jo and S. Imada, J. Phys. Soc. Jpn. 59, 3358 (1990).
37. H. Ogasawara, A. Kotani, and B.T. Thole, Phys. Rev. B 44, 2169 (1991).
38. M.A. Green, Nucl. Instrum. Methods Phys. Res. A 319, 83 (1992).
39. G. van der Laan, M. A. Hoyland, M. Surman, C.F.J. Flipse, and B.T. Thole, Phys. Rev. Lett. 69, 3827 (1992).
40. H. Krakauer, A.J. Freeman, and E. Wimmer, Phys. Rev. B 28, 610 (1983).
41. M. Surman, I. Cragg-Hine, J. Singh, B. Bowler, H.A. Padmore, D. Norman, A.L. Johnson, A. Atrei, W.K. Walter, D.A. King, R. Davis, K.G. Purcell, and G. Thornton, Rev. Sci. Instrum. 63, 1341 (1992).
42. C.T. Chen, Rev. Sci. Instrum. 63, 1229 (1992).
43. A.A. MacDowell, J.B. West, G.N. Greaves, and G. van der Laan, Rev. Sci. Instrum. 59, 843 (1988).
44. G. van der Laan and H.A. Padmore, Nucl. Instrum. Method. Phys. Res. A 291, 225 (1990).
45. A. Smith, unpublished.
46. D.M. Brink and G.R. Satchler, *Angular Momentum* (Oxford University Press, London, 1962).
47. A.P. Yutsis, I.B. Levinson and V.V. Vanagas, *Mathematical Apparatus of the Theory of Angular Momentum* (Israel Program for Scientific Translation, Jerusalem, 1962).
48. D.A. Varshalovich, A.N. Moskalev, and V.K. Khersonskii, *Quantum Theory of Angular Momentum* (World Scientific, Singapore, 1988).
49. B.T. Thole and G. van der Laan, Phys. Rev. Lett. 70, 2499 (1993).
50. G. van der Laan and B.T. Thole, Phys. Rev. Lett. 60, 1977 (1988).
51. B.T. Thole and G. van der Laan, Phys. Rev. B 38, 3158 (1988).
52. P. Carra, B.T. Thole, M. Altarelli, and X. Wang, Phys. Rev. Lett. 70, 694 (1993).
53. C.T. Chen, Y.U. Idzerda, H.J. Lin, N.V. Smith, G. Meigs, E. Chaban, G. Ho, E. Pellegrin, and F. Sette, Phys. Rev. Lett., submitted.
54. M.B. Stearns, in *Magnetic Properties of 3d, 4d, and 5d Elements, Alloys and Compounds*, Ed. K.H. Hellwege and O. Madelung, Landolt-Bornstein, new Series, Vol. III/19a (Springer-Verlag, Berlin, 1986); and references therein.
55. D. Bonnenberg, K.A. Hempel, and H.P.J. Wijn, in *Magnetic Properties of 3d, 4d, and 5d Elements, Alloys and Compounds*, Ed. K.H. Hellwege and O. Madelung, Landolt-Bornstein, new Series, Vol. III/19a (Springer-Verlag, Berlin, 1986); and references therein.
56. G. van der Laan and B.T. Thole, Phys. Rev. B 42, 6670 (1990).
57. G. van der Laan, J. Phys. Soc. Jpn. 63, 2059 (1994).
58. P. Carcia, A. Meinhaldt, and A. Suna, Appl. Phys. Lett. 47, 178 (1985).
59. L. Néel, J. Phys. Rad. 15, 225 (1954).
60. C.T. Chen, Y.U. Idzerda, H.J. Lin, G. Meigs, A. Chaiken, G.A. Prinz, and G.H. Ho, Phys. Rev. B 48, 642 (1993).
61. Y. Wu, J. Stöhr, B.D. Hermsmeier, M. G. Samant, and D. Weller, Phys. Rev. Lett. 69, 2307 (1992).
62. C.F.J. Flipse, J.J. de Vries, A. Partridge, W.J.M. de Jonge, G. van der Laan, M. Surman, F.J.A. den Broeder, and M.T. Johnson, to be published.
63. G.H.O. Daalderop, P.J. Kelly, and F.J.A. den Broeder, Phys. Rev. Lett. 68, 682 (1992).
64. M.T. Johnson, J.J. de Vries, N.W.E. McGee, J. aan de Stegge, and F.J.A. den Broeder, Phys. Rev. Lett. 69, 3575 (1992).
65. F.J.A. den Broeder, H.W. van Kesteren, W. Hoving, and W.B. Zeper, Appl. Phys. Lett. 61, 1468 (1992).
66. B.T. Thole and G. van der Laan, Phys. Rev. B 44, 12424 (1991).
67. G. van der Laan and B.T. Thole, Phys. Rev. B 48, 210 (1993).
68. B.T. Thole and G. van der Laan, Phys. Rev. B 49, 9613 (1994).
69. L.H. Tjeng, C.T. Chen, P. Rudolf, G. Meigs, G. van der Laan, and B.T. Thole, Phys. Rev. B 48, 13378 (1993).

X-RAY MAGNETIC SCATTERING

Andrew J. Rollason

Department of Physics
Keele University
Keele
Staffordshire
ST5 5BG

INTRODUCTION

The study of magnetism and magnetic structures has traditionally been the preserve of neutron scattering where the magnetic interaction cross section is of comparable size to the nuclear scattering cross section. The magnetic interaction of photons with matter is relatively weak at non-relativistic energies and becomes stronger with energy, the amplitude scaling according to $\hbar\omega/mc^2$. However the advent of modern synchrotrons means that this weak cross section is more than compensated by extremely intense fluxes of photons. Over the past 10 years many magnetic studies of materials have therefore been initiated using photons as the probing particles and the special characteristics of synchrotron radiation and the cross section have been exploited to yield exciting new results which cannot readily be provided by neutron scattering. The limitations of neutron scattering include high absorption in specific materials (eg. gadolinium), inability to separate orbital and spin contributions to the magnetic ordering, the need for large samples (cm^3), little possibility of surface studies and low Q resolution with a corresponding loss of information about the extent of magnetic correlations. X-rays can provide an advantageous, alternative tool to study each of these cases. This article describes the development of this field and the role of modern synchrotrons, the principles and current status of magnetic X-ray scattering and reviews some of the major results in both elastic and inelastic studies of antiferromagnets and ferro-/ferrimagnets.

HISTORICAL

The quantum-relativistic Klein-Nishina formula describes the relativistic scattering of a photon by an electron[1]. In the low energy limit this process converges on classical Thomson charge elastic scattering. At higher energies the semi-relativistic cross section may be expanded as a series of energy terms, the higher of which include the interaction between the photon field and the magnetic dipole moment of the electron. For a description of traditional Compton scattering from unpolarised targets (free electrons), it is sufficient to average (integrate) over all polarisation states of the initial (final) photon-electron pair which is the result given by Klein and Nishina for free electrons. This averaging is not essential however. Forty years ago Lipps & Tolhoek[2] provided a detailed analysis of polarisation dependent cross sections for Compton scattering from spin-polarised electrons. Platzman & Tzoar[3] carried this further to the interaction with bound electrons in real solids and suggested the possibilities of observing spin-momentum distributions in ferromagnets and magnetic diffraction peaks from antiferromagnets.

The first evidence of the observability of magnetic ordering by X-rays was produced in ground-breaking experiments by de Bergevin and Brunel[4] on the antiferromagnets NiO and Fe_2O_3. Remarkably this was achieved with a sealed X-ray tube although the predicted signal from the $(^1/_2\ ^1/_2\ ^1/_2)$ and $(^3/_2\ ^3/_2\ ^3/_2)$ NiO magnetic superlattice peaks was only 10^{-8} of the Thomson scattering. The satellite signals disappeared above the Néel point thus confirming their magnetic origin. The same authors later went on to demonstrate charge-magnetic intereference scattering in ferro- and ferrimagnets using both conventional sources and synchrotron radiation[5].

Sakai & Ono[6] produced the first evidence of spin-momentum distributions in solids by Compton scattering circularly polarised γ rays from iron. The circularly polarised γ rays were themselves produced by embedding a cooled (50 mK) Co^{57} isotope source within a magnetised iron foil. Nuclear interactions with the magnetically ordered iron produced a partial circular polarisation of the emitted γ rays. Nowadays there are easier ways to obtain circular polarisation and the opportunity presented by the availability of synchrotrons was taken up by Cooper et al.[7] with a statistically accurate measurement of the spin-momentum distribution for iron using out-of- orbit circular polarised radiation. Although the X-ray magnetic cross sections are normally very weak, the photon fluxes from modern synchrotrons are sufficiently high to allow them to be studied quantitatively with comparable count rates to those achieved in neutron scattering.

SYNCHROTRON RADIATION POLARISATION CHARACTERISTICS

Apart from an enormous increase in flux, synchrotron radiation also has very special polarisation characteristics which can be exploited in magnetic scattering. In the orbital plane the radiation is nearly 100% linearly polarised while above and below the plane there is a significant proportion of opposed circularly polarised radiation. It is clearly necessary therefore to describe synchrotron-based scattering experiments by a cross section which acounts for a mixed polarisation state of the beam. This is achieved by the use of the

Poincarré-Stokes vector $\mathbf{P} = (P_\xi, P_\eta, P_\zeta)$ which is a fictitious vector representing the proportions of orthogonal linear (P_ξ), circular (P_η) and 45° linear (P_ζ) radiation in the beam. If a basis set of complex polarisation vectors ϵ_1 and ϵ_2 is chosen such that ϵ_1 is normal to the (vertical) scattering plane $\mathbf{k} \times \mathbf{k}'$ and ϵ_2 is parallel to that plane, then

$$P_\xi = \frac{|\epsilon_1|^2 - |\epsilon_2|^2}{|\epsilon_1|^2 + |\epsilon_2|^2}$$

$$P_\eta = \frac{Im(\epsilon_1^* \epsilon_2)}{|\epsilon_1|^2 + |\epsilon_2|^2} \tag{1}$$

$$P_\zeta = \frac{Re(\epsilon_1^* \epsilon_2)}{|\epsilon_1|^2 + |\epsilon_2|^2}$$

$P_\xi = 1$ implies the beam is completely linearly polarised perpendicular to the scattering plane (σ-polarisation) and $P_\xi = -1$ implies linear polarisation in the scattering plane (π-polarisation). Ideally $|\mathbf{P}|=1$ implying a 100% polarised beam but in practice most synchrotrons have only limited emissivity and $|\mathbf{P}|$ is limited to 90-98%. For a synchrotron P_η, at 45° to the horizontal orbit, is identically zero.

Figure 1 shows the vertical profile for linear and circular polarised synchrotron radiation from a dipole bending magnet[8]. The solid curves show P_ξ and P_η and the dashed curve is the total flux. The data refer to the SRS at SERC, Daresbury Laboratory with a total polarisation of 97%. Obviously the degree of circular polarisation increases above orbit but at the same time the total flux is decreasing and for experiments that need circular polarisation a compromise position must be chosen, usually about 1 cm above orbit. An alternative to the out-of-orbit configuration is to use a helical wiggler magnet where the in-orbit axis "sees" an oblique (out-of-orbit) electron beam, a device adopted by the Photon Factory in Japan. This produces a much higher flux of circular polarised radiation.

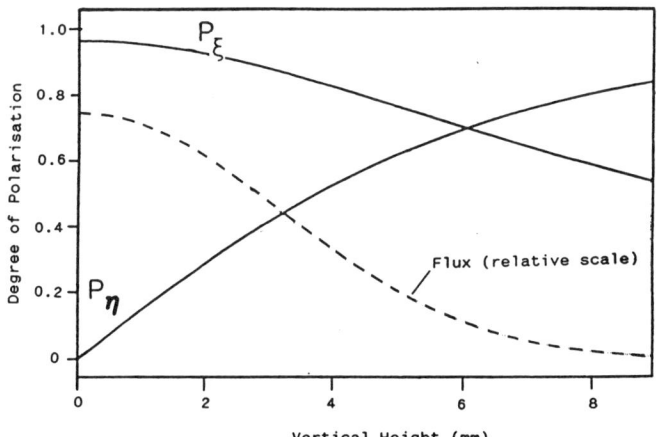

Figure 1. Degree of polarisation as a function of height above the orbital plane. The energy of the incident beam is 12.8 keV. P_ξ and P_η denote the linear and circular polarisation components respectively.

CROSS SECTIONS FOR POLARISED X-RADIATION

Derivations of the cross sections for magnetic scattering have been given in depth by several authors[9-13]. Only general details and phenomenological results which illustrate the subject will be presented here. The magnetic interaction can be simply thought of as the interaction of the photon fields with the magnetic dipole moment of the electron - spin or orbital. At energies sufficiently away from any characteristic excitation of the target ("high energies" in the literature) the amplitude is bilinear with the spin **S** and orbital **L** magnetic moments and the proportionality constants are distinct functions of the photon polarisations and wavevectors (implying geometrically separable contributions). The magnetic scattering can be further distinguished from charge scattering where the polarisation dependence is again different since the former rotates the plane of linear polarisation whereas the latter does not. Circular polarisation introduces an interference between charge and magnetic amplitudes which are otherwise 90° out of phase as does the proximity to a characteristic excitation. At characteristic photon energies a resonance occurs between the scattering and quasi-atomic excitations which can modify the interaction amplitude thus leading to enormously enhanced scattering cross sections. The magnetic cross sections are therefore very closely linked to effects such as spin-dependent absorption, magnetic circular dichroism and anomalous dispersion. These resonant scattering measurements are known as XRES (X-ray Resonant Enhanced Spectroscopy). The technique enables weak magnetic structures to be observed such as in the heavy fermion system URu_2Si_2 (with a total magnetic moment of only 0.04 μ_B) or the critical magnetic scattering above T_N in holmium. Because of the link with characteristic energies, XRES also offers species sensitive scattering which is not possible with neutrons.

Lipps and Tolhoek[2] and Platzman and Tzoar[3] give a relativistic treatment of the scattering of hard X-rays from free electrons. What is required is a low energy treatment of atomic bound electrons including orbital effects. The non-relativistic Hamiltonian for multiple electrons j in a quantised electromagnetic field, describing the bound electron energies, the photon field and the interaction between the two, is

$$H = \Sigma_j \frac{1}{2m}(P_j - \frac{e}{c}A(r_j))^2 + \Sigma_{ij} V(r_{ij}) - e\frac{\hbar}{2mc}\Sigma_j s_j \cdot \nabla \times A(r_j)$$
$$- e\frac{\hbar}{2m^2c^2}\Sigma_j s_j \cdot E(r_j) \times (P_j - \frac{e}{c}A(r_j)) \quad (2)$$
$$+ \Sigma_{k\lambda}\hbar\omega_k(c^+(k\lambda)c(k\lambda) + \frac{1}{2})$$

Scattering occurs in first order for terms in A^2 and second order for terms in A of the interaction part of this Hamiltonian. Blume[10] considers an expansion of the interaction Hamiltonian to second order in energy and from Fermi's Golden Rule derives the equation for the 2nd differential cross section off-resonance. **K** is the momentum transfer **k** - **k'** and

$$\frac{d^2\sigma}{d\Omega dE}\bigg|_{\substack{\lambda \to \lambda' \\ a \to b}} = r_0^2 |\langle b|\Sigma_j e^{iK \cdot r_j}|a\rangle \epsilon \cdot \epsilon'$$
$$- i\frac{\hbar\omega}{mc^2}\langle b|\Sigma_j e^{iK \cdot r_j}M_j|a\rangle|^2 \times \delta(E_a - E_b + \hbar\omega_k - \hbar\omega_{k'}) \quad (3)$$
$$\text{where} \quad M_j = i\frac{K \times p_j}{\hbar\omega^2} \cdot A + s_j B$$

The polarisation factors **A** and **B** are given by

$$A = \epsilon'\times\epsilon \tag{4}$$

$$\text{and } B = \epsilon'\times\epsilon + (\hat{k}'\times\epsilon')(\hat{k}'\cdot\epsilon) - (\hat{k}\times\epsilon)(\hat{k}\cdot\epsilon') - (\hat{k}'\times\epsilon')\times(\hat{k}\times\epsilon)$$

The amplitude in eq(3) consists of two terms - Thomson charge scattering and a purely magnetic scattering term. Examination of the polarisation vectors of the initial and final states reveals that magnetic scattering rotates the plane of polarisation ($\epsilon\times\epsilon'$) while charge scattering remains coplanar ($\epsilon\cdot\epsilon'$). Furthermore the polarisation factors (eq(4)) are geometrically distinct. If eq(1) is evaluated for elastic scattering the cross section more clearly shows its link with the magnetic and charge structures.

$$\frac{d\sigma}{d\Omega} = r_0^2 \left| F(K)\,\epsilon\cdot\epsilon' - i\,\frac{\hbar\omega}{mc^2}\left[\frac{L(K)\cdot A}{2} + S(K)\cdot B\right] \right|^2 \tag{5}$$

where
$$L(K) = \Sigma_j e^{-iK\cdot r_j}\,\frac{i\,K\times P_j}{\hbar k^2}$$

and
$$S(K) = \Sigma_j e^{-iK\cdot r_j}\,\sigma_j\cdot r_j$$

The two magnetic amplitudes are now clearly distinguishable in eq(5) as the orbital magnetisation transform $L(K)$ and the spin transform $S(K)$.

The cross section can be seen to contain contributions from charge (Thomson) scattering, magnetic scattering and an interference term between the two processes. The relative importance of each of these terms is governed primarily by the energy factor $\hbar\omega/mc^2$ which generates the terms in the proportion $1:10^{-6}:10^{-3}$ for soft X-ray energies. Clearly, from this point of view, it is desirable to try to work with the interference term whenever possible. This is not possible for diffraction from antiferromagnets where the magnetic and charge Bragg peaks occur at different points in reciprocal space but is possible for diffraction from ferromagnets as will be discussed later. The interference term though is imaginary and undetectable except as a phase shift unless either the structure factor (charge or magnetic) or a polarisation factor is complex. This can be realised for anomalous scattering near an absorption edge or for a non-centrosymmetric charge distribution or for interactions using circularly polarised light.

When the initial photon beam is a mixed state characterised by the polarisation vector **P** and for a polarisation-insensitive detector the complete cross section away from resonance has been given by Blume and Gibbs[10] in terms of P_ξ, P_η and P_ζ, the complex charge scattering form factor $F = F' + iF''$ and the complex magnetic dipole moment components $L_{123}(K)$ and $S_{123}(K)$ in the axes system with **1** along **k** + **k'**, **2** along **k** × **k'** and **3** along **k** - **k'**. The equation is very long and detailed and will not be reproduced here. It can be be simplified dramatically in most specific applications. As an example we show the cross section for a centro-symmetric magnetic-structure (magnetisation density transform real) expanded to terms of order $\hbar\omega/mc^2$. This includes only the charge and interfence scattering terms, viz.

315

$$\frac{d\sigma}{d\Omega} = \frac{1}{2} r_0^2 |F(K)|^2 [(1+P_\xi) + (1-P_\xi)\cos^2 2\theta_B] \qquad (6)$$

$$-r_0^2 (\frac{\hbar\omega}{mc^2}) \{ [(1+P_\xi) S_2' + \cos 2\theta_B (1-P_\xi)(2\sin^2\theta_B L_2' + S_2')] \times F''\sin 2\theta_B$$

$$+4P_\eta [\cos\theta_B^3 (L_1' + S_1') - \sin\theta_B^2 S_3'] \times F'\sin\theta_B^2$$

$$-4P_\zeta [\sin\theta_B (L_1' + S_1') - \cos\theta_B S_2'] F''\sin\theta_B^3 \cos\theta_B \}$$

The experiments by de Bergevin and Brunel on ferromagnetic iron[5] utilised eq(6). With their unpolarised source the cross section depended only on the anomalous form factor F" and the component of magnetisation (spin only) in the 2 direction, perpendicular to the scattering plane. By reversing the magnetic field they obtained the asymmetric ratio cross section

$$\frac{\Delta I}{I} = 4(\frac{\hbar\omega}{mc^2}) \frac{1+\cos 2\theta_B}{1+\cos^2 2\theta_B} \cdot \frac{F''|S_2|}{|F|^2} \sin 2\theta_B \qquad (7)$$

which has a magnitude of the order of 10^{-3} in agreement with experiment.

If a polarisation sensitive detetor is available, analysis can be made of the final photon beam to distinguish magnetic and charge scattering and orbital / spin contributions. This has been done by Gibbs et al.[12] and Hannon et al.[13] for Ho and is described below.

ANTIFERROMAGNETS

The first system to be studied in detail using X-rays was the spiral antiferromagnet Ho where the structure was already substantially understood from neutron measurements. The promise of X-rays was that the potentially higher Q resolution would enable more precise determination of the magnetic structures. Ho is a spiral antiferromagnet in the temperature ramge 20 K < T < 132 K and develops a ferromagnetic component below 20 K. The magnetic moment rotates by 50° per lattice layer at the Néel temperature and this decreases to 30° at 20 K. The magnetic structure produces magnetic satellites on either side of the charge Bragg peaks at (0 0 21+τ) and the modulation wavevector τ decreases with temperature. The first measurements were made by Gibbs et al[11] using both rotating anode and synchrotron sources. The ratio of the magnetic to charge cross section was predicted by Blume[10] to be of the order 10^{-5} and this was confirmed experimentally. Gibbs followed the development of the magnetic modulation wavevector τ over the entire antiferromagnetic phase and showed that it locks in to a value commensurate with the lattice at T_c. The measurements were repeated later by Bohr et al.[14] with higher resolution to investigate the appearance of a new magnetic feature (called a spin-slip) near the onset of the conical spiral ferromagnetic phase. Their results are presented in fig.1. The precision of these X-ray measurements has since spurred on further high resolution neutron studies which confirm these results.

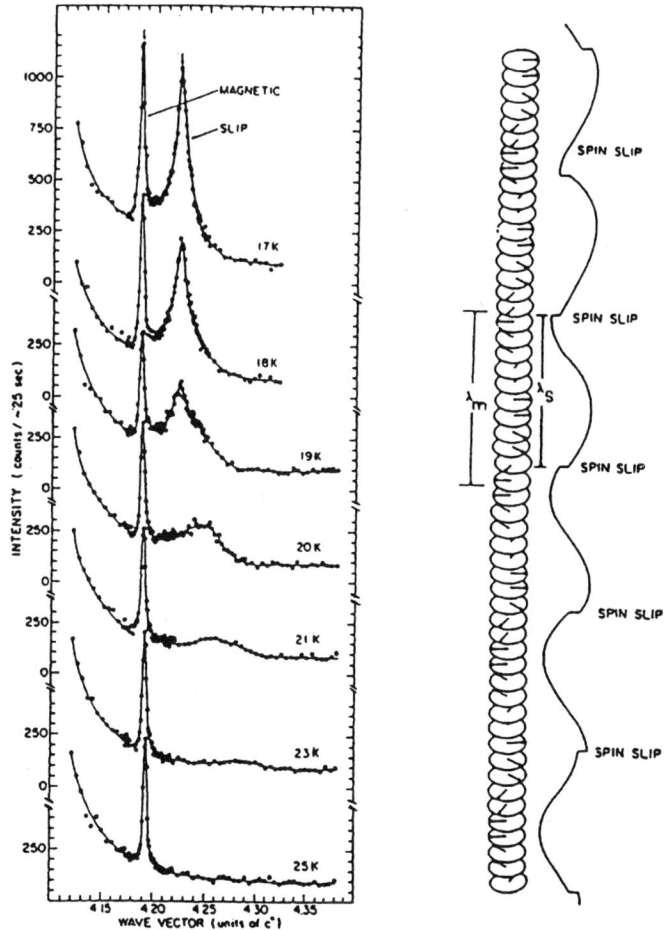

Figure 2. Temperature dependence near the conical-spiral transition of the magnetic satellite reflection (0,0,4+τ) in holmium after X-ray measurements by Bohr et al[14]. An additional magnetic peak is clearly resolved at the spin-slip position due to a distortion from the incipient ferromagnetic component. The real space spiral magnetic structure with the spin-slips is shown at the right.

Non-Resonant High Energy Diffraction

If the photon energy is raised with respect to mc^2 the pure magnetic scattering amplitude in eq(5) increases dramatically with respect to the Thomson term and antiferromagnetic reflections become much more readily observable. This is the motivation for recent measurements on MnF_2 which is antiferromagnetic below $T_N = 67$ K. It was measured at HASYLAB by Lippert et al[15] at an energy of 80 keV, the factor $(\hbar\omega/mc^2)^2$ in the cross section then being 100 times greater than usual. Count rates of 100 per second were obtained for the (300) magnetic reflection with a high signal to background ratio. The higher energies cause some practical problems for the measurements due to low absorption and the closeness together of reflections in angle space (multiple scattering). Diffractometers need to be rather long to achieve adequate resolution and separation of close reflections.

Resonant Diffraction (XRES)

The field which has so far made the most rapid progress is resonant X-ray magnetic diffraction. Many difficulties exist with high energy scattering experiments and the magnetic reflections are always several orders of magnitude smaller than the Thomson peaks. Tuning the photon energy to an absorption edge in an antiferromagnet causes the elastic excitation to include electric multipole resonances which greatly enhance the magnetic scattering amplitude by a factor

$$F_{LM}(E) = \sum_{ab} \frac{P_a P_a(b) \Gamma_x(aMb;EL)}{2(E_a - E_b - E) - i\Gamma} \tag{8}$$

where $\Gamma_x(aMb;EL)$ represents the electric multipole (multipolarity L) transition matrix element from state $|a\rangle$ to $|b\rangle$ and P_a, $P_a(b)$ represent the density of occupied and unpaired empty states respectively. Γ is the energy width of the excited state $|b\rangle$. Ho was again the test system to be studied with the resonance technique[12,13]. At its L_{III} absorption edge (7.847 keV) there is a 3p to 5d electric dipole transition which enhances the cross section by a factor of 50. Weaker electric quadrupole transitions (from 3p to 4f in the case of holmium) are also possible. In other materials much greater resonances can be produced. Uranium Arsenide has an M_{IV} resonant enhancement of the order of 10^6 ! Thus the magnetic scattering peaks can be made as large as the charge peaks themselves and their detection is simplified. To get a large magnetic enhancement requires an elecric dipole transition to a narrow, unpaired electron band. This unfortunately places restrictions on the energy $E=E_a-E_b$ which for many materials means that the wavelength is too long to be Bragg reflected. The energetics are well matched to 4d, 5d and 5f systems but of these only the 5f actinides are of real interest in magnetism. These systems have been studied in detail.

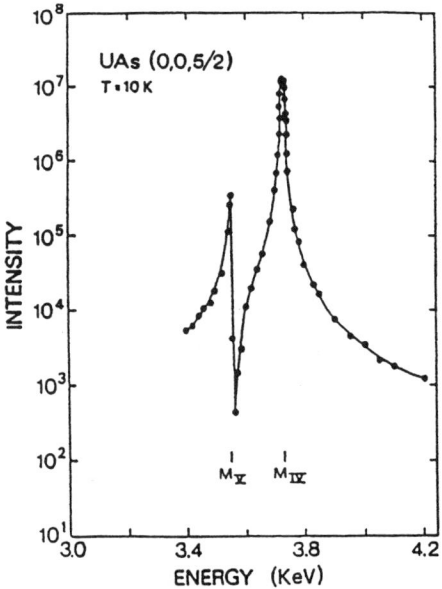

Figure 3. The energy dependence of the the resonant enhancement of the reflected intensity of the (0,0,5/2) magnetic satellite in UAs at the M4 and M5 edges. Note the log scale. Data are taken from McWhan et al[16].

Figure 3 shows the energy dependence of the resonant enhancement of the magnetic satellite $(0,0,{}^5/_2)$ of UAs through the M_{IV} and M_V edges from data by McWhan et al.[16] Two resonances are visible with an anti-resonance mode formed by interference situated between them. The enhancement of about 10^5 is obvious. The widths and relative intensities of the two peaks (branching ratio) are of interest. Qualitative results are made difficult by a rapidly varying absorption factor but data by Tang et al.[17] on a series of uranium compounds UO_2, USb and $U_{0.85}Th_{0.15}Sb$, suggest that the branching ratio decreases as the magnetic sub shell is filled. At a configuration of $5f^7$ the ratio should be unity.

NpAs is one of the most exotic materials to be studied to date. It has a rich magnetic phase diagram and has been extensively studied with polarised neutrons. In the temperature range T=154 K to T_N= 174 K the antiferromagnetic ordering is incommensurate with the rock-salt lattice structure, $\tau \approx 0.233$. The ordering becomes commensurate with $\tau=0.25$ below this range and then below 140 K there is a first order phase transition to a type I ($\tau=1$) antiferromagnet. Figure 4 shows data obtained at the M_{IV} edge resonance (3.85 keV) by Langridge et al.[18] for the temperature development of the $(0,\tau,2)$ magnetic peak in NpAs above the Néel point. In this region it is the critical scattering prior to the onset of magnetic order which is being observed. Clearly lowering the temperature has the effect of narrowing the magnetic diffraction peak. This implies an increase in the long range order of the magnetic structure and it is interesting that the higher resolution X-ray results predict a much longer correlation length than is apparent with established neutron data. This is probably indicative of the near surface sensitivity of the X-ray measurements which sample a surface modified magnetic structure.

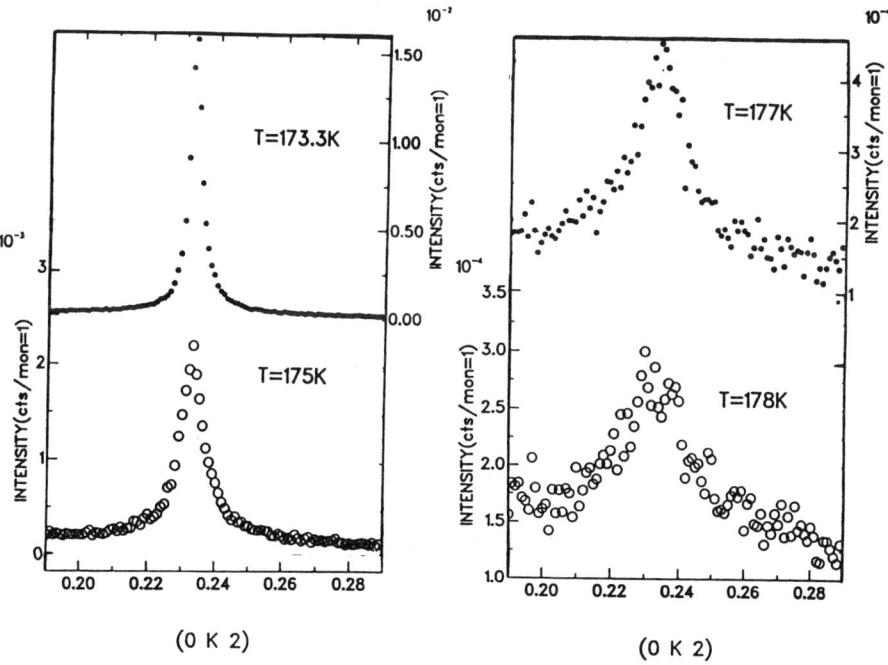

Figure 4. Magnetic X-ray critical scattering for NpAs. The instrumental resolution is 10 times narrower than the reflection width at 178 K and allows the real space correlation lengths to be inferred from the widths with some confidence. They are much longer than suggested by neutron measurements. (Data are taken from Langridge et al.[18])

Polarisation analysis of the (0,0,2l+τ) diffraction peak in measurements at the L_{III} edge of Ho was carried out by Gibbs et al[12-14] to try to distinguish the magnetic scattering from the charge scattering. A graphite analyser crystal was used at 90° scattering to select either the σ-σ or the σ-π channels. It was found that the energy dependence of the enhancement peaked at different energies 3 eV above and below the edge and depended on the polarisation channel and the order of the reflection studied. The scattering below the edge is believed to be due to electric quadrupole transitions and that above due to electric dipole transitions. This is indicative of the extreme sensitivity of the XRES measurement technique.

FERROMAGNETS

The observation of magnetic form factors for ferro- or ferrimagnetic systems is more problematical since both the pure magnetic scattering and the interference terms occur at the same place as the much more intense Thomson charge scattering peaks. The pure magnetic (signal) to charge (noise) ratio would be impossibly small to detect. It is preferred to look at the interference term between charge and magnetic scattering since this is several orders of magnitude larger. Typically for the interference term a maximum relative signal of the order of 0.1 percent is to be expected and thus great demands are still made on the sensitivity and stability of any measurement. Circular polarisation is used to observe the interference.

Equation (6) gives the cross section (to order $\hbar\omega/mc^2$) for diffraction from a ferromagnet with synchrotron radiation of mixed polarisation state. The 45° polarisation term is assumed to be zero for the synchrotron, leaving just linear and circular polarisation contributions. The circular polarisation term can be isolated by reversing the magnetisation direction in the sample and obtaining the asymmetric ratio with the total scattering as denominator. In the absence of any aomalous scattering, the ratio takes the form

$$\frac{\Delta I}{I} = -4\left(\frac{\hbar\omega}{mc^2}\right) P_\eta \frac{[\cos{}^3\theta_B(L+S)\cos\alpha - \sin{}^3\theta_B S \sin\alpha]}{1/2[(1+P_\xi) + (1-P_\xi)\cos{}^2 2\theta_B] |F|^2} F'\sin{}^2\theta_B \qquad (9)$$

where α is the angle in the scattering plane between the magnetisation direction and k_1+k_2.

Particularly for scattering angles near 90° the scale of this quantity is controlled by the polarisation factor $P_\eta / (1+P_\xi)$ and therefore larger ratios will be obtained when the radiation is polarised in the scattering plane ($P_\xi \approx -1$). This implies that the scattering plane must be horizontal (which is an unusual configuration for synchrotron based diffractometers). The higher the degree of linear polarisation, the bigger this polarisation factor will be and the 3rd generation synchrotrons with high emisssivities should have significant advantages for the measurement of this ratio. A change from 90% polarisation to 98% polarisation increases the ratio by a factor 4 !

Figure 5 shows the magnetic form factor obtained for single crystals of Fe(3%Si) by Laundy et al[19] using a white incident beam and energy proportional detector to separate harmonics. The cross section terms have been removed from the data to leave the magnetic-charge form factor ratio as a function of momentum transfer. The data agree with those from earlier neutron measurements[20] and have comparable statistical accuracy. The added advantage of the X-ray technique is that the geometrical terms in eq(9) can be exploited to separate the L and S contributions. Choosing $2\theta_B$ to be 90° and α to be alternately 90° and 0° selects S or L+S contributions respectively. This work is currently under development.

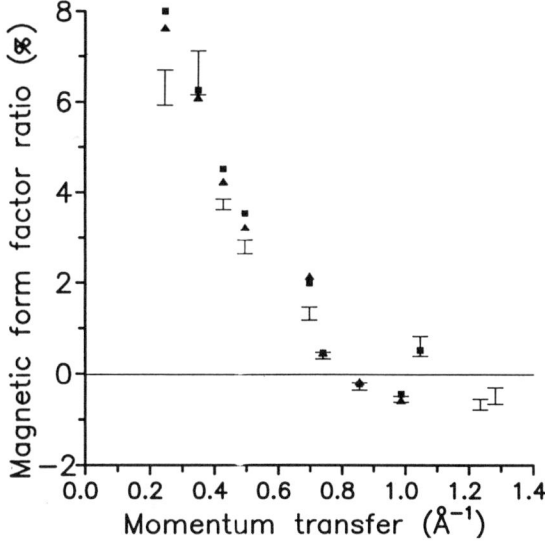

Figure 5. Magnetic form factor ratio for ferromagnetic iron measured with white beam circular polarised radiation. The error bars represent the X-ray results and are in good agreement with the neutron data[20] (squares) and atomic calculations (triangles). Data from Laundy et al.[19]

The stability of these measurements is a fundamental problem to the technique. Attempts to use monochromatic radiation have proved too sensitive to small movements of electron beam, monochromator or sample and are not yet at the 0.1% accuracy level. The situation is helped enormously by using a white beam method where small movements of the apparatus result only in slight shifts of the reflection energies. The current developments in helical wigglers should prove useful since it will be possible to switch the handedness of the circular polarised photons at the insertion device with a constant magnetic field and no other changes at the sample.

COMPTON SCATTERING

The magnetic interaction is instrinsically a relativistic effect as evinced by the factor $(\hbar\omega/mc^2)$ which appears in the cross section. In the relativistic region however the main interaction is not elastic but inelastic - namely Compton scattering and therefore it should not be surprising that the Compton cross section is also sensitive to magnetic ordering of the sample. A particularly unexpected result which has emerged from recent work though is that the Compton interaction ignores any orbital magnetic ordering and is thus sensitive only to unpaired electron spin. First a brief introduction to this unfamiliar field.

The Compton effect is the 2 particle quantum relativistic interaction between a photon and an isolated electron which transfers energy to the electron and softens the scattered photon. The scattered photon also picks up energy via a sort of Doppler process from the initial motion of the electron and thus a study of the second differential cross section as a function of energy yields information about the electronic momenta. In a solid target the electron may still be treated as free provided the requirements of an Impulse Approximation are fulfilled - namely that the energy and momentum transfers should be high compared to characteristic kinematic properties of the target. Under this condition the Born approximation is automatically satisfied and the measured cross section is approximately proportional to a ground state property of the system known as the Compton profile -

$$\frac{d^2\sigma}{d\Omega d\omega} \propto J(p_z) \tag{10}$$

$$\text{where} \quad J(p_z) = \int\int n(p) \, dp_x dp_y$$

$$\text{and} \quad n(p) = \left| \int e^{-i p \cdot r} \psi(r) \, d^3r \right|^2$$

The Compton profile is a measure of the number of electrons having a particular z-component of velocity where z is parallel to the photon scattering vector. For a free electron gas model which closely reproduces the conduction band electrons in alkali metals n(p) is the Fermi distribution function, fully occupied in the 3 dimensions of momentum space up to k_F, and the Compton profile resembles an inverted parabola with the same cut off. If the crystalline potential is increased some of these electrons are scattered into higher order (Umklapp) zones and the momentum density takes on the symmetry (though not the periodicity) of the reciprocal lattice. Conduction electrons remain closely peaked around the origin while core electrons are diffused throughout momentum space and therefore Compton scattering is particularly sensitive to valence effects in solids.

The constant of proportionality in eq(10) depends on the geometry of the scattering process and the polarisation states of the photons. Usually the experiment is performed with unpolarised γ rays and a polarisation-insensitive detector. The constant of proportionality for mildly relativistic electrons is then the well known Klein-Nishina formula which transforms into the Thomson cross section in the elastic (or low energy) limit and the distribution of electron momenta produced by atomic binding is superimposed (within the spirit of the Impulse Approximation) onto the scattering from a free electron.

$$\frac{d^2\sigma}{d\Omega d\omega_2} = \frac{r_0^2}{2}(\frac{\omega_2}{\omega_1})^2(\frac{\omega_1}{\omega_2} + \frac{\omega_2}{\omega_1} - \sin^2\theta) \int\int\int n(p) \, \delta(\omega - \frac{\hbar^2 K^2}{2m} - \frac{K \cdot p}{m}) \frac{d^3p}{(2\pi)^3} \tag{11}$$

If the effect of electron spin is now included in the cross section, Lipps and Tolhoek[2] and Platzman and Tzoar[21] showed that the cross section is enhanced by an additional term depending on the momentum distribution of unpaired spin electrons and the polarisation geometry. To first order in $\hbar\omega/mc^2$ the equation (still within the Impulse Approximation for "free" electrons with spin aligned in the z-direction and containing no orbital angular momentum term) is

$$\frac{d^2\sigma}{d\Omega d\omega} = \iiint A^*A(n(p\uparrow)+n(p\downarrow)) + 2\,Im(A^*B_z)(n(p\uparrow)-n(p\downarrow)) \quad (12)$$
$$\times \delta(\omega - \frac{\hbar^2 K^2}{2m} - \frac{K \cdot p}{m}) \frac{d^3p}{(2\pi)^3}$$

where $A = \epsilon_1 \cdot \epsilon_2$, $\quad q = \hat{k}_1 - \hat{k}_2$

and $\quad B = -\frac{\hbar\omega}{mc^2}\epsilon_1 \cdot \epsilon_2(k_1 \times k_2) - \frac{1}{2}q \cdot q(\epsilon_1 \times \epsilon_2) - q \times (q \times (\epsilon_1 \times \epsilon_2))$

The magnetic term is still imaginary unless the polarisation factors are complex. For circular polarisation of degree P_η and $P_\xi = P_\zeta = 0$

$$\frac{d^2\sigma}{d\Omega d\omega_2} = \frac{1}{2}r_0^2(\frac{\omega_2}{\omega_1})^2 [\iiint (\frac{\omega_1}{\omega_2} + \frac{\omega_2}{\omega_1} - \sin^2\theta)\, n(p) \quad (13)$$
$$- P_\eta(1-\cos\theta)\, S \cdot (k_1 \cos\theta + k_2)\,(n(p\uparrow)-n(p\downarrow))]$$
$$\times \delta(\omega - \frac{\hbar^2 K^2}{2m} - \frac{K \cdot p}{m}) \frac{d^3p}{(2\pi)^3}$$

Again the field flipping technique can be used to isolate this spin contribution. Applying magnetising fields in first one direction and then another and subtracting gives

$$\Delta \frac{d^2\sigma}{d\Omega d\omega_2} = -r_0^2(\frac{\omega_2}{\omega_1})^2 P_\eta(1-\cos\theta)\, S \cdot (k_1 \cos\theta + k_2) \times J_{mag}(p_z) \quad (14)$$

Figure 6 shows an early measurement made on iron by Cooper et al.[7] The profile shows a dip at low momenta-evidence of a negative polarisation density in a diffuse s-band.

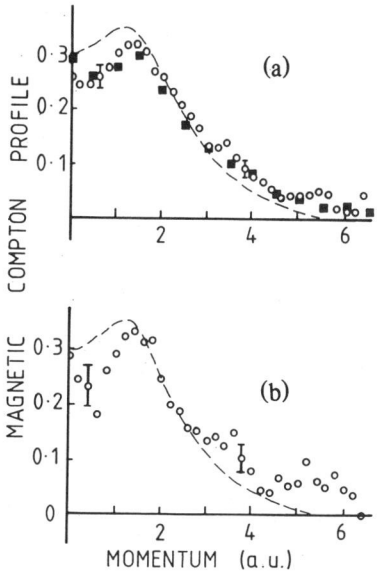

Figure 6. Magnetic Compton profiles for iron measured at (a) 46 keV and (b) 62 keV, showing the dip at the origin due to negative spin polarisation. Open circles: synchrotron data of Cooper et al.[7], squares: radioisotope data at 129 keV of Sakai et al.[22], dashed curves spin-dependent APW calculation[23].

Iron is predominantly a spin system with quenched orbital magnetisation component. HoFe$_2$ on the other hand is a strong L system with a magnetic moment of 10 μ_B. The system is a cubic ferrimagnet with the iron S-moment antiparallel to the Ho L moment. The magnetic cross section for elastic scattering has been proven to be geometrically different for the two moments and several low energy calculations originally considered the Compton cross section to also depend in a similar fashion on L. The L contribution was assumed (see Lovesey[24]) to modify the magnetic term in eq(13) as

$$P_\eta (1-\cos\theta) \left[\frac{1}{2} S \cdot (k_1 \cos\theta + k_2) + L \cdot (k_1 + k_2) \cos^2 \frac{\theta}{2} \right] \quad (15)$$

Judicious choice of the polarisation-geometry should therefore emphasize either S or L contributions and indeed it should have been possible to separate the two contributions. Figure 7 shows the results of several polarisation-inequivalent measurements made by Cooper et al.[25] It is remarkable that no contribution from L magnetisation density can be seen. This is in disagreement with eq(15). Subsequent measurements performed with different apparatus by various groups confirm this finding. A systematic study of the transition metals with a regularly increasing contribution to L magnetisation also showed no differences.

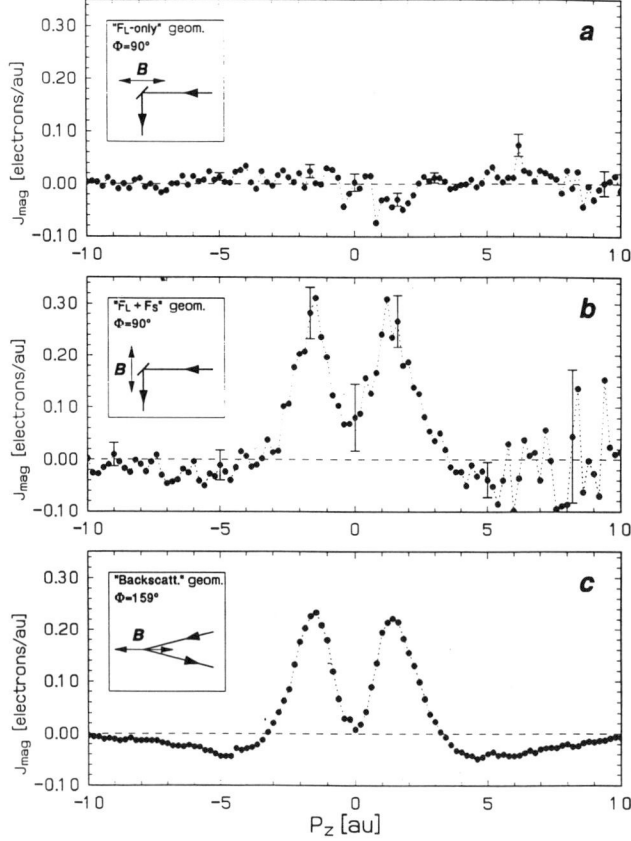

Figure 7. Magnetic Compton profiles of HoFe$_2$ measured in magnetic field configurations of distinct polarisation geometries[25]: (a) geometry chosen to select only L contribution to magnetic scattering (b) L + S geometry and (c) close to backscattering geometry (S only).

The conclusion must be that the cross section is in fact independent of orbital magnetisation! A model for the S contribution alone has to be employed. This is in line with the idea of the Impulse Approximation which being a localised interaction does not allow the concept of an angular motion. Although an approximation in name it allows the prediction of Compton profiles to a very high accuracy. Thus in contradiction of theoretical expectations, Compton scattering definitely provides a means of observing spin magnetism alone.

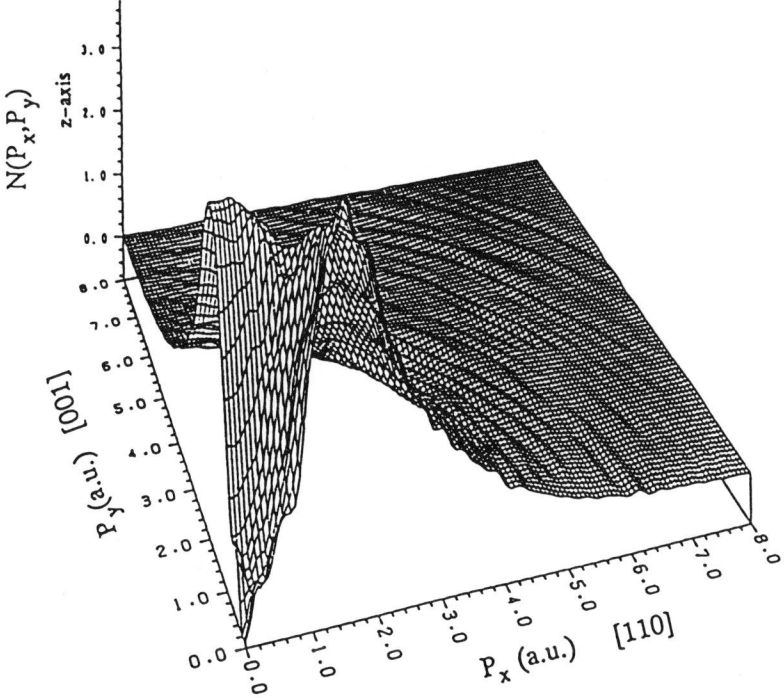

Figure 8. 2D momentum distribution of unpaired spin electrons reconstructed from 14 magnetic Compton profiles[26]. The reconstruction gives the 3D density which has been integrated over the (110) direction to improve statistics. The region of momentum space around the origin clearly shows the size and width of the negative spin polarisation appearing as a dip in the Compton profiles.

Magnetic scattering is a very exciting development in the Compton field and several groups are keenly working on the task of extracting as much detail as possible from the data. One way in which this is done is to reconstruct the 3-dimensional spin-momentum density $n(p\uparrow)$ from a multiplicity of individual profiles measured in characteristic directions of the lattice. This has the beneficial effect of avoiding the planar integration describd by eq(10) and thus improves the resolution and sensitivity of the result. The availability of synchrotrons and circular polarisd radiation means that many ambitious experiments can be undertaken. Figure 8 shows the 2D (integrated over one direction) spin momentum density of Fe(3%Si) reconstructed from 14 separate crystallographically-inequivalent profile measurements made by Sakai et al.[26] at the Photon Factory, Japan. The negative polarisation dip at the origin now clearly dominates the structure.

(γ,eγ) SPECTROSCOPY

(γ,eγ) spectroscopy is the final goal of the pioneering Compton scattering experiments which were carried out in the 1930's. To satisfy the debate about the particulate nature of the momentum conservation in the Compton interaction attempts were made to detect the recoil electron in coincidence with the scattered photon. Although suffering from poor angular resolution due to weak photon beams, these were successful and the recoil direction was determined in agreement with a two-particle conservation of energy and momentum.

The advent of powerful high energy synchrotron beams has enabled these experiments to be repeated but with sufficient angular resolution to show the influence of the intrinsic momentum of the electrons in the target. No momentum is transferred perpendicular to the scattering vector in the Compton interaction so measuring the angle of the recoil electron and knowing the magnitude of the momentum transfer enables the transverse momentum components to be identified. Thus all components are identified if a triple differential cross section is measured[27]

$$\frac{d^4\sigma}{d\omega' dE' d\Omega_\gamma d\Omega_e} = \frac{1}{2} r_0^2 \frac{\omega'}{\omega} \sum_i \frac{p'}{E} \rho_i(p_1) X(p, p') \delta(\omega + E - E')$$

where $X(p, p') = \frac{R}{R'} + \frac{R'}{R} - 2 + 2\sum_{j=1,2} [\epsilon \cdot \epsilon'_j + \frac{(\epsilon \cdot p)(\epsilon'_j \cdot p')}{R} - \frac{(\epsilon'_j \cdot p)(\epsilon \cdot p')}{R'}]^2$

and $R = E\omega - p \cdot k$, $R' = R - \omega\omega'(1 - \cos\theta)$ (16)

The summation over j is performed if the final polarisation states of the photon are not observed. Experiments are carried out using X-rays of medium energy (100 keV to 200 keV) to impart sufficient energy to the electrons for them to escape the target. The electrons are then detected in coincidence with the Compton scattered photons and the angular correlation $I(d\Omega_\gamma d\Omega_e)$ is measured which determines p_x and p_y. If the scattered photon energy is also analysed (as in a Compton measurement) p_z is determined and the cross section is roughly proportional to n(p).

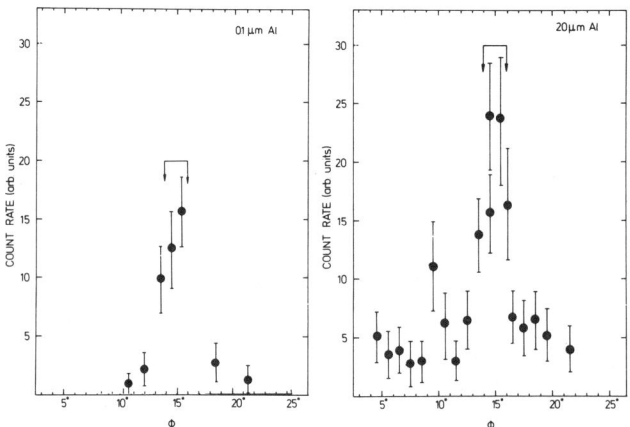

Figure 9. (γ,eγ) angular correlation profiles for polycrystalline Al foils[27] of 0.1 μm and 20 μm thickness measured with 148 keV photons. The arrowed bars represent the angular width of electrons at the Fermi surface.

Measurements were made by the author at HASYLAB of the angular correlation profile from Al and Cu foils which although suffering from poor statistics clearly show the extent of the Fermi surface[28,29]. Figure 9 shows angular correlation data for measurements made of Al foils of widely differing thickness. The width of the central Fermi sphere can be clearly identified in both sets of data although the thicker sample is over 200 electron elastic scattering mean free paths thick! Multiple scattering of the recoil electrons is highly saturated if tight collimation of the experiment is adopted. Figure 10 shows the p_z (photon energy) distribution at the peak of the angular correlation curve ($p_x=p_y=0$). The Fermi surface is again easily identifiable. Thus if statistics can be omproved somewhat, we now have a technique for measuring directly the 3D momentum density n(p) which avoids the planar integration (eq(10)) implicit in traditional Compton scattering. Neither is it necessary to employ reconstruction methods. This implies a much greater sensitivity to valence electrons unhindered by any large background from core states and also allows a much higher (geometry limited) momentum resolution than before. Utilisation of circular polarised radiation should also yield spin momenutm distributions as for Compton Scattering.

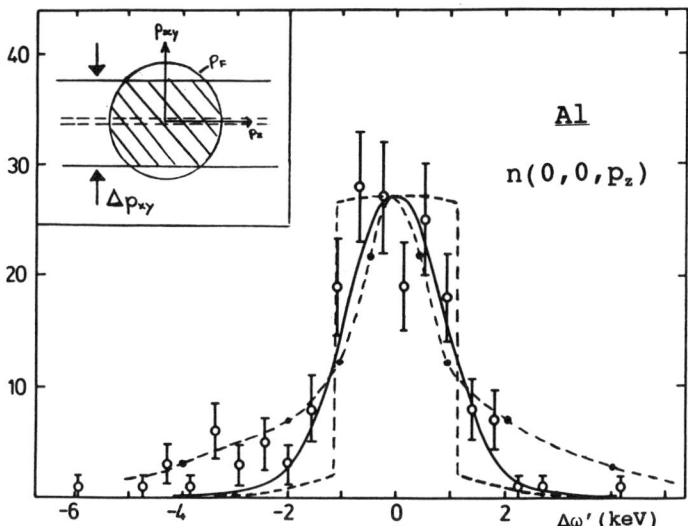

Figure 10. Photon energy distribution at (γ,eγ) angular correlation peak of 1.8 μm Al foil[27]. The inset shows the transverse momentum collimation in comparison to the Fermi sphere. If this is sufficiently narrow the measurement represents the 3D momentum distribution n(p).

To complete the circle we now return to the predictions made by Lipps and Tolhoek[2,3] concerning the production of polarised electron beams by scattering unpolarised gamma radiation from a magnetised target. The authors claim that the probability of a spin-flip occurring during the short interaction time is low and that the degree of polarisation of recoil electrons from a completely spin oriented target is at least 80% for energies up to 250 keV. Thus we have an experimental situation analagous to spin-resolved photo-emission where the initial spin state of the electron can be determined by first ejecting it from the target. Figure 11 shows the degree of polarisation of the recoil electrons from a 100% polarised target with the electron spin orientation prependicular to the photon scattering plane.

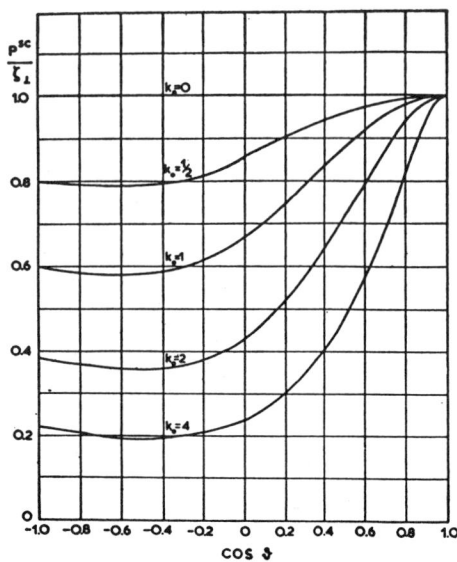

Figure 11. Fractional polarisation of recoil electrons ejected by Compton scattering at various photon energies as a function of angle[2]. The energies are quoted in natural units. For 250 keV photons ($k_0 = 1/2$) the degree of polarisation in backscattering geometry is 80%.

If an efficient way can be found to observe the spin of these recoil electrons this phenomenon can be tied in with the "normal" ($\gamma, e\gamma$) experiment to create a very interesting magnetic spectroscopy technique which samples $n(p,\sigma)$. For a sample of magnetised iron the fraction of electrons which are unpaired is about 8%. Mott detectors typically have a polarisation detection efficiency of 10^{-4} to 10^{-3} and so spin differences of up to 10^{-4} can be anticipated. This low signal could be compensated by employing multi-element photon detectors. We would therefore have an alternative way of observing the identical cross section relevant to magnetic Compton Scattering but would be observing the 3D spin momentum distribution directly with the additional possibility of resolving the binding energies of the scattering electron states.

CONCLUSION

Rapid developments are occurring in the field of X-ray magnetic scattering which are being driven on by the availability of 3rd generation synchrotron sources and new insertion devices on older rings. The higher fluxes and better polarisation characteristics of these sources are enabling more and more sensitivity in the X-ray experiments with the result that completely new spectroscopies are emerging. Comparisons between neutron and X-ray results will be very fruitful, as witness the new measurements of correlation and ordering ranges in the exotic actinide compounds. Inelastic scattering has also been given a boost with its proven sensitivity to spin magnetic order and further results are to be expected when high energy beam lines become more available. The X-ray field has grown dramatically in the past 10 years and the study of spin polarised systems will ensure that this trend continues.

ACKNOWLEDGEMENTS

The author would like to acknowledge the contributions to the work described in this paper made by collaborators at Keele University, Daresbury Laboratory and Hamburg Synchrotron Laboratory. Finally I should like to thank the SERC for financial support for attendance at the SERC Workshop on Polarized Electron/ Polarized Photon Physics.

REFERENCES

1. O. Klein and Y. Nishina in: "The Theory of Photons and Electrons", J. M. Jauch and F. Rohrlich, eds., Addison-Wesley Publishing Co., Reading Mass. (1959)
2. F. W. Lipps and H. A. Tolhoek, Polarization phenomena of electrons and photons, Physica 20:395 (1954)
   H. A. Tolhoek, Electron polarization, theory and experiment, Rev. Mod. Phys. 28:277 (1956)
3. P. M. Platzman and N. Tzoar, Magnetic scatterring of x-rays from elecrons in molecules and solids, Phys.Rev. B 2:3556 (1970)
4. F. de Bergevin and M. Brunel, Observation of magnetic superlattice peaks by x-ray diffraction on an antiferromagnetic NiO crystal, Physics Letters A 39:141 (1972)
5. F. de Bergevin and M. Brunel, Diffraction of x-rays by magnetic materials. I. General formulae and measurements on ferro- and ferrimagnetic compounds,Acta Cryst. A 37:314 (1981)
   M. Brunel and F. de Bergevin, Diffraction of x-rays by magnetic materials. II. Measurements on Antiferromagnetic Fe_2O_3, Acta Cryst. A 37:324 (1981)
   M. Brunel, G. Patrat, F. de Bergevin, F. Rousseaux and M. Lemonnier, Etude de la polarisation circulaire du rayonnement synchrotron dans la gamme des rayons x par diffraction sur un composé ferrimagnetique, Acta Cryst. A 39:84 (1983)
6. N. Sakai & K. Ono, Compton profile due to magnetic electrons in ferromagnetic iron measured with circularly polarised γ rays, Phys. Rev. Lett. 37:351 (1976)
7. M. J. Cooper, D. Laundy, D. A. Cardwell, D. N. Timms, R. S. Holt and G. Clark, Spin-dependent momentum distribution in iron with circularly polarised synchrotron radiation, Phys. Rev. B 34:5984 (1986)
8. D. Laundy, private communication (1991)
9. H. Grotch, E. Kazes, G. Blatt and D. A. Owen, Spin dependent Compton scattering from bound electrons: quasi-relativistic case, Phys. Rev. A 27:243 (1983)
10. M. Blume, Magnetic scattering of x-rays, J. Appl. Phys. 57:3615 (1985)
    M. Blume and D. Gibbs, Polarization dependence of magnetic x-ray scattering, Phys. Rev. B 37:1779 (1988)
11. D. Gibbs, D. E. Moncton and K. L. D'Amico, Magnetic x-ray scattering studies of the rare-earth metal holmium, J. Appl. Phys. 57:3619 (1985)
12. D. Gibbs, D. R. Harshman, E. D. Isaacs, D. B. McWhan, D. Mills and C. Vettier, Polarization and resonance properties of magnetic x-ray scattering in holmium, Phys. Rev. Lett. 61:1241 (1988)
13. J. P. Hannon, G. T. Trammell, M. Blume and D. Gibbs, X-ray resonance exchange scattering, Phys. Rev. Lett. 61:1245 (1988)
14. J. Bohr, D. Gibbs, D. E. Moncton and K. L. d'Amico, Magnetic structure of holmium, Physica A 140:349 (1986)
15. M. Lippert, Th. Brückel, Th. Köhler, J. R. Schneider and W. Jauch, Bulk magnetic scattering of high energy synchrotron radiation, HASYLAB Annual Report, 621 (1993)

16. D. B. McWhan, C. Vettier, E. D. Isaacs, G. E. Ice, D. P. Siddons, J. B. Hastings, C. Peters and O. Vogt, Magnetic x-ray scattering study of uranium arsenide, Phys. Rev. B 42:6007 (1990)
17. C. C. Tang, W. G. Stirling, G. H. Lander, D. Gibbs, W. Herzog, P. Carra, B. T. Thole, K. Mattenberger and O. Vogt, Resonant magnetic x-ray scattering in a series of uranium compounds, Phys. Rev. B 46:5287 (1992)
18. S. Langridge, W. G. Stirling, G. H. Lander, J. Rebizant, J. C. Spirlet, D. Gibbs and O. Vogt, X-ray study of the critical magnetic scattering in NpAs, Europhys. Lett. 25:137 (1994)
19. D. Laundy, S. P. Collins and A. J. Rollason, Magnetic x-ray diffraction from ferromagnetic iron, J. Phys. Condens. Matter. 3:369 (1991)

    S. P. Collins, D. Laundy and A. J. Rollason, Magnetic form factors of ferromagnetic iron by x-ray diffraction, Phil. Mag. B 65:37 (1992)
20. C. G. Shull and Y. Yamada, Magnetic electron configuration in iron, J. Phys. Soc. Japan B 17:1
21. P. M. Platzman and N. Tzoar, Inelastic magnetic x-ray scattering, J. Appl. Phys. 57:3623 (1985)
22. N. Sakai, O. Terashima and H. Sekizawa, Nucl. Instrum. Methods 221:419 (1984)
23. S. Wakoh and Y. Kubo, Spin-dependent momentum distribution in iron, J. Mag. Mag. Materials, 5:202 (1977)
24. S. W. Lovesey, Theory of magnetic photon scattering: total and Compton cross sections for circularly polarised hard x-rays, Physica Scripta 44:51 (1991)
25. M. J. Cooper, E. B. Zukowski, S. P. Collins, D. N. Timms, F. Itoh and H. Sakurai, Does magnetic Compton scattering only measure spin magnetisation? J. Phys. Cond. Matter L (1992)
26. N. Sakai, Latest results in Compton profile studies, 9th International Conference on Positron Annihilation, Szombathely, Hungary (1991)
27. F. Bell, Th. Tschentscher, J. R. Schneider and A. J. Rollason, The triple differential cross section for deep inelastic photon scattering: a (γ,γ') experiment, J. Phys. B 24:L533 (1991)
28. J. R. Schneider, F. Bell, Th. Tschentscher and A.J. Rollason, $(\gamma,e\gamma)$ spectroscopy: a new technique to determine elecron momentum densities of solids, Rev. Sci. Instrum. 63:1119 (1992)
29 F. Bell, Th. Tschentscher, J. R. Schneider and A. J. Rollason, The electron momentum density of copper studied by $(\gamma,e\gamma)$ spectroscopy, J. Phys.: Condens. Matter 3:5587 (1991)

SPIN-POLARISED SCANNING TUNNELLING MICROSCOPY AND RELEVANT TECHNIQUES- A SURVEY OF PRESENT STATUS

T-H Shen

Department of Physics
The University of Leeds
Leeds LS2 9JT
U.K.

INTRODUCTION

Spin-polarised STM and related techniques have emerged amid a general background where the advance of materials research is demanding ever more powerful analytical tools. In the field of thin film magnetism, the ability to grow epitaxial metallic films to a high degree of perfection has generated fascinating new topics. For example, there has been considerable research activities in the magnetic behaviour of multilayer structures made of thin metallic layers with a thickness of a few atoms. These multilayers exhibit a number of interesting effects such as oscillations in the interlayer coupling and giant magnetoresistance (MR) [1-5]. Because of their novel behaviour, these thin films may find their way to potential technological applications: for instance, magnetic-field detection, for which a large MR at small fields is desirable. Epitaxial magnetic films grown on suitable substrates have also led to the study of films with crystallographic structures not commonly found in nature as well as the study of magnetism of reduced dimensions. As an important subject in both fundamental research and data storage applications, the observation of magnetic domain structures is now being performed on an ever-decreasing lateral scale. From the point of view of technological applications, as data recording density increases such that the domain size in magnetic thin film media is shrinking to submicron level, the structures and characterisation of domain boundaries are becoming more important. It is also evident that the ultimate goal of surface magnetic studies would involve establishing the electron spin polarisation and the electronic state of each surface atom with atomic resolution.

A number of experimental techniques has been developed over the past decade or two to study surface magnetism, such as spin-polarised field emission, photoemission, magnetic circular dichroism by means of X-ray absorption and X-ray fluorescence, spin-polarised LEED, Lorentz microscopy, and scanning electron microscopy with polarisation analysis (SEMPA), of which the SEMPA technique[6] offers at present a spatial resolution of about 20nm. (For review articles concerning some of these techniques, please refer to those of Wohlecke and Borstel[7], Feder[8], Heinzmann and Schonhense[9], Meier[10], Kirschner[11], Heinzmann[12] and, Oepen and Kirschner[13] as well as other articles mentioned in this volume.)

On the other hand, since its invention by Binnig and Rohrer some ten years ago scanning tunnelling microscopy (STM)[14] has proven to be an extremely useful instrument for research in a vast range of areas and topics. Scanning probe microscopy, as it is now generally known because of a variety of interactions being employed in addition to electron tunnelling, provides an immensely powerful tool for studying and modifying surfaces at the nanometer to atomic level. For example, magnetic force microscopy has now been well established in the study of magnetic domains with a resolution limited by the long range of the magnetic interaction, whilst Faraday rotation has been demonstrated with near field optical microscopy[15].

There are a number of recent examples illustrating that the detection of electron-spin-polarisation effects in STM has received increasing attention. In their beautiful experiments Manassen et al reported the observation of individual paramagnetic spins in oxidised silicon surfaces[16]. Johnson and Clarke reported the design and preliminary results from a prototype spin-polarised STM operated in air employing a magnetisation modulation at the sample[17]. In ultra high vacuum, spin-polarised scanning tunnelling microscopy has been successfully demonstrated by Wiesendanger and co-workers using ferromagnetic tips to probe antiferromagnetic sample surfaces[18]. Recent work on Fe_3O_4 appeared to suggest that spin states imaging might be realised at the atomic level[19]. Further advances of spin-polarised STM are represented by the work of two other independent research groups with completely different approaches. The first was reported by Alvarado and Renaud[20]. They showed that the spin polarisation of the tunnelling electrons from a ferromagnetic tip into a GaAs sample could be detected by measuring the circularity of the light emitted by the radiative recombination of electrons in the GaAs and from their results they were able to estimate the degree of polarisation for the tunnel electrons from a Ni tip. The second piece of work, by Sueoka and co-authors[21], demonstrated the possibility of observing spin-polarised tunnelling current using STM with optically pumped GaAs, a somewhat inverted process of the first case.

In this brief survey, we shall describe in some detail the various approaches employed by these research groups in their pioneering work on spin-polarised STM as well as other relevant techniques and provide some assessment of the significance of these advances. Some possible future developments of the technique will also be discussed.

THE TECHNIQUE OF SPIN-POLARISED STM

The observation of spin-polarised tunnelling process requires a number of conditions; (i) that electrons emitted from initial states are to some degree spin-polarised, (ii) that the process of tunnelling through a barrier will usually have a negligible effect on the spin states of the electrons, and (iii) that the final states available in the sample for tunnelling are also to some degree spin-polarised. Experimentally it was shown in the 70's that field emitted electrons within a fraction of an electron volt of the Fermi surface from magnetic tips were spin-polarised[22]. For W-EuS junction field emitters, the polarisation was found[23,24] to be over 80%. Tunnelling measurements of metal-insulator-superconductor (MIS) junctions in a magnetic field as well as metal-insulator-metal (MIM) junctions with ferromagnetic electrodes also demonstrated the effect of spin-polarised tunnelling[25,26]. Theoretical analysis of such spin-polarised tunnelling has also been reported for planar junction geometry[27]. Also, the presence of small quantities of magnetic impurities or other localised spin centres in bulk tunnel junction has been known to significantly affect tunnelling characteristics[28-31]. This dependence of tunnel current on the spin centres has lead to the recent development of a paramagnetic resonance type of technique using STM[16]. As for spin-polarised STM measurement, in order to investigate the spin polarisation as well as surface magnetic structure of the specimen, it will be necessary in some way to reverse the spin-polarisation state of the tip or sample so that the 'topographic' contribution to the tunnelling current can be separated from the spin-polarised tunnelling effect.

The Concept

The concept of spin-polarised STM has been discussed by a number of authors[17,18]. In a 'conventional' STM, the tunnel current is proportional to the local density of states (LDOS) of the sample, which is a function of both in-plane coordinates (x,y) along the surface and the out-of-plane coordinate z. If a constant tunnel current is maintained at a fixed bias by adjusting the z-distance whilst the tip scans across the surface, the constant LDOS surface at a given energy is mapped out. One can also set z constant during an x-y scan to obtain the LDOS as a function of (x,y) at a given z. In both cases, we can obtain the 'topological' information of the surfaces on an atomic scale. For instance, the localised bonding nature of Si gives rise to a highly corrugated surface topography on an atomic scale, whilst as for Au, because of the relatively free-electron-like behaviour of the states near the Fermi energy, the surface topography obtained by STM looks somewhat smoother. By varying the bias between the tip and sample, it is possible to probe the energy dependence of the LDOS, hence providing a spectroscopic technique as well for the study of electronic structures at the surfaces.

For spin-polarised STM, a magnetic tip is used. For a ferromagnetic metal such as nickel, the exchange splitting of the spin-up and spin-down sub-bands implies that the majority spin sub-band has few states available at the Fermi energy[17]. Therefore the electronic states close to the Fermi level are to a large degree spin-polarised. As the process of tunnelling through a non-magnetic insulating barrier will have a negligible influence on the polarisation states of the tunnel electrons, these spin-polarised tunnel electrons will now act as spin-sensitive probes of the spin-dependent LDOS of the sample surfaces. However the following subtle aspect must to be recognised. In a spin-polarised STM, as the tunnel electrons of one particular spin state will map the LDOS of a corresponding spin-subband near the Fermi energy, the average tunnel current for the two opposing states of spin polarisation is proportional to the average LDOS of both spin-up and spin down sub-bands. The difference between them will reflect the difference in the LDOS between the two corresponding sub-bands. The former is in fact the same as that of the conventional STM and can be used as a measure of the topography at the surfaces. The latter contains information about the magnetic structure at the surfaces. Since in a typical spin-polarised STM measurement, both surface topography and spin-polarised effect are mixed together, in order to separate the two contributions, it is necessary to measure the two spin-polarisation states in some way. The spin-polarised STM concept can be grossly summarised in a 'three-step' model; namely, the magnetic tips provide a source of spin-polarised electrons, the electrons tunnel through the barrier with little change to their polarisation states, and the spin-polarised electrons tunnelling process will be sensitive to the final LDOS of the sample of a particular spin sub-band. Whilst such a highly simplified account of spin-polarised STM appears to work well in an intuitive explanation of the experimental observations, we note that a detailed theoretical model for spin-polarised STM is still lacking. This of course will not detract from the usefulness of the method as a powerful experimental technique in many applications and is well in line with the general order of development in the SPM field. However a better theoretical understanding is no doubt extremely important, a point well worthy of emphasising.

The Prototype Operated in Air

Johnson and Clarke reported their design and preliminary results of a prototype of spin-polarised STM operated in air[17]. It is well known that surface magnetism is adversely influenced by surface contamination. However, the authors quoted earlier work on interfacial charge-spin coupling involving the use of ferromagnetic films to spin polarise electric currents injected into a bulk normal metal[32]. The polarised electrons diffused a characteristic distance, known as spin depth, of considerable length before losing their polarisation. These experiments showed that electrons within 1 to 10meV of the Fermi energy injected from a ferromagnetic material across a dirty metal interface are spin-polarised with the magnitude of the polarisation being about 0.06 for an alloy of nickel and iron, supporting the idea that operation in air may be possible.

To separate the spin-dependent contribution from the non-spin topographic one, Johnson and Clarke employed a technique in which the magnetisation of the sample is modulated such that it alternated between parallel and anti-parallel to the magnetisation of the tip. The modulation technique resulted in two quantities being determined simultaneously. The first is the average tunnelling current, which is proportional to the sum of the spin-up and spin-down LDOS at the sample surface. The second is the amplitude of the modulated tunnel current due to the magnetisation alteration of the tip or sample. This amplitude is proportional to the difference between the LDOS of the two spin sub-bands. The effective spin polarisation of the tunnel junction is then the ratio of the two quantities. The arrangement is sensitive to in-plane magnetisation. A prototype of spin-polarised STM was constructed. To obtain a better control of magnetisation and minimise stray external fields, ferromagnetic samples were made in the form of a torus. A single crystal of Ni was used as a tunnelling tip. The Ni tip was magnetised to saturation by two small pieces of magnet completed to form a torus with an iron yoke. Silver films were used as sample controls and Au tips as tip controls. Therefore the prototype essentially consisted of a ferromagnetic tip with saturated magnetisation from which spin-polarised electrons tunnel into a ferromagnetic sample. When the sample magnetisation is alternated between parallel and anti-parallel to that of the tip, a modulation of the tunnel current was measured at the same frequency. This signal satisfies criteria developed to characterise spin-polarised tunnelling. Spurious signals were also observed in the experiments and their probable causes identified.

The values of the polarisation determined in their experiments were typically two to six times larger than that of MIM tunnelling. The difference was explained in terms of spin randomisation in the oxide layer during the tunnelling process. A number of problems have been identified and some remedies suggested for the development of a spin-polarised STM to operate in vacuum using a similar modulation technique. These include thermal drift, magnetostrictive effects, and relative displacements between tip and sample driven by stray magnetic fields.

The work reported by Johnson and Clarke appeared to have shown that spin-polarised STM with magnetisation modulation was achievable even when the system was operated in air. Nevertheless a system operated in air will have very limited use in the study of surface magnetism as surface contamination can alter the behaviour drastically. Consequently, operation in ultra high vacuum (UHV) comes in as a natural requirement. The relative complexity of the prototype design as a result of the need for modulation makes radical improvement in the design necessary for any system operated in a UHV condition.

The UHV Experiment

Wiesendanger and co-workers took a somewhat different approach and successfully conducted the first spin-polarised STM experiment in ultra high vacuum[18]. Their experiments relied on the use of separate magnetic and non-magnetic tips to distinguish the effect of spin-polarised tunnelling from the non-magnetic morphological contribution of the surfaces. With such an approach, a surface with well defined characteristic magnetic domain structure is required as a testing system. They chose the Cr(001) surface for the following reason. Recent self-consistent total-energy calculations showed that topological antiferromagnetism separated by single steps is the energetically favourable structure of the surface[33]. This result is compatible with both the absence of magnetisation from spin-resolved photoemission[34] and the existence of spin-split surface states observed using energy- and angle-resolved photoemission[35]. The former is due to the cancellation between oppositely magnetised terraces within the area illuminated by the incoming photon beam and the latter due to majority- and minority-spin states inside each ferromagnetic terrace. This topological antiferromagnetism with terraces alternately magnetised in opposite directions provides an ideal test structure for spin-polarised STM experiments and minimises the long range magnetic interaction between a ferromagnetic tip and the sample. Ferromagnetic CrO_2 films were chosen to make magnetic tips. It has been shown from spin-resolved photoemission that CrO_2 films exhibited a spin polarisation of nearly 100% for binding energies near 2 eV below the Fermi level[36]. These films also have a large remanent magnetisation which makes the tip more likely to be in a single domain. Because of the in-plane magnetisation of the film and the shape anisotropy of the tip, these tips have a preferred magnetisation direction perpendicular to the sample surface in the experiment. Therefore the arrangement is sensitive to out-of-plane magnetisation of the samples. Electrochemically etched tungsten tips were used to perform the surface topographical imaging. As the determination of the step heights is important for the analysis of the results, they calibrated the STM scanning unit using Si (111) -7×7 and Si (001) -2×1 surfaces.

Their topographic results with a W-tip showed that the Cr(001) surface has terraces separated predominantly by monatomic steps, consistent with the model of topological antiferromagnetic order mentioned earlier. The monatomic step-height was found to be 0.149±0.008 nm, in good agreement with half of the cubic unit-cell height of 0.144 nm for bcc Cr. With a CrO_2 tip, the STM images of the Cr(001) surfaces showed qualitatively the structure of terraces separated by monatomic steps. Two different results were found with the CrO_2 tips compared to tungsten tips. The first, from I-V characteristics, showed the appearance of typical semiconductor-vacuum-metal tunnelling instead of that of metal-vacuum-metal tunnelling and established the need for a positive sample bias of at least 2 V in order to obtain a stable tunnelling current of 1 nA, consistent with results of spin-resolved photoemission work[36]. The second unusual result was that there existed a periodic alternation of the measured monatomic step heights between 0.16 nm and 0.12 nm compared with the mean single step-height value of 0.144 nm. Furthermore such a periodic alternation has never been found in topographic STM images with a tungsten tip. This periodic alternation of the monatomic step-height values was interpreted by them as being due to an additional contribution from the spin-polarised tunnelling.

If a terrace has the same direction of magnetisation as the apex of the CrO_2 tip, the tunnelling current will increase due to a contribution from the spin-polarised tunnelling[26]. Since the STM is operated at constant current, this increase of the tunnel current leads to a corresponding increase in the z-direction from the mean distance between the tip and the sample surface. Similarly if a terrace has the opposite direction of magnetisation, the tunnelling current will be decreased, resulting in a decrease of the tip-sample distance. The values of the measured single step height therefore alternate between a larger and a smaller value from the mean step height value. As the distance is adjusted to maintain a constant current, the distance dependence of the tunnel current will need to be known to determine the effective polarisation of the tunnel junction. Taking an explicit relationship between the total tunnelling current and the effective spin polarisation, the changes in the tip-sample distance and, the mean local tunnelling barrier height, measurable from the slope of the log current vs distance, the authors were able to determine the effective polarisation of the tunnelling junction. The derived value for the spin-polarisation was (20±10) %.

The spin-polarised STM work of Wiesendanger *et al* is based on a detailed analysis of the difference between STM measurements of magnetic and non-magnetic tips. Although their experiments have established clearly the spin-polarised tunnelling effect with STM, this determination was based on only one type of measurement. For those who are more sceptical and cautious, a second independent type of measurement performed *in situ* on the same system would appear to be far more convincing. Unfortunately to date such experiments have yet to be reported. Nevertheless, the effect of spin polarised tunnelling for a STM with a magnetic tip has been observed by other independent research groups with entirely difference approaches. We shall briefly discuss these experiments in the subsequent sections.

Circularly Polarised Light Emission with STM

Independent evidence for spin-polarised tunnelling with STM comes from very interesting work by Alvarado and Renaud[20], who demonstrated vacuum tunnelling of spin-polarised electrons from a ferromagnetic Ni tip into a GaAs(110) surface. The experiment is based on the measurement of the circular polarisation of the recombination luminescence excited by the tunnel electrons. The p-type doped GaAs, with relatively large spin-orbit-spilt valence bands, acted as a useful spin detector[37]. A high doping level of the GaAs was chosen to position the Fermi level close to the uppermost of the spin-orbit-split valence bands, making possible the radiative recombination of minority charge carriers injected at low energies. The advantage of the technique is that it allows the observation of spin-polarised tunnelling effect without the interference of topographic features of the sample under study and does not depend critically on the tunnelling current or on its fluctuation. The light emitted from the radiative recombination excited by the tunnel electrons passes through a Pockels cell and then a linear polariser. The intensities for $\pm\lambda/4$ retardation are then measured by a photomultiplier and the degree of circular polarisation determined. The circular polarisation of the emitted photons is proportional to the degree of spin polarisation of the recombining minority carriers at the fundamental band gap. As the spin-orbit splitting at the valence band of GaAs is large, the recombination happens at the uppermost valence band. The degree of circular polarisation was shown[38] to be equal to the product of the spin-polarisation detection sensitivity, a quantity dependent upon the electron lifetime at the bottom of the conduction band as well as the spin relaxation time[39], the initial spin polarisation of the tunnel electrons, and $\cos\theta$ where θ is the angle between the electron-spin-polarisation vector and the direction of light propagation towards the optical detector.

In their experiment, Ni tips were made by electrochemically etching polycrystalline wire followed by cleaning in UHV by Ne-ion bombardment and subsequent annealing. The tip was magnetised *in situ* using an electromagnet and the measurements were taken with the tips magnetised remanently. The orientation of the magnetisation at the front end of the tip was determined by stray magnetic field measurement. Again because of the shape anisotropy, the magnetisation of the micron-sized region at the front end of the tip is preferentially oriented along the tip's axis[40]. During the experiment, they were able to flip this magnetic domain by 180°. In order to determine the spin polarisation, the circular polarisation of the luminescence was measured for both polarities of the magnetising electromagnet.

Their main results showed that the measured spin polarisation for a Ni tip depended on the electron injection energy; i.e., the bias voltage between the tip and sample. The values of spin polarisation of the electrons were found to be largest at lower electron injection energies (E_k) above the bottom of the conduction band. The value of the spin polarisation was estimated to be (-31±5.6) % for E_k=0.3 eV, where the magnitude is an estimated lower limit. The negative sign indicated dominant minority electron injection. Whilst these values of electron spin polarisation of Ni are of particular interests in comparing them with similar data obtained by other techniques, the central importance of this work by Alvarado and Renaud is that it provides us with an unambiguous independent observation of spin-polarised electron tunnelling from a ferromagnet with STM.

Optical Pumping and Spin-Polarised Tunnelling

The experiment by Sueoka and co-workers[21] is to some degree an inverse process of that with circular polarised light emission mentioned previously and is another independent experiment on spin-polarised tunnelling. In their experiment, the possibility of observing spin-polarised tunnelling current using an STM with optically pumped GaAs was demonstrated. They used a GaAs thin-film sample pumped by circularly polarised light and a ferromagnetic polycrystalline Ni tip. When a GaAs crystal is illuminated by circularly polarised photons with energy near the band-gap energy at the centre of the Brillouin zone spin-polarised carriers are excited into the conduction band with maximum spin polarisation of the carriers along the wave vector of the incident light and with an expected value of 50%. This occupancy of spin-polarised electrons in the conduction band will have the maximum influence on the spin-polarised electron tunnelling from the magnetic tip when the tip magnetisation vector is parallel or anti-parallel to the spin polarisation vector of these electrons in the conduction band. Very thin GaAs samples were used in their experiment and were optically pumped with the incident light from the back of the samples along the tip axis. Tips were made of Ni wires pre-etched to a small constriction broken off to get a fresh surface and then magnetised prior to measurements. The actual tunnelling experiments were carried out in air.

They found that the tunnelling current was perturbed by the modulation of the power and polarisation of the pumping light. The corresponding variation of the tunnel current arises due to three dominant effects; namely, the thermal expansion of the tip or sample, the change in excited carrier concentration in the GaAs sample and the spin-polarised tunnelling. The first two effects can be eliminated with improved power stability and better alignment of the optics. These effects also have a characteristic dependence on the bias voltage in a manner considerably different from that of spin-polarised tunnelling. Well adjusted pumping optics enabled the separate detection of the spin-polarised signal, which depended upon the circular polarisation of the incident photons and the tip magnetisation. A non-magnetic Pt tip was also used as a comparison. The spin polarisation for the tunnel electrons was estimated from the experimental result. Sueoka and co-workers considered their estimated value of -43% for the spin polarisation of the tunnel electrons might contain a large uncertainty because of the large errors in the estimation of spin relaxation time and the excited carrier polarisation. However the sign of the polarisation was considered meaningful. Their work appeared to be in qualitative agreement with that of Alvarado and Renaud[20]. With regard to the contrast to experiments performed in UHV, where the adverse effects of surface contamination to spin-polarised tunnelling were noted[20,41], Sueoka *et al* suggested that their detected signal may result from a wide-area contribution of the sample and tip to the detriment of atomic resolution. The averaged signal was hence detected over a much larger area. It was also suggested in their work that, if it is possible to interchange the role of the tip and sample, a spin-polarised STM with a nonmagnetic tip may be realised. It would appear that such an STM would allow the spin polarisation of the tunnel electrons from the tip to be altered readily, an important step towards the separation of surface topography and spin-polarised tunnelling during a single imaging.

Other Advances

Since these earlier experiments, further progress in spin-polarised STM study has continued. With regard to the experiment on Cr (001) surfaces by Wiesendanger et al[18], numerical modelling has been applied to address the following issues. Assuming a simple spin-dependent tunnelling probability, the effective spin-polarisation (P) of a CrO_2- vacuum-Cr(001) junction was calculated using published density-of-states calculations[42]. It was found that in general, P was characteristic for a special junction at a specific bias voltage. The change of bias can alter the value as well as the sign of the effective spin-polarisation. The P value obtained from the experiment[18] at +2.5 V sample bias agreed well with the calculation. The effect of magnetic forces was also considered[42], as one might argue that the influence of magnetic forces could also lead to the observed alternation of the values of monatomic step-height. The calculation showed that the effect of magnetic dipole forces acting between the ferromagnetic CrO_2 tip and the antiferromagnetic Cr sample can be neglected in the experiment. The influence of magnetic exchange forces was also small with this particular experimental arrangement.

Considerable efforts have been made in the fabrication and characterisation of magnetic tips suitable for spin-polarised STM experiments as CrO_2 tips tend to be difficult to fabricate[41]. A range of antiferromagnetic materials with Néel temperatures above room temperature has been examined as possible candidates for STM tips, including anti-ferromagnetic alloys of Mn, such as MnPt, MnNi, and single element antiferromagnets[43], such as Cr. Ferromagnetic materials such as Fe, Ni, Co have also been considered[41]. As most of the materials tend to be easily oxidised in air and oxidation as well as contamination in air influence spin-polarised STM studies in an uncontrolled way, *in situ* tip preparation under UHV condition will be desirable. Wiesendanger and co-workers reported the use of an *in situ* tip preparation technique which reproducibly yields clean and microscopically sharp tips[41]. With their technique, metal wires are first electrochemically etched to form a constriction of 20-100 μm. The top part and the bottom part are fixed on separate tip holders and the two parts pulled apart under UHV condition. The cleanness of the tips at the sharp end is then in principle limited by only the impurity level of the bulk. The probes appear to have a well defined micro tip at the end which is independent of size of the starting constriction, and atomic resolution can be routinely achieved using these tips with Si(111) 7×7 as testing surfaces. The use of single crystal Fe whiskers was also considered[41].

Ferromagnetic Fe tips have been used in the study of Magnetite (Fe_3O_4) (001) surfaces in conjunction with non-magnetic tungsten tips[19]. The work demonstrated a selective imaging of the different magnetic ions Fe^{2+} and Fe^{3+}, suggesting that magnetic imaging can be realised at the atomic level. However, the nature and origin of such atomic contrast are still unclear, although a spin-polarised tunnelling effect might be one probable cause.

SOME RELEVANT STM BASED TECHNIQUES

Concurrently with the study of spin-polarised tunnelling effect by STM, other closely related techniques based on STM have also been developed over the past few years. In order to provide a wider perspective of this rapidly advancing and exciting field, we shall mention below three powerful new techniques in which spin and magnetism play an important role.

Paramagnetic Spin Resonance with STM

The presence of small quantities of magnetic impurities or other localised spin centres at a tunnel junction has been known to affect tunnelling characteristics considerably[28-31]. Therefore, it may be expected that such effects should be observable with STM. In particular, in a constant magnetic field, a localised spin will precess at the classical Larmor frequency around the direction of the field, so a modulation of the tunnelling current at the same frequency would be expected when the tip is at the vicinity of the localised spin centre. By measuring the rf component of the tunnelling current, Manassen and co-workers demonstrated in their remarkable experiment that indeed such paramagnetic resonance can be observed[16]. Furthermore, with the STM as a local probe, they were able to detect what was believed to be the precession of an individual spin. Their pioneering experiment was carried out with a tungsten tip on thermally oxidised Si surfaces. More detailed experimental studies and theoretical interpretations have been reported in their subsequent work[44-46]. The success of the experiment opens up new possibilities for the development of electron spin resonance on an atomic scale.

Secondary Spin-polarised Electron Emission with STM

STM has also been utilised in another novel way in the experiment reported by Allenspach and Bischof[47]. An STM was operated in the field emission mode to provide a narrow electron beam injected on the magnetic surface and the spin polarisation of the emitted secondary electrons was monitored. As their first result, a plot of spin polarisation of secondary electrons vs applied magnetic field showed a hysteresis loop for an Fe-based metallic glass. Their results demonstrated that the low-energy secondary electrons from a ferromagnet produced by an electron beam extracted from an STM tip operated in the field emission mode were able to escape from the sample and that the secondary electrons were spin polarised. They also showed the importance of using correctly shaped STM tips and the feasibility of applying the technique for high resolution magnetic domain imaging.

Magnetic Circular Dichroism with STM-excited Fluorescence

Very recently de Parga and Alvarado reported the observation of magnetic circular dichroism in cobalt films with STM-excited fluorescence[48]. For experiments involving excitations by photon radiation[49-54], magnetic circular dichroism (MCD) arises from the interplay between the magnetic exchange and the spin-orbit interaction which provides coupling in a given transition between the electron spin of the valence-band states and the spin of the photon involved. Such a technique allows electronic excitations of magnetic materials to be studied without an explicit analysis of electron spin-polarisation. The same effect is expected for the inverse process; that is, radiative recombination of charge carriers involving spin-non-degenerate valence states should also display non-zero circular polarisation if observed in the direction of the magnetisation of the sample. In their experiment, the circular polarisation of radiative recombination of thin Co films was studied with electrons injected from a tungsten STM tip operated both in the tunnelling mode and field emission mode. Their results showed clearly that MCD was observed. Other interesting results include the observation of different MCD behaviour between the tunnelling and field emission mode of electron injection.

SUMMARY AND FUTURE PROSPECTS

The research work published so far indicates that STM based spin-polarised electron tunnelling effects are certainly measurable by experiment. Such observations have been illustrated by a number of independent research groups using different experimental techniques. We have mentioned in the preceding sections the key work in this area; in particular, the spin-polarised STM prototype operated in air with modulation of sample magnetisation, the high spatial resolution UHV work on an anti-ferromagnetic Cr surface, the examination of tunnelling from Ni tips employing circular light emission and the optical pumping techniques, all of which showed the effects of spin-polarised tunnelling with STM. These experimental studies also demonstrated other interesting points that may be useful in terms of future applications of these techniques. With their special experimental arrangement, in-plane magnetisation[17] appeared to be probed by Johnson and Clarke, whilst out-of-plane magnetisation was probed by Wiesendanger and co-workers[18] with sharper ferromagnetic CrO_2 tips. In the work by Alvarado and Renaud[20], it was reported that they were able to flip the magnetisation of the tip apex by 180° during the experiment with an electromagnet operated *in-situ*. If such a change of the magnetisation can be achieved during an experiment without any adverse effect on the sample, the method will offer a useful way to separate the magnetic and topographic information. As to the work reported by Sueoka and colleagues[21], if suitable GaAs tips can be fabricated and optically pumped by circularly polarised light, thus reversing the roles of the tip and the sample in their experiment, these tips will then be potentially useful for spin-polarised STM experiments with the polarisation state of the electrons controlled optically. We note that these publications provided separate confirmation of spin-polarised electron tunnelling using STM with different techniques or different approaches. Although *in situ* studies with combined techniques are still lacking at present, these experiments have established beyond doubt the feasibility and the potential of spin-polarised electron tunnelling with STM.

An obvious application of the spin-polarised STM technique would be in the area of magnetic domain imaging, especially in the study of magnetic domain boundary structures, taking advantage of the high lateral resolution generally associated with SPM techniques. For effective applications of this sort, efficient methods of separating the magnetic structural information from the topographic counterpart will need to be derived. The need to separate the two contributions, however, is not specific to spin-polarised STM based techniques. For instance, such an issue also exists in high resolution work of magnetic force microscopy (MFM). In the experiments described earlier, a number of techniques had already been used for this purpose; for instance, by the comparison of images taken by both magnetic and non-magnetic tips. As far as imaging of magnetic domain structure is concerned, there have already been a number of techniques available to offer different degrees of spatial resolution, such as, scanning magneto optical Kerr effect, SEMPA and MFM. The advantage of a spin-polarised STM based technique would be its higher attainable resolution, with a promise of ultimate resolution on an atomic scale.

A further important aspect of the spin-polarised STM based techniques appears to be the ability to undertake spectroscopic studies. By varying the bias across the tunnel junction, some information regarding the energy dependence of the spin-dependent LDOS can be obtained. Thus it is possible to provide a probe to study the spin-polarised electronic states on a local scale, a subject that cannot be readily addressed by other techniques such as MFM.

Another significant aspect of spin-polarised STM may lie in the fact that such techniques are likely to provide a spin-polarised micro-probe for many future potential applications. For instance, a spin-polarised version of the ballistic electron emission microscopy (BEEM) may offer the ability to study spin-polarised hot electron transport across thin magnetic films on a microscopic scale, a subject of close relevance to our present research programme at Leeds.

Many future developments are likely. There is a clear need to have a better understanding, both experimentally and theoretically, of the processes involved in spin-polarised tunnelling with STM. A careful examination of other magnetic effects which are likely to arise through the use of magnetic tips will also be required. New techniques for the separation of surface topography and spin-polarised effects will need to be sought; e.g., a possible reversal of tip magnetisation directions during successive scanning, and reversing the spin polarisation direction of tunnelling electrons possibly by using optically pumped GaAs tips mentioned above. The application of spin-polarised STM and related techniques to novel magnetic systems such as magnetic superlattices will also become an important aspect. Such applications are likely to include planar imaging of domain boundary structures, cross-sectional imaging of magnetic superlattices, and spin-polarised electron transport.

To conclude on a note of optimism, perhaps it is fair to say that STM based spin-sensitive techniques have great potential, are likely to break new ground in polarised photon and polarised electron physics, and may even lead us to a new era in which magnetism can be studied and magnetic technological applications made on an atomic scale.

ACKNOWLEDGEMENT

The author wishes to express his sincere thanks to Professor D. Greig for his encouragement, useful discussions and a critical reading of the manuscript. Thanks are also due to Dr. C.G.H Walker for his assistance in the preparation of the manuscript.

REFERENCES

1. M.N.Baibich, J.M.Broto, A.Fert, F.Nguyen van Dau, F.Petroff, P.Etienne, G.Creuzet, A.Friederich, and J.Chazelas *Phys. Rev. Lett.* 61: 2472 (1988)
2. G. Binasch, P.Grunberg, F.Saurenbach, and W.Zinn, *Phys. Rev. Lett.* 64: 2304 (1990)
3. A.E.Berkowitz, J.R.Mitchell, M.J.Carey, A.P.Young, S.Zhang, F.E.Spada, F.T.Parker, A.Hutten, and G.Thomas *Phys. Rev. Lett.* 68: 3745 (1992)
4. S.S.P.Parkin *Phys. Rev. Lett.* 71: 1641 (1993)
5. M.J.Hall, B.J.Hickey, M.A.Howson, M.J.Walker, J.Xu, D.Greig and N.Wiser *Phys. Rev. B* 47: 12785 (1993)
6. R. Allenspach, *Phys. World* 7 (No 3): 44 (1994)
7. M. Woehlecke and G. Borstel, "Optical Orientation", F. Meier and B.P. Zacharchenya, eds., Modern Problems in Condensed Matter Sciences, Vol 8 (North-Holland, Amsterdam, 1984).
8. R. Feder, "Polarised Electrons in Surface Physics", R. Feder, ed., (World Scientific, Singapore, 1985).
9. U. Heinzmann and G. Schönhense, "Polarised Electrons in Surface Physics", R. Feder, ed., (World Scientific, Singapore, 1985).
10. F. Meier, "Polarised Electrons in Surface Physics", R. Feder, ed., (World Scientific, Singapore, 1985).
11. J. Kirschner, "Polarised Electrons at Surfaces", Springer Tracts in Modern Physics 106, (Springer, Berlin, 1985).
12. U. Heinzmann, "Photoemission and Absorption Spectroscopy of Solids and Interfaces with Synchrotron Radiation", M. Campagna and R. Rosei, eds, (North-Holland, Amsterdam, 1990).
13. H.P. Oepen and J. Kirshner, *Scanning Microscopy* 5: 1 (1991)
14. G. Binnig and H. Rohrer, *Helv. Phys. Acta* 55: 726 (1982)
15. C. Schoenenberger and S.F. Alvarado, *Z. Phys B* 80: 373 (1990); D. Rugar, H.J. Mamin, P. Guethner, S.E. Lambert, J.E. Stern, I. McFadyen and T. Yogi, *J. Appl. Phys.* 68: 1169 (1990); J.K. Trautman, E. Betzig, J.S. Weiner, D.J. DiGiovanni, T.D. Harris, F. Hellman and E.M. Gyorgy, *J. Appl. Phys.* 71: 4659 (1992).
16. Y. Manassen, R.J. Hamers, J.E. Demuth and A.J. Castellano, Jr., *Phys. Rev. Lett.* 67: 2531 (1989).
17. M. Johnson and J. Clarke, *J. Appl. Phys.* 67: 6141 (1990).
18. R. Wiesendanger, H.-J. Guntherodt, G. Guntherodt, R.J. Gambino and R. Ruf, *Phys. Rev. Lett.* 65: 247 (1990).
19. R. Wiesendanger, I.V. Shvets, D. Burgler, G. Tarrach, H.-J. Guntherodt, J.M.D. Coey and S. Graser, *Science* 255: 583 (1992).
20. S.F. Alvarado and P. Renaud, *Phys. Rev. Lett.* 68: 1387 (1992).
21. K. Sueoka, K. Mukasa and K. Hayakawa, *Jpn. J. Appl. Phys. Part I* 32: 2889 (1993).
22. G. Chrobok, M. Hofmann, G. Regenfus and R. Sizmann, *Phys. Rev. B* 15: 429 (1977).
23. N. Muller, W. Eckstein, W. Heiland and W. Zinn *Phys. Rev. Lett.* 29: 1651 (1972).
24. E. Kisker, G. Baum, A.H. Mahan, W. Raith and B. Reihl, *Phys. Rev. B* 18: 2256 (1978).

25. R. Meservey, P.M. Tedrow and P. Fulde, *Phys. Rev. Lett.* 18: 1270 (1970). P.M. Tedrow and R. Meservey, *Phys. Rev. Lett.* 26: 192 (1971); R. Meservey, D. Paraskevopoulos and P.M. Tedrow, *Phys. Rev. Lett.* 37: 858 (1976).
26. M. Julliere, *Phys. Lett.* 54A: 225 (1975); S. Maekawa and U. Gafvert, *IEEE Trans. Magn.* MAG-18: 707 (1982).
27. J.C. Slonczewski, *Phys. Rev. B* 39: 6995 (1989).
28. L.Y.L. Shen and J.M. Rowell, *Phys. Rev.* 165: 566 (1968).
29. J. Appelbaum, *Phys. Rev. Lett.* 17: 91 (1966).
30. E.L. Wolf, "Principles of Electron Tunnelling Spectroscopy" (Oxford Univ. Press, London, 1985).
31. A.L. Belyanin, A.R. Kessel and V.A. Zhikharev, *J. Phys. C* 49: 57 (1982).
32. M. Johnson and R.H. Silsbee, *Phys. Rev. Lett.* 55: 1790 (1985).
33. S. Blugel, D. Pescia and P.H. Dederichs, *Phys. Rev. B* 39: 1392 (1989).
34. F. Meier, D. Pescia and T. Schviber, *Phys. Rev. Lett.* 48: 645 (1982).
35. L.E. Klebanoff, S.W. Robey, G. Liu and D.A. Shirley, *Phys. Rev. B.* 30: 1048 (1984); L.E. Klebanoff, R.H. Victora, L.M. Falicov and D.A. Shirley, *Phys. Rev. B* 32: 1997 (1985).
36. K.P. Kamper, W. Schmitt, G. Guntherodt, R.J. Gambino and R. Ruf, *Phys. Rev. Lett.* 59: 2788 (1987).
37. H.C. Siegmann, *Europhys. News* 14: 9 (1983).
38. D.T. Pierce and F. Meier, *Phys. Rev. B.* 13: 5484 (1976).
39. G.E. Pikus and A.N. Titkov, in "Optical Orientation", F. Meier and B.P. Zacharchenya, eds., Modern Problems in Condensed Matter Sciences, Vol 8 (North-Holland, Amsterdam, 1984) p133.
40. Ch. Schnonenberger and S.F. Alvarado, *Z. Phys. B* 80: 373 (1990)
41 R. Wiesendanger, D. Burgler, G. Tarrach, U. Hartmann, H.-J. Guntherodt, I.V. Shvets and J.M.D. Coey, *Appl. Phys.* A53: 349 (1991).
42. R. Wiesendanger, D. Burgler, G. Tarrach, A. Wadas, D. Brodbeck, H.-J. Guntherodt, G. Guntherodt, R.J. Gambino and R. Ruf, *J. Vac. Sci. Technol. B* 9: 519 (1991).
43. I.V. Shvets, R. Wiesendanger, D. Burgler, G. Tarrach, H.-J. Guntherodt and J.M.D. Coey, *J. Appl. Phys.* 71: 5489 (1992).
44. D. Shachal and Y. Manassen, *Phys. Rev. B* 44: 11528 (1991).
45. D. Shachal and Y. Manassen, *Phys. Rev. B* 46: 4795 (1992).
46. Y. Manassen, E. Ter-Ovanesyan, D. Shachal and S. Richter, *Phys. Rev. B* 48: 4887 (1993).
47. R. Allenspach and A. Bischof, *Appl. Phys. Lett.* 54: 587 (1989).
48. A.L.V. de Parga and S.F. Alvarado, *Phys. Rev. Lett.* 72: 3726 (1994)
49. G. Schutz, W. Wagner, W. Wilhelm, P. Kienle, R. Zeller, R. Frahm and G. Materlik, *Phys. Rev. Lett.* 58: 737 (1987).
50. G. van der Laan, B.T. Thole, G.A. Sawatzky, J.B. Goedkoop, J.C. Fuggle, J.-M. Esteva, R. Karnatak, J.P. Remeika and H.A. Dabkowska, *Phys. Rev. B* 34: 6529 (1986).
51. C.F. Hague, J.-M. Mariot, P. Strange, P.J. Durham and B.L. Gyorffy, *Phys. Rev. B* 48: 3560 (1993).
52. C.M. Schneider, M.S. Hammond, P. Schuster, A. Cebollada, R. Miranda and J. Kirschner, *Phys. Rev. B* 44: 12066 (1991).
53. J. Bansmann, M. Getzlaff, C. Westphal, F. Fegel and G. Schonhense, *Surf. Sci.* 269/270: 622 (1992).
54. L.M. Baumgarten, C.M. Schneider, H. Petersen, F. Schafers and J. Kirschner, *Phys. Rev. Lett.* 65: 492 (1990)

Series Publications

Below is a chronological listing of all the published volumes in the *Physics of Atoms and Molecules* series.

ELECTRON AND PHOTON INTERACTIONS WITH ATOMS
Edited by H. Kleinpoppen and M. R. C. McDowell

ATOM–MOLECULE COLLISION THEORY: A Guide for the Experimentalist
Edited by Richard B. Bernstein

COHERENCE AND CORRELATION IN ATOMIC COLLISIONS
Edited by H. Kleinpoppen and J. F. Williams

VARIATIONAL METHODS IN ELECTRON–ATOM SCATTERING THEORY
R. K. Nesbet

DENSITY MATRIX THEORY AND APPLICATIONS
Karl Blum

INNER-SHELL AND X-RAY PHYSICS OF ATOMS AND SOLIDS
Edited by Derek J. Fabian, Hans Kleinpoppen, and Lewis M. Watson

INTRODUCTION TO THE THEORY OF LASER–ATOM INTERACTIONS
Marvin H. Mittleman

ATOMS IN ASTROPHYSICS
Edited by P. G. Burke, W. B. Eissner, D. G. Hummer, and I. C. Percival

ELECTRON–ATOM AND ELECTRON–MOLECULE COLLISIONS
Edited by Juergen Hinze

ELECTRON–MOLECULE COLLISIONS
Edited by Isao Shimamura and Kazuo Takayanagi

ISOTOPE SHIFTS IN ATOMIC SPECTRA
W. H. King

AUTOIONIZATION: Recent Developments and Applications
Edited by Aaron Temkin

ATOMIC INNER-SHELL PHYSICS
Edited by Bernd Crasemann

COLLISIONS OF ELECTRONS WITH ATOMS AND MOLECULES
G. F. Drukarev

THEORY OF MULTIPHOTON PROCESSES
Farhad H. M. Faisal

PROGRESS IN ATOMIC SPECTROSCOPY, Parts A, B, C, and D
Edited by W. Hanle, H. Kleinpoppen, and H. J. Beyer

RECENT STUDIES IN ATOMIC AND MOLECULAR PROCESS
Edited by Arthur E. Kingston

QUANTUM MECHANICS VERSUS LOCAL REALISM: The Einstein, Podolsky, and Rosen Paradox
Edited by Franco Selleri

ZERO-RANGE POTENTIALS AND THEIR APPLICATIONS IN ATOMIC PHYSICS
Yu. N. Demkov and V. N. Ostrovskii

COHERENCE IN ATOMIC COLLISION PHYSICS
Edited by H. J. Beyer, K. Blum. and R. Hippler

ELECTRON–MOLECULE SCATTERING AND PHOTOIONIZATION
Edited by P. G. Burke and J. B. West

ATOMIC SPECTRA AND COLLISIONS IN EXTERNAL FIELDS
Edited by K. T. Taylor, M. H. Nayfeh, and C. W. Clark

ATOMIC PHOTOEFFECT
M. Ya. Amusia

MOLECULAR PROCESSES IN SPACE
Edited by Tsutomu, Isao Shimamura, Mikio Shimizu, and Yukikazu Itikawa

THE HANLE EFFECT AND LEVEL CROSSING SPECTROSCOPY
Edited by Giovanni Moruzzi and Franco Strumia

ATOMS AND LIGHT: INTERACTIONS
John N. Dodd

INTRODUCTION TO THE THEORY OF X-RAY AND ELECTRONIC SPECTRA OF FREE ATOMS
Romas Karazija

POLARIZATION BREMSSTRAHLUNG
Edited by V. N. Tsytovich and I. M. Oiringel

INTRODUCTION TO THE THEORY OF LASER–ATOM INTERACTIONS (Second Edition)
Marvin H. Mittleman

ATOMIC INNER-SHELL PHYSICS
Edited by Bernd Crasemann

COLLISIONS OF ELECTRONS WITH ATOMS AND MOLECULES
G. F. Drukarev

THEORY OF MULTIPHOTON PROCESSES
Farhad H. M. Faisal

PROGRESS IN ATOMIC SPECTROSCOPY, Parts A, B, C, and D
Edited by W. Hanle, H. Kleinpoppen, and H. J. Beyer

RECENT STUDIES IN ATOMIC AND MOLECULAR PROCESS
Edited by Arthur E. Kingston

QUANTUM MECHANICS VERSUS LOCAL REALISM: The Einstein, Podolsky, and Rosen Paradox
Edited by Franco Selleri

ZERO-RANGE POTENTIALS AND THEIR APPLICATIONS IN ATOMIC PHYSICS
Yu. N. Demkov and V. N. Ostrovskii

COHERENCE IN ATOMIC COLLISION PHYSICS
Edited by H. J. Beyer, K. Blum. and R. Hippler

ELECTRON–MOLECULE SCATTERING AND PHOTOIONIZATION
Edited by P. G. Burke and J. B. West

ATOMIC SPECTRA AND COLLISIONS IN EXTERNAL FIELDS
Edited by K. T. Taylor, M. H. Nayfeh, and C. W. Clark

ATOMIC PHOTOEFFECT
M. Ya. Amusia

MOLECULAR PROCESSES IN SPACE
Edited by Tsutomu, Isao Shimamura, Mikio Shimizu, and Yukikazu Itikawa

THE HANLE EFFECT AND LEVEL CROSSING SPECTROSCOPY
Edited by Giovanni Moruzzi and Franco Strumia

ATOMS AND LIGHT: INTERACTIONS
John N. Dodd

INTRODUCTION TO THE THEORY OF X-RAY AND ELECTRONIC
SPECTRA OF FREE ATOMS
Romas Karazija

POLARIZATION BREMSSTRAHLUNG
Edited by V. N. Tsytovich and I. M. Oiringel

INTRODUCTION TO THE THEORY OF LASER–ATOM INTERACTIONS
(Second Edition)
Marvin H. Mittleman

ATOMIC INNER-SHELL PHYSICS
Edited by Bernd Crasemann

COLLISIONS OF ELECTRONS WITH ATOMS AND MOLECULES
G. F. Drukarev

THEORY OF MULTIPHOTON PROCESSES
Farhad H. M. Faisal

PROGRESS IN ATOMIC SPECTROSCOPY, Parts A, B, C, and D
Edited by W. Hanle, H. Kleinpoppen, and H. J. Beyer

RECENT STUDIES IN ATOMIC AND MOLECULAR PROCESS
Edited by Arthur E. Kingston

QUANTUM MECHANICS VERSUS LOCAL REALISM: The Einstein,
Podolsky, and Rosen Paradox
Edited by Franco Selleri

ZERO-RANGE POTENTIALS AND THEIR APPLICATIONS IN ATOMIC
PHYSICS
Yu. N. Demkov and V. N. Ostrovskii

COHERENCE IN ATOMIC COLLISION PHYSICS
Edited by H. J. Beyer, K. Blum, and R. Hippler

ELECTRON–MOLECULE SCATTERING AND PHOTOIONIZATION
Edited by P. G. Burke and J. B. West

ATOMIC SPECTRA AND COLLISIONS IN EXTERNAL FIELDS
Edited by K. T. Taylor, M. H. Nayfeh, and C. W. Clark

ATOMIC PHOTOEFFECT
M. Ya. Amusia

MOLECULAR PROCESSES IN SPACE
Edited by Tsutomu, Isao Shimamura, Mikio Shimizu, and Yukikazu Itikawa

THE HANLE EFFECT AND LEVEL CROSSING SPECTROSCOPY
Edited by Giovanni Moruzzi and Franco Strumia

ATOMS AND LIGHT: INTERACTIONS
John N. Dodd

INTRODUCTION TO THE THEORY OF X-RAY AND ELECTRONIC
SPECTRA OF FREE ATOMS
Romas Karazija

POLARIZATION BREMSSTRAHLUNG
Edited by V. N. Tsytovich and I. M. Oiringel

INTRODUCTION TO THE THEORY OF LASER–ATOM INTERACTIONS
(Second Edition)
Marvin H. Mittleman

POLARIZED ELECTRON/POLARIZED PHOTON PHYSICS
Edited by H. Kleinpoppen and W. R. Newell

INDEX

Above threshold ionisation, 177
Achiral, 218
Actinides, 296
Alignment angles γ, δ, ε, 76
Alignment parameter, 227
Alkanes
    standing, 220
    lying, 220
Amplitude, 312, 315, 316, 318, 319
Angle-resolved photoemission, 336
Angular correlation, 232
Annihilation Radiation, 172
Angular symmetry, 227
Anisotropy, 280, 282
Antiferromagnets, 312, 313, 316, 317
Atomic beam, 83
Atoms, 2
Auger electrons, 11
Auger peaks, 154
Autoionization, 225, 248

Band gap, 143
Beamline, 297
Bethe-Heitler formula, 12
Born-Ochkur approximation, 266
Bose condensation, 175
Branching ratio, 307
Bremsstrahlung, 11, 12
Brillouin zone, 109, 142, 336

Calibration, 117
Centrifugal barrier, 248
Chiral, 197
Chiral effects in electron scattering, 204
Chirality, 175
    time-even and time-odd, 206
Chopper, 298
Circular dichroism, 302
Circular polarization, 302, 313-315, 321, 324
Close coupling
    eigenstate, 39
    pseudostate, 37, 39
CO_2 laser, 183
Coherence lengths
    of two-photon radiation, 187
    of photon wavepackets, 187

Coherent excitation of atoms by
    photon excitation, 5
Coherent nature of the excitation process, 5
Collisions
    polarized electrons, 1
    polarized photons, 1
Complete atomic collision experiments, 3, 5
Complete descriptions, 17
Complete experiment, 2, 7, 10, 81, 160
Complete scattering experiments, 261
Compton, 312, 313, 323, 325-330
Coplanar geometry, 52
Correlation
    electron-photon, 62
    angular, 63
    method, 63
    polarisation, 63
    the n^1P states of helium, 64
    the 3^3P states of helium, 66
    the 3^1D state of helium, 67
    the 3^3D state of helium, 70
    the ^2P states of hydrogen and sodium, 71
Correlation measurements
    angular, 81
    polarization, 81
Coulomb
    direct, 15
    exchange amplitude, 7
    exchange interaction, 13, 166
    explosion, 177
Coulomb direct amplitude, 7
Coulomb direct interaction, 13
Cross section, 312, 313, 316, 317, 319, 321-328, 330
Crossed-beam experiment, 8
Curie temperature, 279

Daresbury SRS, 225
Density matrix, 199
Depolarisation, 88, 202
Dielectric tensor, 270
Diffraction, 313, 316, 318-321
Dipole
    approximation, 181
    moment, 313, 315, 317
Direction of magnetisation, 336

Doppler
  broadening, 249
  laser cooling, 13
  pedestal, 249
Doppler-free spectroscopy, 249
Double excitation, 248, 254

Earth alkaline atoms, 81
Elastic scattering, 313, 316, 325, 328
Electric dipole, 296
Electron
  atom collision processes, 3
  atom collisions, 14
  atomic hydrogen, 46
  beam, 199
  gun, 135
  longitudinal, 197
  molecule scattering, 197
  optic dichroism, 218
  optics, 1, 150
  photon correlation
    photon coincidence technique, 159
    photon coincidence, 165
  polarimetry, 113
  scattering, 13
  scattering technique, 215
  spectrometer, 226
  spectroscopy, 225
  spectroscopy with polarization analysis, 147
  spin polarised, 140
  spin polarization, 113, 300
  transverse, 197
Electrostatic spherical sector, 137
Elwert-Haug theory, 12
Epitaxial, 281
Exchange, 23, 197
  amplitude, 25
  asymmetry, 23
  effects, 262
  in electron-atom scattering, 23
  interaction, 15, 296

Faraday effect
  high-field, 240
  rotation, 235
  spectroscopy, 238
$FeB_{13.5}Si_{3.5}C_2$, 153
Fermi, 321, 326, 327
  edge, 101
  energy, 332
  level, 143
Ferrimagnets, 312, 313
Ferro-/ferrimagnets, 312
Ferromagnetic metals, 270
Ferromagnets, 313, 316, 321
Feynman diagrams
  free-free scattering, 182
  diagrams for SEPE, 182
Fixed polarisation, 220
Floquet theory, 181
Fluorescence, 300

spectroscopy, 225
Form factor, 317, 322
Free electron wave function, 181
Free-free scattering, 177, 178
Fresnel coefficients, 272
Fundamental spectra, 300

GaAlAs laser diode, 136
GaAs cathode, 109
  photocathodes, 29, 135, 138
Gallium arsenide source, 109
Garnet, 296
Giant resonances, 225
Goedkoop filter, 296
Grating, 297

Hamiltonian
  spinless, 53
  Dirac, 53
  Breit-Pauli, 54
Hanle effect, 5
Hartree-Fock, 251
Helicity, 172, 298
Hexapole magnet, 160
Hysteresis, 278

Impulse Approximation, 323, 324, 325
Inelastic scattering, 330
Inner shell excitation, 248, 252
Instrumental asymmetry, 97
Interference, 315, 316, 320, 321
  amplitude, 7
  asymmetry, 24, 33
Ionization asymmetry, 9
Isolated core excitation, 249, 250
Isotropic spectrum, 302

K-matrix methods, 258
Kerr
  effect, 295
  polarisation, 273
  rotation, 279
Lamb shift, 188
Langmuir-Taylor detector, 160, 162
Laser assisted collisions, 179
Laue diffraction, 142
LEED pattern, 142
  polarimeter, 116
Lifetime spectrum, 172
Linear dichroism, 302
Loops
  paramagnetic, 276
  ferromagnetic, 276
Lorentz force, 271
Lu-Fano plots, 253
Lyman-α, 192

Magnetic
  circular dichroism, 236
  correlations, 312
  dipole operator, 307

films, 276
 interaction, 312, 315, 323
 multilayer, 304
 order, 320, 330
 ordering, 312, 313, 323
 phase transition, 275
 processes at surfaces, 147
 structure, 317, 318, 320
 sub-cross sections, 3
 thin film, 332
 X-ray dichroism, 295
 X-ray scattering, 312
Magnetism, 312, 319, 326
 bulk, 174
 surface, 174
Magneto anisotropic energy, 306
Magneto-optic
 Kerr effect, 269, 281, 288
 measurements, 279
Magneto-optical rotation, 236
Magnetometery, 274
Magnetometery techniques, 275
Magnetoresistance, 332
Many-body, 296
Micro-Mott polarimeter, 102
Model of coherent impact excitation, 4
Moderators, 170
MOKE
 effect, 283
 loops, 283
Molecules, 197
Monochromator, 297
Mott detector, 140, 149
 polarimeter, 95
 scattering, 24
MPI, 177
Multi-photon, 178
Multilayer structures, 332
MXD, 295
Neutron scattering, 31, 313
Neutrons, 315, 320
Ni, 142
Non-chiral, 210

Obscuration plate, 299
Optical pumping, 29
Optically pumped GaAs, 333, 338
Optically pumped sodium atoms, 8
Orbital, 312-317, 323-326
Orbital magnetization, 302
Orientation of magnetisation, 271
Oriented molecules, 203, 209, 210, 215, 218
Oriented target, 262

P-I loops, 152
Parity, 189
 allowed transitions, 263
 forbidden transitions, 264
Particle-photon coincidence experiments, 3
Particle-photon coincidences, 3

Pauli matrices, 205
Perpendicular magnetocrystalline anisotropy, 306
Photoattachment, 51
Photoelectron spectra, 231
Photoemission of electrons
 from solids, 17
 adsorbates, 17
 from atoms and solids, 17
Photoionization, 14
 of rate-gas atoms, 18
Photoionization and fluorescence of calcium and strontium, 225
Photoionization spectra of the alkaline earth atoms, 225
Photon-particle coincidence experiments, 2
Photons, 312, 323, 324, 328, 329
Physical processes
 quantum mechanically, 1
 complete, 1
PMA, 306
Pockels cell, 31, 136, 142
Polar Kerr magnetometery, 289
Polarimeters, 117
Polarimetry instrumentation survey, 121
Polarisation, 199
 of the fluorescent radiation, 232
Polarisation-dependent effect, 184
Polarised electrons, 197, 219
Polarization
 reversals, 31
Polarization correlations
 electron impact excitation, 81
Polarization spectroscopy, 249
Polarization Stokes parameter, 163, 164
Polarization vector, 184
Polarized
 atoms, 1, 27
 electrons, 1, 8, 17, 27
 one-electron atoms, 23
Polarized electron source
 solid, 107
 gaseous, 107
Polarized electron spectroscopy, 157
Polarized particles
 electrons, 1
 positrons, 1
 atoms, 1
Polarized photon-particle coincidences, 11
Polarized photons
 with surfaces and solid state targets, 17
Polarized positrons, 169
Polarized sodium atoms, 16
Pondermotive potential U_p, 179
Positron polarimeter, 172
Positron scattering, 13
Positronium
 para-state, 171
 ortho-state, 171
Positronium, 169
Positrons, 2, 170

Potassium atoms, 159

Q resolution, 312, 317
Quadratic Zeeman effect, 242
Quadruple moment, 302
    photons, 188
    transitions, 296
Quanta, 181
Quantum mechanically complete and optimum information, 3

R-Matrix, 64, 202
    approximation, 40
    intermediate energy, 40
R-Matrix method, 181
Ramsauer-Townsend effect, 5, 10
Rare earths, 296
Reduced matrix element, 300
Reflection symmetry, 63
Relativistic electron-atom scattering, 26
Resonance, 315, 316, 319, 320
    enhancement, 319, 320
    trapping, 82
    photoemission, 308
Retarding field
    analyzer, 140
    Mott detector, 149
Retarding polarimeters, 118
RHEED, 288
Rydberg electron, 251
Rydberg states, 239
Satellite structure, 307
Scanning electron microscopy with polarisation analysis, 147, 332
Scattering asymmetry, 150
Scattering matrix, 205
SEMPA, 148, 153
Sherman function, 97, 100, 105, 149
Simultaneous electron photon excitation, 177, 178, 181
Singlet amplitude, 7
Singlet interactions, 15
SMOKE, 149
Soft x-ray, 297
Spatially oriented, 210
Spectroscopic interactions
    with surfaces, 1
    with solids, 1
Spherical sector, 138
Spin, 312, 313, 315-318, 323-327, 329, 330
    asymmetry, 37, 50, 55
    analysis in (e,2e), 52
Spin analysis
    of the scattering, 8
    polarized electrons by polarized atoms, 8
Spin asymmetries
    elastic scattering, 42
    for atomic hydrogen, 41
    for electron-sodium scattering, 43
    in Bremsstrahlung, 12
    in (e,2e), 57
    inelastic scattering, 42

Spin change, 262
Spin effects, 1
    in electron-atom scattering, 37
    atomic collision processes, 159
Spin flip, 199, 262, 263
    process, 10
    amplitude, 10
Spin integrated photoemission Ni(110), 100
Spin orbit asymmetry, 24
Spin orbit interaction, 13, 197, 199, 270, 296
    in electron-atom Bremsstrahlung, 12
Spin parallel-antiparallel asymmetries
    in (e,2e)-collisions, 11
Spin polarized STM, 332, 333
Spin polarized tunnelling, 338
Spin polarization, 81, 140, 337
Spin polarized electron scattering
    in condensed matter, 261
    electrons, 2, 76, 159
    electron scattering, 218
    inverse photoemission, 133
    positrons, 175
    sodium, 159
    secondary electron emission, 147
    Auger electron spectroscopy, 147
    Auger peaks, 155
    x-ray photoemission spectroscopy, 147
Spin polarized electrons
    from photoionization of atoms, 19
Spin polarized atom, 14
Spin reactions, 3
Spin resolved photoemission, 297, 336
    Iron on Cu(100), 104
    Ni(110), 104
    Xe, 104
Spin triplet systems, 15
Spin vectors, 142
Spinless Hamiltonian, 38
Stern-Gerlach magnet, 30
Stimulated Bremmstrahlung, 178
STM
    paramagnetic spin resonance, 339
    secondary spin polarised electron emission, 340
    magnetic circular dichroism, 340
STM measurements
    magnetic, 337
    non-magnetic tips, 337
Stokes parameters, 4, 84, 159, 166, 190
Strained lattice cathodes, 112
Substrates, 281
Sum rules, 302
Superconducting magnet, 300
Superelastic electron scattering, 82
Superelastic scattering, 83
Surface magnetism, 147, 149, 332
Synchrotron radiation, 295, 312-314, 321
    electron polarimetry, 95
    Faraday rotation, 235
Synchrotrons, 312-314, 322, 327

T Matrix, 181, 198, 202
Thompson scattering, 313, 316, 319, 321, 324

Threshold ionization, 48
Time-inverse process, 6
Topological antiferromagnetism, 336
Transfer, 178
Transition metals, 296
Transition probability, 300
Trilayer structures, 286
Triple differential cross section, 51
    amplitude, 7
    interactions, 15
    to singlet transition, 203
Tunnelling current, 336
Two-photon
    transitions, 248
    selection rules, 248
Two-photon decay
    characteristics, 187
Two-photon excitation, 177

Ultimate analysis, 166
Ultra thin film, 147, 277
Unoriented chiral molecules, 210

Vector magnetometry, 283
Vector polarisation, 4
Virtual 'off-shell', 181
Virtual interactions, 177
Volkov solution , 180
VSM, 300

Wannier threshold law, 37, 51
Wavepacket
    photon, 195
    description, 195
Wavepacket envelope
    autocorrelation, 192

X-rays, 312, 313, 315, 317, 328
XRES, 315, 319, 320

Young's interference, 13

Zeeman Spectroscopy, 239
Zeeman states, 30